Emerging Infectious Diseases and the Threat to Occupational Health in the U.S. and Canada

PUBLIC ADMINISTRATION AND PUBLIC POLICY

A Comprehensive Publication Program

Executive Editor

JACK RABIN
Professor of Public Administration and Public Policy
School of Public Affairs
The Capital College
The Pennsylvania State University—Harrisburg
Middletown, Pennsylvania

Assistant to the Executive Editor
T. Aaron Wachhaus, Jr.

Available Electronically

Emerging Infectious Diseases and the Threat to Occupational Health in the U.S. and Canada

edited by
William Charney
Healthcare Safety Consulting
Seattle, Washington

Taylor & Francis
Taylor & Francis Group
Boca Raton   London   New York

CRC is an imprint of the Taylor & Francis Group,
an informa business

CRC Press
Taylor & Francis Group
6000 Broken Sound Parkway NW, Suite 300
Boca Raton, FL 33487-2742

International Standard Book Number-10: 0-8493-4637-1 (Hardcover)
International Standard Book Number-13: 978-0-8493-4637-8 (Hardcover)
Library of Congress Card Number 2006040779

Library of Congress Cataloging-in-Publication Data

Charney, William, 1947-
   Emerging infectious diseases and the threat to occupational health in the U.S. and Canada / by William Charney.
      p. cm. -- (Public administration and public policy ; 120)
   Includes bibliographical references (p.).
   ISBN 0-8493-4637-1 (alk. paper)
   1. Emerging infectious diseases--United States. 2. Emerging infectious diseases--Canada. 3. Medicine, Industrial--United States. 4. Medicine, Industrial--Canada. I. Title. II. Series.

RA643.5.C43 2006
362.196'9--dc22   2006040779

Visit the Taylor & Francis Web site at
http://www.taylorandfrancis.com

and the CRC Press Web site at
http://www.crcpress.com

Dedication

This book is dedicated to E.D. whose inspiration kept my well from running dry.

This book is also dedicated to all the selfless health care workers who put themselves in harms way, shift after shift, to protect public health.

May this volume bring some respite to their daily exposures and help protect them in their time of need.

Preface

To me, the most extraordinary example of poor preparedness took place in the last week of September 2005. The South had just begun the long recovery from the devastation of Hurricane Katrina, and was bracing for the anticipated onslaught of Hurricane Rita. After Katrina, local, state, and federal governments had been widely criticized regarding their disaster management and were fearful of more missteps. Repercussions were widespread: gas and oil prices around the country surged because of current compromises of fuel supplies from the hard-hit Gulf of Mexico, and fears for the future.

My home state of Georgia was not likely to be directly affected by Hurricane Rita. But what did our governor do? Just before close of business on Friday, he announced that all the state's public schools would be closed on Monday and Tuesday to save fuel from school bus transportation. Apparently our energy planning is so inadequate that the most responsible reaction to a potential hurricane in another state was to deprive the Georgia's children of an education, and require their parents to take days off from work. Weren't there alternatives? Perhaps free public transportation for a few days, which might even have encouraged commuters to stick with the habit and leave their automobiles permanently at home. Or it would have been an excellent time for a bold new conservation program, or an agenda to make Georgia a leader in energy efficiency or alternative energy development. What about a telecommuting initiative or proposing the establishment of a new institute on climatology? No, instead our children's future was the first priority to be compromised; a metaphor of grave concern for larger decisions.

The federal government's response was no better. After the devastation of New Orleans, the government was handed an opportunity to rebuild a model city. They could have developed an exemplary public transportation system for commuters and for mass evacuations during emergencies;

crafted a plan to regenerate the wetlands that would help to buffer hurricanes and create habitat for wildlife; and developed other responsible environmental policies that could reduce future risk and enhance environmental preservation. Instead, President Bush acknowledged that Katrina was "not a normal hurricane," but failed to acknowledge that we should pay attention to the science that has been developed regarding global warming and its effect on hurricane severity. Instead the government proposed bypassing laws protecting the environment and fair wages, gave enormous no-bid contracts to large corporate donors, discouraged media portrayals of governmental errors, and bemoaned this devastating "act of God."

So what major policy link did the president make to Hurricane Katrina? He referred, once again, to the oft-cited tragedies of September 11th, 2001, stating how the hurricane demonstrated that Americans (unlike terrorists) "value human life."

But what lessons about preparedness and priorities did 9/11 really teach us? On that date, 3,400 people died because of four intentional plane crashes, because they were in the wrong place at the wrong time. Among the consequences of these deaths was a major redefinition and redirection of the role of government in and funding for public health. Certainly, governments must protect their citizens, so it is appropriate to address possible future threats, and indeed could prove essential. However, there is an immediate, real threat which we know will kill enormous numbers of Americans if we do not change our strategy, and that is the redirection of funds away from basic public health services to bioterrorism (BT) prevention.

What problems do basic public health services try to address, and why is diversion of resources away from them of concern? Using annual national data on mortality from various risk factors and diseases, I calculated that approximately 6,620 Americans were likely to have died on September 11th, 2001 from the major sources of mortality that many basic public health services work to address; 3,166 of these deaths were attributable to leading preventable risk factors (e.g., diet, inactivity, alcohol, etc.).[1]

The importance of these numbers is not just in their size, which is considerable, but their predictability. A similar volume of deaths from these same causes took place, not just on September 11th, 2001, but on September 12th, 2001, and on every day since then.

Concerns about disproportionately funding BT versus other public health functions have been building for some time: as early as December 2001 the American Medical Association resolved that the general enhancement of state and local public health agencies should be among our nation's highest priorities, and should be built, not eroded, by BT responses. Many thought that the Bush administration's smallpox vaccination plan was a misguided redirection of public health funds for BT, and it was successfully thwarted. Initial smallpox vaccination cost estimates ranged from $600 million to $1 billion[2] and plans for vaccination and treatment of smallpox, anthrax, and botulism were projected to exceed $6 billion over the following decade.[2] But concerns about an inadequate science base for this initiative, and concerns about it being a distraction from more fundamental public health needs helped to redirect this effort. The Association of State and Territorial Health Officials called smallpox immunization "the ultimate unfunded federal mandate," and the National Association of County and City Health Officials also expressed concern that efforts to combat smallpox and other potential BT threats would divert resources from current, pressing needs.[2] Even the CDC's own *Morbidity and Mortality Weekly Report* documented the difficulty for state health departments to allocate "the necessary time and resources for the pre-event smallpox vaccination program."[3]

But while smallpox immunization efforts are no longer a major focus for public health departments, spending on preparedness that is specific to BT is still magnified beyond what the extent of current threats might logically prescribe. Due to this funding, state health departments increased the number of epidemiology workers doing infectious disease and terrorism preparedness 132% between 2001 and 2003.[4] But concurrent with this increase in BT funding and mandates, 66% of health departments had problems allocating time for general planning, and 55% had problems establishing even basic disease surveillance systems.[5]

These observations are not intended to diminish the tragedy of September 11th or of Hurricane Rita. If our government wishes to appropriate substantial funds to prevent potential future threats to our security, this may well be justifiable. But public health funding for current threats should not be compromised; we should simultaneously try to prevent and to prepare for catastrophes that are caused by destructive individuals and those that are caused by destructive societies. We must recognize that a highly predictable tragedy is happening daily, that we already have available many strategies to help reduce the numbers of deaths from these predictable causes, and that more people will die unless we ensure that protecting the population against these routine, predictable causes of death remains a top priority. Let us not make Americans wonder if they must be in the right place at the right time if they want to stay healthy. Let us not have one more American die because of September 11, 2001.

Erica Frank, M.D., M.P.H.

Contributors

George Avery, Ph.D., M.P.A.

Dr. Elizabeth Bryce

Erna Bujna

Hillel W. Cohen, M.P.H., Dr.P.H.

Christina M. Coyle, M.D., M.S.

Stephen J. Derman

Michael J. Earls

Kim Elliott, M.A.

Erica Frank, M.D., M.P.H.

Robert Gould

Michael R. Grey, M.D., M.P.H.

Ted Haines, M.D., M.Sc.

Margaret A. Hamburg, M.D.

Jeanette Harris, R.N.

Shelley A. Hearne, Dr.P.H.

John H. Lange

Jeffrey Levi, Ph.D.

Nora Maher, M.Sc./Occupational
Hygiene

Giuseppe Mastrangelo

Lisa McCaskell

Mark Nicas, Ph.D., M.P.H., C.I.H.

Laura M. Segal, M.A.

Victor Sidel

Bernadette Stringer, Ph.D., R.N.

Barb Wahl

Dr. Annalee Yassi

Prologue

A pestilence isn't a thing made to man's measure; therefore we tell ourselves that pestilence is a mere bogy of the mind, a bad dream that will pass away. But it doesn't always pass away, and, from one bad dream to another, it is men who pass away, and the humanists first of all, because they haven't taken their precautions. Albert Camus, *The Plague*

Emerging Infectious Diseases and the Threat to Occupational Health in the U.S. and Canada is a relevant and topical reminder that what Camus wrote about remains as true today as it did in the fictional mid-twentieth-century north African town of Oran. Novelists and historians alike have shown us that millennial dawns are often accompanied by rising anxiety and fear in many Western societies. Not infrequently such fears and anxieties have been driven by very real threats — such as epidemic diseases or wars — that, in turn, trigger well-intentioned responses with ambiguous or clearly adverse impacts on individual liberties and the public welfare. Anyone who watches network news, listens to radio and television talk shows, or visits with work colleagues by the water cooler cannot help but be aware that globalization of infectious diseases has followed fast on the heels of worldwide economic and cultural globalization. Whether we are talking about new infectious threats, such as SARS, mad cow disease, or avian influenza; more familiar public health threats such as HIV/AIDS or tuberculosis; or the frightening possibility of genetically modified or weaponized "classic" infectious diseases, such as smallpox or anthrax, today we are more aware collectively of the need for a resilient and effective public health system. It has been many decades since the public health system has received as much attention as it has in the aftermath of the September 11th terrorist attacks, but as Charney and colleagues demonstrate in *Emerging Infectious Diseases*, bioterrorism may be the least of our worries. The fact of the matter is that our medical and public health communities are not nearly as prepared as they must be to adequately respond to existing, emerging, and potential threats to our nation's health and social welfare. The U.S. experience with Hurricane Katrina in September 2005 is simply the latest in a series of events that underscores this point, albeit in heart-wrenching detail, for all the world to see.

The reasons for this state of affairs are complex, diverse, and not easily remediable. Indeed, it is a topic worthy of a book unto itself, and a number of prominent organizations — such as the Institute of Medicine, the National Institute for Occupational Safety and Health (NIOSH), and the Robert Wood Johnson Foundation, to name a few — have published scholarly monographs that address the current state of readiness of the nation's health care and public health infrastructure. Taken as a whole, they are sobering reading. The Institute of Medicine's landmark 1988 report, *The Future of Public Health*, explored fully the disarray in the public health community, and anticipated the now widely recognized need to overhaul the education and training of public health and medical professionals, ineffective risk assessment and risk communication strategies, and outdated emergency and disaster management planning and procedures.[1,2,3,4]

While knowledgeable observers disagree as to fundamental or contributory causes for our present circumstances, several stand out as worthy of particular attention, largely because they offer helpful guidance as to what direction we might move in if we are to begin to redress these problems. To their credit, the contributors to *Emerging Infectious Diseases* address each of these issues cogently and forthrightly.

First, changes in health care financing and economics, in particular the managed care revolution, have seriously depleted the nation's ability to respond to any substantial surge in hospitalization, regardless of the proximate cause. Public hospitals were once a mainstay in most major metropolitan areas and the institution of last resort for those communities' poorest citizens. Today, they have either closed or struggle to maintain solvency in an increasingly competitive health care environment. Private and not-for-profit hospitals, too, have responded to these changes by downsizing bed capacity to the point where many now operate at near capacity year round. These same institutions struggle when even mild wintertime influenza epidemics cause demand for hospital beds to exceed capacity, there is emergency room gridlock, and often precipitous discharge of sick patients from hospitals. It is a problem facing urban, suburban, and rural hospitals alike, and it is more than just a matter of bed capacity. Many health care institutions face chronic problems with adequate staffing, training, and equipment. Hospitals are not alone in this regard. Nursing home and intermediate care facilities, home care programs, community health centers, and even physicians' offices are little more prepared to handle any surge in demand. Consequently, patients discharged from the hospital to make room for mass casualties or surges caused by infectious disease outbreaks may find that there is simply no room at the inn. Put simply, any sudden increase in demand for medical services — whether caused by routine influenza cases, or more worrisome still, natural or manmade disasters — pressure our health care system beyond its ability to respond. Addressing the problem of inadequate surge capacity will require creative planning and cooperation at the local, state, and national level and as certainly will demand coordination, planning, and funding by public and private means.

Second, the historical schism between the health care and public health communities must be bridged. The training of health care workers needs to prepare them for their new responsibilities as frontline workers in the nation's public health defense. Is it too much to expect in this day and age that all health care organizations and health care workers will be aware of key public health contacts in their community? Is it too much to expect the integration of basic principles of occupational health and safety into our daily practices? The lack of awareness of and attention to such basic principles of infection control as hand washing serve as a reminder that we have a long way to go.

While health care professionals must be prepared to step outside their comfortable clinical role and into a broader role of public health provider, the converse is true for those working in public health. One of the critical areas in need of attention in terms of public health education and training is preparing public health workers to be more cognizant of their responsibilities in communicating with their medical colleagues and the communities they serve. As an occupational and environmental health consultant to a state health department for many years, I observed that public health officials need to be better prepared to respond quickly, accurately, and reassuringly to legitimate concerns on the part of the public and health care workers. The 2003 SARS epidemic in Canada is instructive. At the time and in retrospect, both medical and public health workers recognized that communication was too slow, too ambiguous, and lacked the credibility needed to manage the crisis.[5]

Third, there needs to be a consistent and unarguable commitment to protecting the health and safety of public health and health care workers in dealing with natural and manmade disasters. Few of us would argue that our local firefighters deserve to have the proper training and equipment to perform their duties as emergency responders. How ironic, then, it was that during the first hours and days following the collapse of the World Trade Center opinions expressed on an occupational/environmental health listserv implied that individuals who suggested that emergency responders clawing through the rubble needed to be protected against asbestos, irritants, and other respirable dusts were at best unpatriotic, and possibly traitorous. As the dust has settled and the trauma receded into our memories, it is clear that the best interests of the heroic men and women who responded to the disaster were not served by the failure to provide proper training and equipment. The September 11th experience drove important attitudinal shifts within emergency response community itself: specifically, the emergency response and disaster management systems now recognize the necessity of integrating health and safety awareness into training and preparedness efforts and the need for knowledgeable and readily available personnel with the sort of skills needed to secure the public's health when medical and public health systems are being stressed maximally. These issues form the foundation of multiple chapters in *Emerging Infectious Diseases,* covering topics from protecting occupational health and safety during naturally occurring diseases (Charney), ventilation controls for emerging diseases (Derman), the public health problems and emerging diseases (George Avery), respirators for emerging diseases (Lange, Nicas, and Yassi), natural disease pandemics vs. the bioterror model (H. Cohen), what went wrong during the SARs epidemic in Canada (Bunja and McCaskell), influenza pandemics and public health readiness (Cohen et al.), occupational health vs. public health and infection control, where the boundaries are emerging during disease epidemics (Maher), and Bernadette Stringer, "Hospital Cleaners and House-keepers: The Frontline Workers in Emerging Diseases."

Finally, communication between the Centers for Disease Control, state health departments, hospitals, emergency response systems, and community physicians remains a weak link, if somewhat strengthened by the investment of significant resources over the last few years.

Following the diagnosis of a case of cutaneous anthrax in New York City in 2001, a public information hotline established in the wake of the September 11th attacks was bombarded by over 15,000 calls in a single day. Estimates are that between 50 and 200 individuals will seek medical care following an "event" for every individual actually exposed. This bald fact underscores the need for advance preparation and ready access to timely information in real time. Communication strategies are central to all preparedness efforts to date and there is still far to go.

Despite enormous sums of money being channeled to bioterrorism-related preparedness efforts in the aftermath of September 11th, this policy has not been without its share of detractors. Indeed, the Katrina experience has given legitimacy to critics who have argued — many quite consistently and reasonably — that the diversion of public health funding toward bioterrorism has undermined rather than strengthened our nation's public health system. The national smallpox vaccination program — that in the interest of self-disclosure, I should note began with the vaccination of four

physicians by my occupational medicine group at the University of Connecticut Health Center — was also criticized roundly for being hastily
prepared, founded on ambiguous evidence, and slow to address legitimate
medicolegal concerns about liability, employee benefits, and workers' compensation and disability claims for those willing to be vaccinated. Many
contributors to *Emerging Infectious Diseases* would agree with these assessments and find fault with pillars of our nation's public health system,
including the Centers for Disease Control and the U.S. Department of Health
and Human Services. The stakes are very high and certainly there is room
for open and fair-minded debate on policy matters as critical as these.

As we consider how to move forward from this point, we should bear
in mind Santayana's admonition that "Those who cannot learn from history
are doomed to repeat it." An honest appraisal of our experiences with
natural (e.g., Katrina, SARS) and manmade (September 11th) disasters will
identify existing limitations and opportunities to improve so that such
missteps can be minimized in the future. Unfortunately, history has other
lessons to teach us as well. Based on past experience, it is not unreasonable to conclude that preparedness and training are necessary but not
sufficient to ensure a rational, transparent, and well-coordinated response
to each and every public health threat. As a practicing clinician, I know
that better systems can and should be implemented to limit adverse
outcomes in patients who commit themselves to our care. At the same
time, medicine is an inexact science and a flawed art. Not all medical
errors are preventable. The analogy holds true with regards to public
health. It is easy to see chaos and discoordination in situations where
fluidity, lack of readily accessible information, and uncertainty are — if
not irreducible — at least to some degree inevitable. This is not an
argument for public health nihilism, rather it is an appeal for thoughtful
critique, remediation when possible, and charity toward those individuals
and organizations charged with a very difficult task.

There are reasons for optimism as well. In the aftermath of the
September 11th attacks, strides have been made in communications,
training, and preparedness. Few disagree with the necessity of rebuilding
our public health infrastructure, expanding the available pool of broadly
trained public health professionals, addressing inadequacies in the surge
capacity of our health care system, or providing adequate training to health
care workers to meet contemporary medical and public health threats. It
is easy to overlook that this attitudinal sea change has been driven to an
important degree by legitimate recognition that multiple threats, from
bioterrorism to global infectious disease pandemics, can no longer be
ignored. Our public health and health care systems are better prepared
than they were only a short while ago, even if we admit that we are not
nearly as prepared as we ought to be and that competing priorities

invariably generate tension and divisiveness among key, and I would add well-intentioned, stakeholders.

Historians of medicine and public health describe a process where the often negative early public response to newly emerging diseases typically, if slowly, gives way to a chronic disease model that offers more opportunities for scientifically based clinical management, reassurance, and legal protections. Such was the case with HIV/AIDS. In the early 1980s when the disease first gained notoriety, medical knowledge on basic issues such as risk factors, disease transmission, and treatment were rudimentary. Not unexpectedly, both the public and the health care community were anxious, a situation that too often was detrimental to victims of the disease. As research addressed many of these uncertainties, protective strategies, and modified medical practices emerged. Today, HIV/AIDS is a chronic disease with correspondingly less fear and anxiety attached to it.

The history of public health has also taught us that public health threats typically raise uncomfortable questions about the limits of the law, civil liberties, and ethics. As with the HIV/AIDS epidemic, contemporary public health threats — whether they be newly emerging infectious diseases, such as avian flu, or bioterrorism — have engendered their share of ethical and legal questions. To extend this analogy into the domain of bioterrorism and emerging diseases, it might be accurate to say that we are in the first phase of threat awareness. There is a great deal we do not know, or at least know with sufficient clarity to help us improve our preparedness with the level of confidence that is needed — and that will be possible in time. Put differently, the experience of those to whom we naturally turn for advice and guidance, such as military, medical, or public health leaders, is more limited than we care to acknowledge.

Emerging Infectious Diseases makes abundantly clear that the underfunding and understaffing of the nation's public health infrastructure and workforce are no longer tenable in today's world. If we are willing, we have sufficient science and technology to guide us and much can be accomplished if political, medical, and public health leaders are willing to engage collaboratively in dispassionate analysis, open debate, and fair-minded criticism.

Years from now, global public health threats will be as immediate as they are today. It is to be hoped that, as with other medical and public health threats, we will by then have accomplished the research needed to lessen current uncertainties and engaged in the constructive debate needed to integrate our nation's health care and public health systems. Perhaps then, public health threats will no longer be seen in isolation, but as a condition of the modern world that requires vision, planning, and funding to achieve the security to which we all aspire.

Michael R. Grey

Introduction

William Charney, DOH

"If disease is an expression of individual life under unfavorable conditions, then epidemics must be indicative of mass disturbances of mass life," Rudolf Verchow

"Conditions ripe for flu disaster," *Seattle Times,* February 6th, 2005

"Canada stockpiles drugs to combat global flu pandemic," *Vancouver Sun,* February 4th, 2005

"Fatal plague outbreak feared in Congo," *Seattle Times*, February 19th, 2005

"Stalking a deadly virus, battling a town's fears," *New York Times* April 17th, 2005

"Bird flu virus mutation could spread worldwide," *Seattle Post Intelligencer,* February 22nd, 2005

"Lack of health insurance in the U.S. will kill more people than Katrina," Krugman, *New York Times*, September 18th, 2005

"Bird flu threat: Think globally, prepare locally," *Seattle Times,* April 15th, 2005

Naturally occurring emerging infectious diseases pose an immense threat to populations worldwide.[1] Almost every week this topic of the threat of pandemics makes the headlines of major newspapers, including but not limited to the *New York Times,* as shown in the quotes above. Threat analysts are reporting almost weekly of the possibility and inevitability of a dangerous pathogen reaching our shores.[2] In North America, where we like to believe that we are protected by our science and technology, a dangerous ambivalence has somehow taken hold. Since 1993[3] scientific texts have been warning and then urging healthcare facilities to step up their response capabilities for the potential of a virulent, naturally occurring, airborne transmissible organism. Most of these warnings have been ignored. Former Secretary of Health and Human Services[4] Tommy Thompson said, upon being purged from the Bush administration, that what worried him most was the threat of a human flu pandemic. "This is really a huge bomb that could adversely impact on the healthcare of the world." And according to Davis, in an article in *The Nation*, despite this knowledge the Department of Health and Human Services allocated more funds for "abstinence education" than for the development of an avian flu vaccine that might save millions of lives.

This text concentrates on one vital theme: the importance of a critical analysis of existing protocols and systems to protect the healthcare community during a naturally occurring infectious disease outbreak — more appropriately called the *occupational health* outcome.

One example is a quote from Robert Webster, a respected influenza researcher, of St. Jude Hospital in Memphis: "If a pandemic happened today, hospital facilities would be overwhelmed and understaffed because many medical personnel would be afflicted with the disease."[5] Another example also cited by Davis, is that under the Democrats and the Republicans, Washington has looked the other way as local health departments have lost funding and crucial "surge capacity" has been eroded in the wake of the HMO revolution.[6]

This book is designed to be a critical analysis. We will show among other things that the bioterror template does not necessarily bleed over to the naturally occurring infection paradigm either in training models or preparation (see Chapter 7). And despite some similarities, being prepared for one does not mean we are prepared for the other. The billions that have been provided after 9/11 for the bioterror preparedness do not mean that they represent money well spent for the naturally occurring pathogen response. Confusing the two can lead to dangerous myths that can leave us unprepared. In Chapter 8 by Cohen, Gould, and Sidel it is stated that, "massive campaigns focusing on bioterrorism preparedness have had adverse health consequences and have resulted in the diversion of essential public health personnel, facilities, and other resources from urgent, real public health needs."

Occupational Health Paradigms

In the occupational health/protecting healthcare workers arena, problems still seem to abound. Hospital design parameters do not provide for enough negative pressure isolation rooms, either for patient care, triage, emergency trauma rooms, radiology, or for high risk aerosolized procedures such as bronchoscopies. A *limited* amount of isolation capacity (see Chapter 6 by Derman) will not adequately defend healthcare systems against transmissions, especially in a patient surge situation. This volume will also analyze the problems and deficiencies of healthcare workers who are not adequately trained either in respirator protection for airborne transmission or decontamination for surface removable contaminants, as well as first responder vehicles which are not designed to protect firefighters or EMT personnel from cross-transmission. Triage area ventilation systems are not controlled for transmissions. There is still controversy about types of personnel protective equipment, especially types of respirators (see, Chapters 3, 4, and 5). Nonclinical departments, such as diagnostic imaging or housekeeping have not been adequately prepared to deal with virulently infectious patients, and healthcare facilities are still not cleaning adequately to defend against pathogens[7] (see Stringer, Chapter 11). Healthcare systems are not being tested for their preparedness for naturally occurring pandemic scenarios. Regulatory agencies that set guidelines and rules sometimes do not reflect current scientific literature on isolation and respiratory protection.

Avoiding the so-called Black Death syndrome, the fourteenth-century's pandemic, is going to take putting the problems that exist today in our healthcare facilities on the radar screen and in many instances changing the business as usual criteria. Our healthcare systems are not set up to receive large populations of infectious patients, either through design of the facilities or the way healthcare is administered. Codes for mechanical systems and pressure differentials would not apply. Mixing of infectious patients with noninfectious patients would not apply. Low level and inexpensive personnel protective equipment, now supplied, would in most cases not apply, especially during clinical procedures involving aerosolization. Current respirators now considered generally acceptable for protection against infectious agents would not apply. Training of healthcare workers at present levels of readiness would not apply, as current training models would be inadequate to meet the severity of the toxicity (training levels of healthcare workers were criticized during the SARS outbreak in Canada (see Bunja and McCaskell, Chapter 1). Community buildings may have to be used, and to date most communities have not scouted or prepared community buildings for large influxes of infectious patients.

Studies of cross-infection for contagious airborne diseases (influenza, measles, TB, for example) have found that placing patients in single rooms is safer than housing them in multibed spaces, which means current hospital designs might not apply.[8] Severe Acute Respiratory Syndrome (SARS) outbreaks in Asia and Canada dramatically highlighted the shortcomings of multibed rooms for controlling or preventing infections both for patients and healthcare workers. SARS is transmitted by droplets that can be airborne over limited areas. Approximately 75% of SARS cases in Toronto resulted from exposure in hospital settings.[9] The pervasiveness in American and Canadian hospitals of multibed spaces in emergency departments and wards will severely impact infection control measures during an outbreak.[10] Quarantine models from state to state would have to be made more enforceable, while implementation models for large scale quarantines have yet to be tested.

Chapters 1 and 2 on SARS in Canada shine bright lights on the holes in the acute care responding systems and should be taken as messengers/harbingers of important information for the American healthcare community and the protection paradigms for occupational health outcomes. And despite the role that nosocomial infection transmissions have played in educating about airborne transmission to patients, protection of healthcare responders to potential infections has lagged. The classic studies of Riley[11] were very important to the comprehension of airborne transmission of tuberculosis in a healthcare setting. Charney's work developing a portable negative pressure unit to cheaply convert hospital rooms to negative pressure and air-scrubbing through HEPA filtration is another example of an occupational health response to an emerging pathogen.[12] However, the totality of occupational protection against emerging infectious disease has not appeared on the radar screen with the intensity needed to protect this population of workers.

SARS in Canada accounted for an occupational transmission rate of 43% and in Hong Kong and China accounted for an occupational exposure rate of 20%. With this particular coronavirus, mortality for healthcare workers remained relatively low due to the lower toxicity and infectious virulence of the virus, not to excellent protection standards for healthcare workers. In fact in Toronto, more money was spent hiring the Rolling Stones ($1 million) to promote tourism during the outbreak than was spent on protecting or training healthcare workers. The Canadian experience listed a number of factors that increased transmission to healthcare workers: A brief list follows:[14]

1. Lack of healthcare worker training in decontamination procedures.
2. Protocols from the relevant regulatory agencies that changed almost on an hourly basis, confusing healthcare workers and their responses.
3. Confusion as to the effectiveness of respirator selection and fit-testing protocols.
4. Lack of timely protocols for aerosol-producing clinical procedures.
5. Lack of training for first responders.

6. Questions about isolation and negative pressure, especially in triage areas.
7. Lack of timely protocols for airborne protection, especially in the early stages of the epidemic as the virus was labeled a "surface removal contaminant."
8. Following protocols but still seeing occupational transmission.

Air flow in healthcare settings is ill-prepared for containing transmissions (see Chapter 8). A substantial number of viruses, bacteria, and fungi are capable of spread via the airborne route in hospitals.[15] Among the common exanthems, the evidence in support of airborne transmission is quite strong with respect to varicella zoster and measles.[16] Rubella may also spread through the airborne route. There is a strong base of evidence of airborne transmission of respiratory syncytial virus and adenoviruses in pediatric wards.[17] Hoffman and Dixon, 1977,[18] reported on the transmission of influenza viruses in hospital settings through airborne routes; and the strongest evidence of airborne transmission of influenza is a well-documented outbreak that occurred on a commercial aircraft.[19] All types of viruses can be spread throughout hospitals by airborne transmission. Even SARs, a coronavirus, which was mistakenly considered only a "surface removable" contamination, was found to allow airborne transmission as well.[20] The Marburg virus, now occurring in parts of Africa, has an airborne component exposure.[21] There is evidence that certain enteric viruses may be transmitted through the air. Sawyer[22] reported on a case of a viral-like gastroenteritis that occurred in a Toronto, Ontario hospital in 1985 where 635 hospital personnel were affected and the investigators found no common food or water source and believed contamination was through the airborne route.

There is a looming sense that healthcare facilities would not be prepared for a surging population of victims or that cross-contamination and cross-transmissions could be prevented. SARS actually projected all the difficulties in protecting healthcare workers from a natural emerging disease. From the perspectives of building design, patient flow, air flow parameters, disinfection principles for surface removable or airborne transmissions, personal protective equipment, and most importantly, healthcare worker training, the United States and Canada are underdeveloped and unprepared according to many experts. The occupational health dynamic is often the last item on the agenda when emerging disease is discussed. In an op-ed piece in the *New York Times* written by Barack Obama and Richard Lugar entitled, "Grounding a Pandemic," there was not one word mentioned about how to protect against transmission to healthcare workers.[23] This complacency is unsafe. We are taking for granted that our healthcare systems are going to be able to deal with thousands of sick and dying people, when in fact at the current level of preparedness they will be overwhelmed and chaos is quite predictable. Just from a standpoint of healthcare worker protection technology, the national community and guideline agencies have not adopted a respiratory standard that seems acceptable to protect against airborne transmission, or provided healthcare facilities with enough acceptable respiratory protection equipment or models. Air scrubbers with HEPA or ULPA filters, that could scrub the air of viruses and bacteria and that would be an important ingredient to add protection factors in many healthcare rooms and spaces, are not currently required or used substantially.

Healthcare workers are substantially under-trained for emerging infections to level of risk. This was apparent in Canada during the SARs outbreak and is alluded to in Chapter 1 by Bunja and McCaskell. Constant cross-contamination for surface removable transmission was a problem, as well as a lack of knowledge about respirators.

Until the problems discussed in this section are admitted and addressed the healthcare worker is at increased risk, thereby putting community populations at greater risk. Risk assessment analysis stresses that all parts of the exposure whole be working intelligently together for positive outcomes. We are not there yet.

Public Health Paradigms

This book adds a chapter (Avery 7) on the public health system's ability to respond to a potential pandemic. Avery in Chapter 7 points out how little money is being allocated, despite the fact that there are new and emerging infectious diseases that will pose a global health threat to the U.S. These diseases will endanger U.S. citizens at home and abroad.[24]

- Twenty well-known diseases, including TB, malaria, and cholera have reemerged since 1973, often in more virulent and drug-resistant forms.
- At least thirty previously unknown disease agents have been identified since 1973, including HIV, Ebola, hepatitis C, and Nipah virus for which no cures are available.[25]
- Newer diseases, such as H5N1 (bird flu) and Marburg virus, have emerged that are beginning to mutate and jump from animals to humans.
- Annual infectious disease rates in the United States have nearly doubled to some 170,000 annually after reaching an historic low in 1980.[26]

- Influenza now kills some 30,000 Americans annually and epidemiologists generally agree that it is not a question of whether, but when, the next killer pandemic will occur.

Avery also reminds us in Chapter 7 that the American public health system is compromised by several deficiencies; namely, a shortage of personnel, communication problems, and time lags. He states: "since the 1960s the United States, like much of the rest of the world, has seen a decline in the ability of the public health system to address the threat of infectious disease." It has also been shown through the lens of analysis of Katrina, that public health responses have been severely compromised, from the nonfunding of levees to protect the city of New Orleans to actual response and communications between federal, state, and city responders. Avery points out that 18% of public health laboratory positions are vacant and over 40% of public health epidemiologists lack training in the field.

Many of the agencies needed to respond during a natural public health disaster have suffered in recent and past years from problems that range from cronyism to severe budget cuts. FEMA hired Mike Brown, who had little or no disaster response experience. FEMA had become known as the "turkey farm" where high level positions were filled with political appointees. The Environmental Protection Agency, needed now more than ever in New Orleans, has been crippled by cronyism. The agency has seen an exodus of experienced officials due to both Democrat and Republican administrations' refusals to enforce environmental regulations. In an interview with the British newspaper *The Independent* on September 10th, 2005, Hugh Kaufman, a senior policy analyst with the EPA, complained of severe budget cuts and inept political hacks in key positions. The Food and Drug Administration also has been accused of coziness with the drug companies and the agency's head of women's health issues resigned due to "politics" over "health" in the delay of approving Plan B, the morning after pill.

The current Bush administration's increasing focus on terrorism to the exclusion of natural disasters has been a concern for some time. A recent report by the Government Accountability Office showed that "almost 3 out of every 4 grant dollars appropriated to the Department of Homeland Security for first responders in fiscal year 2005 were for 3 primary programs that had explicit focus on terrorism."[27] More than $2 billion in grant money is available to local governments looking to improve the way they respond to terrorist attacks but only $180 million is available under the grant program for natural disasters or pandemics. The Bush administration has even proposed cutting that to $170 million even though the National

Emergency Medicine Association (NEMA) had identified a $264 million national shortfall in natural disaster funding.

Katrina, like SARS, has put a spotlight on the flaws of the public health response systems that include city, state, and federal response agencies and also healthcare delivery systems. "Confusion, Desperation Reigned at New Orleans City's Hospitals" read the headlines in the *Seattle Post Intelligencer* on September 14th, 2005. Evacuation of the infirm and sick did not take place in a timely lifesaving manner. Hospital backup generators failed as electrical grids went off line. Police communications systems failed. And even after three days, food and water supplies were not entering the city. Toxic waste issues were evident and overwhelming to underfunded agencies. And probably most important, there was no plan to evacuate, feed, and house the 130,000 residents of New Orleans who live below the poverty line, drawing a class line in the sand of our public health readiness.

Cohen and Coyle, in Chapter 10 on influenza speculate on whether the next flu pandemic can be stopped, and mention several obstacles that have to be overcome in "order for there to be any reasonable chance...." Surveillance, they mention, has to be improved to identify the earliest index cases, communication and information systems need to be upgraded, and vaccine models must become truly effective and tested, and be stored in quantities to be effective. These are only some among many obstacles. Since we live in a superaccelerated world, where germs travel as fast as supersonic jets, international cooperation, on a scale so far unmatched, would need to be developed.

National Agencies Response Paradigms

I am somewhat perplexed at the Centers for Disease Control (CDC), Health Canada, and the World Health Organization's inability to be more cognizant of the occupational health effects and protections necessary to assure healthcare workers' protections during the latest SARS outbreak. Their lack of preparedness on the occupational health front is not reassuring for the next potential pandemic. The CDC *Guidelines*[28] for SARS became a questionable model of scientific inquiry. CDC protocols during the outbreak changed on almost a daily basis (see Bunja and McCaskell Chapter 1),[29] confusing healthcare workers and creating a climate of uncertainty, especially on the issues of transmission and protection. John Lange and Giuseppe Mastrangelo (Chapter 3) show that the respiratory requirements within the SARS *Guidelines* were a serious departure from the science of respiratory protection for the protection of healthcare workers. The recommendation of a paper respirator, N95, with leakage factors of 10% at the face seal and 5% at the filter, to protect healthcare workers from an exposure that has a 15% mortality rate, where airborne transmission was not ruled out (and later ruled in),[30] does not follow any of the OSHA rules for respirator selection for serious toxicity (see Chapter 3), especially when there was no dose/response relationship known (see Nicas, Chapter 4). This nonscientific attitude within the CDC continues in the position statement on plague (see letter sent to the CDC director Gerberding from the AFL-CIO, Appendix A) where surgical masks would be allowed as the respirators of choice for healthcare workers responding to plague and that the fit-testing regulation could be waived, this despite the fact that the literature cites a 43% fatality rate using this method[31] and that the recommendation contradicts regulatory safeguards and contradicts peer review science.

Effect of Globalization and Global Warming

Globalization and the global economy have made it easier for diseases to spread from one country to the next. One can travel anywhere in the world now within 24 hours and transport pathogens. The latest dengue epidemic in El Salvador was spread from Vietnam via Cuban workers, then to nearby islands in the Caribbean, on to the South American continent, and into Central America. But developing countries are not the only ones affected. When West Nile virus appeared in the United States in 2003, health officials said 59 people in the New York City area were hospitalized. Since then federal researchers estimate about 1,400 cases have been treated. C. Everett Koop has written that we have achieved the "globalization of disease."

Bird flu is only one of the six emerging global pandemics. They are: Super TB, H5N1 (bird flu), super staph, SARS, super malaria, and HIV. HIV alone has mutated and has gone from 2 to 400 strains in only twenty years. Influenza is justifiably feared (see Chapter 10 by Cohen and Coyle). In 1918 and 1919, 40 to 50 million people (2–3% of the world's population) died.[32] Subsequent influenza pandemics occurred in 1957 and 1968.

Since December 2004, pneumonic plague has resulted in 300 suspected cases and at least 26 deaths in eastern Congo. This is the largest plague outbreak since 1920. Hong Kong flu, which swept across the Pacific Rim in 1968, reaching the United States in the same year, killed an estimated 34,000 Americans in six months. Asian flu claimed 70,000 American lives and a million worldwide. Spanish flu, which occurred in 1918, swept across the trenches in World War I and accounted for half the GI deaths. By some estimates this flu infected at least a billion people worldwide, killing 20 million. Most victims were healthy adults aged 20 to 50. There was an outbreak of Marburg virus in Angola in 2005 and this dangerous virus is transmissible from person to person.

Global warming, as verified by 48 Nobel Prize winners, is another ingredient in the rapid growth of new bacteria and viruses. As global temperatures rise, conditions improve for pathogens to emerge. Global warming has also been associated with the intensity of hurricanes hitting the Gulf Coast this year, as the warming Gulf of Mexico feeds the ferocity of the hurricanes as they travel over water.

Class

The response to Katrina displayed the class bias that exists in the United States. Mexico has already warned that the next flu pandemic will affect the poorer countries disproportionately, and affect the global response capacity. It has long been argued that first world riches were not creating a more level playing field for the third world, that the rising tides were not raising all ships. Warnings have been issued for years by epidemiologists, demographers, and political scientists that if more was not done to bridge the gap between rich and poor countries that the imbalances would affect the "global health." Even the report issued by the National Intelligence Council for the Central Intelligence Agency in January 2000 warned:

> new and reemerging infectious diseases will pose a rising global health threat and will complicate U.S. global security over the next 20 years. These diseases will endanger U.S. citizens at home and abroad and exacerbate social and political instability in key countries and regions in which the United States has significant interests.[33]

This report goes on to say,

> development of an effective global surveillance and response system probably is at least a decade away owing to inadequate coordination and funding at the international level and lack of capacity and funds in many developing countries. The gap between the rich and poorer countries in the availability and quality of health care is widening ... compromising response.[34]

Class is one of the most misunderstood and denied causes for world-wide infections and also response capabilities. Today, with globalization and the rapid ways in which microbes can travel, the planet is shrinking.

What happens in Bangladesh could affect New York City within hours. A class approach to finding solutions to poverty in the third world is essential to protect the first world. Many diseases, such as TB, are poverty driven. Antibiotic resistance in one country could lead to an epidemic in another. Poverty and class divergence are the petri dishes for emerging diseases. A long term strategy needs to be developed to combat poverty and the class differences in the world and in nation-states. This would be the best strategy to protecting the citizens of the world against pandemics.

Are Hospitals or Pubic Health Systems Ready? A Conclusion

There seems to be accumulating evidence that our healthcare institutions will not be prepared for a pandemic scenario, and that our public health agencies, underfunded and understaffed will have multiple problems responding to a pandemic scenario. In New Orleans, during Katrina, hundreds of patients died in hospitals and nursing homes despite some heroic efforts by healthcare workers. There were failures in both healthcare and public health responses on all levels. Katrina is surely the canary, as was SARs, to our need to invest and solidify all aspects of preparedness, both for hospitals and public health, if we are to protect the public health during a surge pandemic.

As many authors point out (e.g., Avery, Cohen, Gould, Frank), necessary funding for public health has been diverted to terrorism disproportionate to risk. This is the political disconnect that Katrina has made us observe. As Frank points out 3,400 people died on September 11th, but over 5,200 people a day die from natural specific diseases that are preventable.

Unless we integrate this information into the body politic, insisting that the domestic pubic health problems are funded and repaired, the United States is at great risk. The breakdown of the levee systems in New Orleans was predicted based on integrated information, and the breakdown of the public health system was predictable based on the defunding of the systems. Tommy Thompson's Health and Human Resources budget for public health, was underfunded year after year, and cut by large percentages by Congress during his period as health secretary. In America you get what you pay for. If we militarize space when one third of children in this country go to bed hungry we have made a disconnect. If we spend $2 to 7 billion a month on a questionable foreign war and occupation when there are millions of Americans without health insurance, we have made a disconnect. If spending on abstinence training outpaces spending on public health, we have made another disconnect that leaves us underserved and quite vulnerable during times of either disasters or pandemics. Since 2001 the country has been waiting for a finished Pandemic Plan and at this writing it is still not completed, let alone tested, and components verified for effectiveness.

In late 2005 Bush appointed Stewart Simonson as the "Bird Flu Czar." Simonson has turned up as number seven on the *New Republic*'s list of 15 Bush administration hacks. Simonson is a lawyer from Amtrak with, according to sources, very little public health experience, and according to the *Washington Post*,[35] a spotty record of nonsuccess at Project Bioshield, a program designed to speed the manufacture of crucial vaccines and antidotes. One blogger, the author of Lonewacko, said acerbically that Simonson is obviously qualified if we have an outbreak of litigation.

Russian roulette belongs in the gambling casinos, not in the public health arena. There is too much at stake.

Contents

TWO REPORTS FROM THE FRONT LINE: TWO NURSES UNION OFFICIALS REPORT ON SAFETY ISSUES DURING THE SARS CRISIS IN CANADA

Chapter 1

Presentation to the Commission to Investigate the Introduction and Spread of Severe Acute Respiratory Syndrome (SARS)

Erna Bujna and Lisa McCaskell

CONTENTS

Health and safety are important in any workplace. But in a healthcare environment, they are doubly important. If workers are not protected from health and safety hazards, patients and the public are not protected either. It's that simple. If workers are not told how to protect themselves, they cannot do so. If unions are left out of the process, they cannot play a role in helping our members get the information they need. Healthcare workers and their patients died as a result of the SARS crisis. (March 2003, SARS I.)

The Ontario Public Services Employees Union (OPSEU) represents 113,000 members, including more than 28,000 healthcare workers. Of these, approximately 15,000 work in hospitals, most of them members of regulated health professions, such as respiratory therapists, x-ray technologists, laboratory technologists, physiotherapists, occupational therapists, diagnostic imaging technologists, speech therapists, and many others. OPSEU members are also cleaners, office and clerical workers, and other nonregulated healthcare workers.

The Ontario Nurses Association (ONA) is the union that represents 48,000 registered nurses and allied health professionals working in a variety of settings across Ontario. ONA represents about 21,500 members in the regional municipalities of Durham, York, Peel, Halton, and the City of Toronto, the region most directly affected by SARS.

The Workplace Safety and Insurance Board has received 160 claims for compensation from healthcare workers who actually exhibited symptoms of SARS and another 98 from healthcare workers who were exposed, but did not develop symptoms.

Both OPSEU and ONA members were either directly or indirectly affected by SARS. Both unions had many members who contracted SARS. Two ONA members died as a result of workplace SARS exposure.

In our positions, both authors educate and respond to our members' health and safety concerns. During SARS, our workload increased significantly. We were inundated with calls for help from our members. We were asked to review the directives and to provide advice and guidance. We did this in hazard alerts and advice on the union websites and in correspondence with union representatives. Representatives of both OPSEU and ONA participated in biweekly teleconferences with the Ontario Hospital Association, the Ministry of Health, and the Ministry of Labour.

During the crisis we heard many frightening stories from staff and from members who risked their own safety and health in order to care for their patients. Many of those workers believe that more could have been done to protect them.

SARS, along with other virulent diseases, *is* a health and safety matter, one that we believe was handled poorly by many employers, by the Ministry of Health, and by the Ministry of Labour. Every worker in the province of Ontario has a right to a safe workplace.

Although both workplace parties have an obligation to ensure that work is done safely, employers have the greatest responsibility under the Occupational Health and Safety Act (OHSA). Section 25 of the Act states that employers must alert workers to workplace hazards and that they must take all precautions reasonable in the circumstances to protect workers from those hazards.

The Ministry of Labour enforces the OHSA. Under Section 54 of the Act, the Ministry has the power to enter any workplace at any time without warrant or notice. If the Ministry's inspector finds that an employer is violating the OHSA, the inspector can issue orders for the protection of workers and the employer can be prosecuted.

For years there have been reports about the high rates of injury and illness among healthcare workers. Despite these reports, we have witnessed a lack of enforcement and the unwillingness of the Ministry of Labour to exercise its powers under the OHSA to deal with health and safety problems within the healthcare industry. If SARS had suddenly plagued private industry, we believe the Ministry of Labour's response would have been much different. In our experience, at least since the mid-1990s, the healthcare sector has been a low enforcement priority of the Ministry. Why is this? Perhaps it is because healthcare workers are caregivers — their focus is on patient care not on protecting or caring for themselves. In the past, they have rarely refused unsafe work or even complained about unsafe conditions. They work in a building (the hospital), which is thought to be the safest place to take someone who is ill or injured, so why wouldn't it also be the safest place to work. It should be! But it isn't. In industry, for example, construction workers know their work is really dangerous. Health and safety issues are real to them and are something they know they must be concerned about. Why is it that in health care, the industry with one of the highest rates of injury and illness, workers, employers, and the public do not have this same awareness?

Why Are Hospitals Vectors of Contagion, Not Havens of Health?

During SARS, healthcare workers realized again that their work is really dangerous, and they realized that their hospitals wouldn't protect them. The outbreak highlighted for ONA and OPSEU again how critical the Occupational Health and Safety Act is in providing a safe environment for our members.

We will focus here on two of three main areas: the directives, and the Occupational Health and Safety Act, in particular, the roles of the joint health and safety committees and the Ministry of Labour. The third area, infection control policies and procedures, which we only touch on, is addressed in greater detail at www.ontarionurses.ca.gov and www.ona.org.[1] Our intent is to highlight areas where the directives, poor health and safety practices, and lack of enforcement may have contributed to the spread of SARS.

Please note that our written submission contains more detail and examples than we have included here. Also, we have not included our preliminary recommendations today — they are all included in the written submission that you have requested from us.

The Directives

The first document addressed to the medical community about SARS that we have found is a March 18, 2001 "Letter to All Physicians in Ontario." This letter contained a warning about the possibility of SARS coming to Canada. Most importantly, the letter set out infection control measures and advice to healthcare workers about how to protect themselves. As far as we know, none of the information about protection of healthcare workers was communicated to workers in any healthcare facility. Why would critical information pertaining to the protection of healthcare workers and infection control practices be sent only to physicians?

Nine days later, on March 27th, the first hospital directive was issued to all acute care hospitals in the province. This first directive required staff only in the emergency departments of the Greater Toronto Area (GTA) and Simcoe County Hospitals to wear N95 masks and other protective gear. Workers in the rest of the hospital were not required to take any special precautions to protect themselves. This distinction between what protection was recommended for which groups of workers in the same facilities arose again and again throughout the crisis. Both unions were constantly trying to establish which workers in which areas were required to wear what personal protective equipment and why.

Directives subsequent to the March 27th document came fast and furious, targeting acute care facilities most often, but also giving direction to long-term care facilities, community care access centers, home care workers, and physicians' offices.

We have grouped the problems with the directives into seven main areas.

Lack of Transparency

For convenience, we are treating the Provincial Operations Centre (POC) as the source of all the directives, although it was never clear whether final authority for the directives lay with the POC or the Ministry of Health. The occasional document was even issued directly by the Ministry of Public Safety and Security. The directives were always posted on a Ministry of Health website using Ministry letterhead, but they were signed by the Commissioner of Public Security and the Commissioner of Public Health and Chief Medical Officer of Health. The relationship between the Ministry and the POC was not made clear to us.

In the early days of the crisis, both unions had difficulty getting access to the directives at all. Although OPSEU and ONA were involved in teleconferences discussing the directives, it was not until April 7th, almost two weeks after the first directive was released, that both unions gained access to what was called the Ministry of Health "dark site." This is where the directives were posted.

Until then, both unions had relied on contacts within the Ontario Hospital Association (OHA) or from union members to provide them with the directives that were governing the work and the safety needs of healthcare workers. Even when both unions were issued the password to access the Health Ministry site, OPSEU and ONA were warned in writing that "the site is not intended for the general public and is password protected to provide access to healthcare providers/associations only."

To date, OPSEU and ONA are not sure who exactly was working at the Provincial Operations Centre, how they were chosen, or what their roles were. This question was raised numerous times at the OHA tele-conferences. To date both unions still do not know. Most importantly, ONA and OPSEU did not know the background and expertise of the people who were drafting the directives that directed the daily work of healthcare workers.

The OHA teleconferences were often frustrating for union representatives who did not obtain answers to health and safety questions in a timely fashion either from representatives from the ministries or the OHA. At the end of April, teleconferences were reduced to once per week. OPSEU and ONA continued to press the OHA's representative (Vice-President, Human Resources, Management Services) and the Ministry of Health representatives to answer questions.

By the end of SARS I (June, 2003), both unions still had no answers to some of the basic questions, such as an explanation of the POC process, and never really knew if OPSEU and ONA's concerns were heard by the POC. If a change was made to a directive that appeared to address one of the unions' concerns, the unions learned of it only when reviewing the new directive.

Some time in June 2003, the Ministry began to post the directives on its public site. To date, ONA and OPSEU do not understand why the content of the directives was considered to be top secret and not a public document until June. When the directives were changed — either strengthened or relaxed — because there was no rationale offered, and because OPSEU and ONA did not know the process being used to determine the changes, the unions' confidence in the directives was diminished. At the teleconference meetings, both unions repeatedly sought clarification and explanations for the changes, especially when protective measures for workers were reduced. Our union representatives requested that their concerns be taken back to the POC for explanations. It was OPSEU and ONA's position that the directives should always err on the side of safety. Neither union received answers to questions about relaxing the directives. For example, if workers throughout a facility are required to wear certain personal protective equipment one day, and the next day only workers in the emergency department are required to wear it, and there is no explanation or rationale offered, it is difficult to be confident that every precaution is being taken to protect the health of our members. This lack of transparency led many of the members to speculate and raise concerns to both unions as to whether political interference, because of feared loss of tourism or shortages of equipment, had led to the changes; or whether in fact there were good epidemiological reasons to explain the decisions.

In summary, the two unions were not privy to the makeup and processes of the Provincial Operations Centre; the creation of the directives took place behind closed doors; and union input, questions, and suggestions about the directives were seldom recognized. Consequently, neither union could be confident that the directives would adequately protect the health and safety of our members.

Directives Were Incomplete

There were notable gaps in the directives that in the opinion of both unions and individual workers could lead to absurd and possibly dangerous results. In some instances, the directives were just confusing. Some workers who had been exposed to SARS were put on what was termed "working quarantine" and were allowed to work, although they were confined to their homes during time off. No official attempts were made to accommodate pregnant workers. And during the first month of the crisis, directives offered remarkably little detail to assist employers and workers to implement them.

The following are examples of gaps in the directives:

1. Transportation problems for health care workers on "working quarantine": Healthcare workers on working quarantine were still using public transportation. In order to prevent the possibility of further exposure of the public, participants of the OHA teleconferences asked that the Ministry of Health address this in its directives.
2. Screeners: Directives were not clear as to the protective equipment that screeners should wear in any facility, creating much confusion and anxiety.
3. Pregnant workers: There was no information in any of the directives to address concerns raised by pregnant workers about the health effects of wearing the N95 respirators, or about exposure to Ribavirin, one of the drugs being used to treat SARS. Workers, pregnant or not, agreed that wearing the N95 masks for any length of time caused increased fatigue, probably because of decreased oxygen intake (the mask restricts breathing) and increased carbon dioxide levels (the mask restricts successful exhalation because as you exhale, air containing carbon dioxide is trapped in the mask and then is breathed in again). For pregnant workers, breathing is already affected by the pressure of the growing fetus on the diaphragm. The interference in their breathing caused by the mask led to extreme fatigue.

The following examples were communicated to OPSEU: Pregnant workers, in some cases, asked to be accommodated into work areas where they would not be required to wear respirators for an entire shift. Some employers may have accommodated workers; some refused. In one case, a manager suggested that a pregnant worker "try a surgical mask" and return to work. In another case, pregnant lab workers were advised they did not need to wear respirators if they had no patient contact. This direction was later rescinded.

The other issue that worried pregnant workers was use of the drug Ribavirin, which is known to cause birth defects and is contraindicated in pregnancy. Workers were concerned that if they became ill with SARS, they would not be offered Ribavirin. They were also worried about mixing and administering the drug. This concern affected all healthcare workers who either were or could become pregnant.

Detailed Directives at Last

On April 20th, almost a month into the crisis, detailed direction was given for the first time on matters such as air supply to SARS units and patient rooms and procedures such as applying and removing personal protective equipment, minimizing patient contact during patient care activities, and cleaning. Four days later, a revised directive was released that contained even more detail.

These directives offered the first concrete evidence that the POC had begun to recognize that employers, supervisors, and workers did not understand how to implement previous directives. A good example of the kind of detail finally provided was the directive on housekeeping and cleaning measures. Until these directives were issued, there had been no direction to ensure that adequate cleaning was being performed to protect patients and workers from infection. Some of the details in these directives addressed what is not known about SARS, such as how long the virus lives on hard surfaces. However, some details are standard cleaning routines that should have been applied when dealing with any droplet-borne infectious illness.

The First Directives for High Risk Procedures

Between April 15th and 21st, nine health care workers at Sunnybrook and Women's Hospital were diagnosed with SARS following exposure to a SARS patient during a long and complex medical intervention. About a week later, the Provincial Operations Centre released directives to address the exposures that may take place during procedures that can produce airborne respiratory secretions carrying SARS. The U.S. Centers for Disease Control published its first SARS-related document concerning aerosol-generating procedures on patients March 20th, more than a month earlier.

Importantly, SARS is primarily a respiratory infection, often requiring procedures that generate airborne respiratory secretions. Why were these directives issued more than a month after the SARS emergency was declared and after nine healthcare workers were infected during such a procedure?

Late Directives for Fit Testing and Respirator Program

The *Regulation for Health Care and Residential Facilities* (under the OHSA) at Section 10(1),(2), mandates that:

> A worker who is required by his or her employer or by this Regulation to wear or use any protective clothing, equipment or device shall be instructed and trained in its care, use and limitations before wearing or using it for the first time and at regular intervals thereafter and the worker shall participate in such instruction and training.

It also states: "Personal protective equipment that is to be provided, worn or used shall be a proper fit." This was the law in the healthcare sector for 10 years before the SARS crisis. It appears that some hospitals in Ontario did not apply this part of the law or that some employers and supervisors simply did not know it was the law. We know of only one hospital with a respiratory protection program before SARS. And throughout the SARS crisis, we heard of many cases where hospitals appeared either to ignore the directives on fit testing, or to be unaware of them.

One example of the confusion about fit testing is found in a June 2003 memo from the Director, Infection Prevention and Control at the University Health Network in Toronto, which states:

"Canadian regulations have never required fit-testing in the healthcare setting." In the opinion of the unions, this statement contradicts the requirement in the health care regulation that requires employers to ensure that "Personal protective equipment that is to be provided, worn or used shall be a proper fit."

The Occupational Health and Safety Act requires an employer, when appointing a supervisor, to appoint a competent person for the purposes of the Act. The section of the memo just cited caused both unions to question the state of institutional knowledge of the Act and the ability to apply those requirements in respect to the appointment of competent supervisors.

The Ministry of Labour's role is to enforce regulations under the Act. In the case of respirators, the Ministry uses as its enforceable standard, the 2002 Canadian Standard Association's Z94.4 "Standard, Selection, Use and Care of Respirators" which requires all Canadian workers to pass a fit test before wearing a respirator. Until the SARS crisis, neither union could find any evidence of Ministry of Labour attempts to proactively ensure compliance with this regulation in the healthcare sector.

The lack of preexisting respiratory programs may have placed workers' health at risk when the crisis hit. Respirator programs provide guidance on issues such as: who assists with fit testing; where respirators are obtained; the life of masks; how to determine if they are soiled or damaged; donning and doffing; maintenance and storage; and what to do if a properly fitting mask cannot be found.

The March 29th directive to acute care hospitals noted the need for respirator fit testing, but didn't say how to achieve it, who could assist, where respirators were to be found, the life of masks, how to determine if they are soiled or damaged, how to store, how to put on and take off, or what to do if properly fitting masks cannot be found for workers.

Directives two days later (March 31st) two GTA long-term care facilities and community care access centers repeated the requirement for fit testing. *That is the last mention of fit testing until a May 2nd communiqué* that listed mask suppliers who also provided fit-testing services. It did not emphasize fit testing, nor requirements in the health care regulations and the CSA Standard; it simply said, "Studies document that proper fit-testing enhances the effectiveness of masks. Through fit-testing, employees can learn which type of mask best fits their facial features."

Although the Health Ministry began to stress fit testing in May, it was much later before workplaces started to implement it. Finally, fit testing began, sporadically, due to union complaints and a nurse's June 6th work refusal. The Ministry of Labour ordered that nurses be fit tested before being required to work in a workplace that required respiratory protection. Later in June, 840 workers at Mount Sinai were fit tested. Yet OPSEU workers at Lakeridge Hospital were *still* being ordered into high-risk areas in June without being fit tested. Some staff were told they could work in high-risk areas without a fit test if they didn't move their heads. At Baycrest, a June 12th memo to cost center managers said that fit testing would begin "in the near future." Bridgepoint trained their fit-testing trainers on July 3rd and planned to complete phase 1 of their fit-testing program July 31st.

Directives Did Not Address Workers Other than Nurses and Doctors, Nor Other Potentially Vulnerable Workplaces

While nurses reported the directives to be vague, confusing, contradictory, and nonspecific, the directives at least acknowledged their work. In contrast, many other healthcare workers found nothing at all in the directives to guide them in their specialized work. This was especially true for staff doing critical diagnostic and treatment functions; and those in frontline clerical jobs in emergency, admitting, and critical care.

Not until April 20th and 24th were detailed directives released on how to safely enter and exit a SARS patient's room. There was even direction on safely removing specimens from a patient room. These directives gave clearer direction on cleaning equipment inside patient rooms. Other than saying that every effort should be taken to avoid sharing equipment, and that disinfection protocols for shared equipment had to be written by infection control, no other guidance was given.

Consider, for example, an x-ray technologist required at times to perform x-rays with a portable x-ray machine that is moved from one patient room to the next. Chest x-rays are one of the critical tools used to diagnose SARS. Some patients had to have daily chest x-rays. Portable x-ray equipment was considered safer, since suspect and probable SARS patients would not have to be transported through the hospital to the x-ray department. But how were technologists to ensure their machines did not become contaminated and carry infections from one area to the next?

The machine should be disinfected after each use, but there was no directive to explain how this should be done, or if indeed portable x-rays were safer at all. Nor was there any consideration, with cleaning protocols added to already heavy workloads, what this meant for productivity and workers' health.

On the other hand, if suspect and probable SARS patients are brought to the x-ray department, how will the department be kept uncontaminated and safe for other patients and workers? The cleaning protocols for patient rooms in the April 24th directive are elaborate and time consuming. What is to be done in other areas of the hospital where patients travel for tests? One assumes that all hard surfaces that patients could touch or cough or breath on during their time in another department should be disinfected, but the directives were silent.

Another critical and much ignored area was respiratory therapy. These workers, known as RTs work in different areas of the hospital performing diagnostic tests and treat patients with respiratory problems in a variety of areas of the hospital. Their work includes intubating (inserting an airway) for patients who cannot breathe for themselves, maintaining mechanical ventilation for these patients, suctioning respiratory secretions, taking special blood samples, and assisting with cardiopulmonary resuscitation.

Only after nine healthcare workers had contracted SARS following a prolonged attempt to intubate a patient at Sunnybrook Hospital, was a directive finally released on May 1st that gave guidance for "High-Risk Procedures in Critical Care Areas during a SARS Outbreak."

Laboratory workers were also ignored in the POC directives. OPSEU could find no mention of any special precautions recommended to laboratory technologists when working with blood, sputum, or other samples from probable or suspect SARS cases. Also omitted from the hospital directives were clerical workers in emergency and admitting departments and throughout hospitals. At least five OPSEU clerical workers contracted SARS in the workplace. One of the first indications OPSEU had of this problem with the directives was the April 1st directive to all acute care hospitals which replaced two previous directives (March 27th and 29th). On March 29th *all staff in any part of the hospital* in the GTA and Simcoe County acute care hospitals, were required to wear N95 masks. On April 1st, a new directive to all acute care hospitals in the province required staff to wear N95 masks when caring for or entering the room of a SARS patient, and when in *direct* contact with patients in intensive and critical care units or emergency departments.

The term *direct contact* was never defined. At first glance that would appear to remove the requirement for clerical workers in critical care areas to wear respirators. However, those workers are often a meter or closer to patients as they take information and assist them in a variety of ways. This directive added to the confusion.

Employers Interpreted Directives Differently

Throughout the healthcare sector employers interpreted the directives and communicated how they were to be implemented to their employees. While this process is understandable, in many cases that we became aware of it was not acceptable. At times both unions had concerns that certain interpretations of the directives may have placed the health and safety of our members at greater risk.

The Provincial Operations Centre was aware of these problems and issued at least two notices on June 3rd and 7th, advising hospitals that compliance with the directives was mandatory and that they were not to be breached or modified. We are unaware of any other actions taken by the Provincial Operations Centre to address this problem.

Interpretations Placed on Directives by Individual Facilities

Humber River Regional Hospital: On Friday, March 28th, OPSEU issued a hazard alert reflecting the March 27th POC directive to all acute care hospitals. In it, OPSEU quoted the directive, stating among other things: "All staff in GTA and Simcoe County hospital emergency departments and clinics to wear protective clothing (gloves, gown, eye protection and mask N95 or equivalent)."

Later that same day, OPSEU received a copy of *SARS Update #3* distributed by the director of employee and labour relations at the Humber.

It stated that: "Provincial officials have advised us that N95 masks are now required *only* for staff and physicians involved in the care of patients in isolation. For all others, including clinics and emergency department, surgical masks are acceptable protection."

When OPSEU followed up with this director to find the source of the information that resulted in this contradiction of existing provincial directives, she said the direction had come from a meeting of something called the west cluster management group that was associated with the "emergency management office." It was unclear whether this group was connected to the Provincial Operations Centre or who the group members were. The director could provide no name or phone number. We assume that the hospital update was later changed to reflect the POC directives, although we received no formal notification of that.

- *St. Michaels Hospital:* A June 5th e-mail stated that it was not necessary for any staff and physicians, other than those identified in the same e-mail, to be fit tested, despite the fact that the May 31st directives indicated otherwise.
- *Mount Sinai:* On June 2nd and 3rd, staff in labor and delivery asked management to allow them to wear personal protective equipment as indicated in the May 31st directive. Management told them that there was minimal risk and therefore personal protective equipment was not required. Nurses who ignored management and persisted in wearing their personal protective equipment as required by the directives, reported being laughed at. The day following their request, a medical student on their unit went home exhibiting classic SARS symptoms following an earlier exposure. The labor and delivery nurses felt that they, their families, and the public were exposed unnecessarily to SARS as a result.

Directives Were Confusing, Changing Rapidly, the Changes Were Neither Highlighted nor Explained

It was our experience that the directives were often revised in substantive ways with no explanation or warning. The Ministry of Public Safety and Security recognized this in an April 3rd letter to Ontario healthcare facilities, which attempted to reassure the parties that the changes were based on "updated, evolving information." Despite POC recognition that frequent changes were of concern to the hospitals, there seemed to be no attempt to broadcast to all stakeholders when new directives were released. The changes to the directives, made from one day to the next, were not highlighted or communicated in any way, at least not to the unions representing healthcare workers. Both authors were forced to regularly check the Ministry of Health "dark site" to see if new directives had been posted, and then to go over them in excruciating detail to try to understand what changes had been made and to speculate as to the reasons for those changes.

The following are a few examples of rapidly changing directives: On March 29th the Ministry of Health issued directives that clearly addressed precautions for all hospital staff. It stated among other things: "For *all staff* when in any part of the hospital: Use frequent hand washing techniques. Use an N95 (or equivalent) mask (ensure mask is fit-tested)." It also clearly outlined additional precautions for staff who visit patient care units, for staff having direct patient contact, information about the reuse of masks and gowns and when they must be disposed of and replaced. It also provided direction about infection control methods to follow after each patient contact.

On April 1st and 3rd new directives were issued to all Ontario acute care hospitals that reduced the precautions outlined in the March 29th directive. The new directives stated: "However, the routine use of gowns, gloves and masks is *not required* provided the patient is not in respiratory isolation." (This directive was developed just after we learned at the OHA meeting on April 1st that the masks were in short supply.) ONA and OPSEU wondered whether employer concerns about not being able to adequately supply masks to all staff may have influenced the Provincial Operations Centre to change its directives.

On May 13th, a "new normal" directive was issued. It was confusing. This directive only briefly referenced the Occupational Health and Safety Act, and stated that fit testing should be initiated immediately; however, it appeared to identify only high-risk areas as requiring fit testing.

On May 31st, the POC released a directive that appeared to offer better protection to workers. However, we believe that the content of this directive should have been issued immediately at the start of SARS II. Both unions question why this important directive was not issued earlier.

On June 16th, a new directive for acute care facilities reduced the number of areas where health care workers were required to wear personal protective equipment. (This followed the first ever SARS health-care worker work refusal.) Both unions questioned whether the relax-ation of the May 31st directive was an attempt by the Provincial Operations Centre to avoid further work refusals about personal protec-tive equipment.

Concerns Other than Worker Health and Safety and Public Safety

At times, decisions were made that raised questions as to whether concerns other than worker health and safety and public safety may have influenced the content of the directives. The following is one example that raised suspicions at OPSEU and ONA that the directives and approaches to SARS were not always based entirely on scientific and epidemiological evidence. (www.ontarionurses.ca.gov).

Sunnybrook and Women's Hospital

A June 4th e-mail from the president and CEO stated that the hospital had an issue with "the appropriate use of full droplet precautions where necessary as opposed to a blanket application of this directive in every area of the hospital." The memo went on to say:

> We have consulted GTA teaching hospitals and they agree that the directives need further interpretation. To try and correct this situation, we have sent our very own [doctors] to work with the Ministry of Health and other infection control practitioners today to revise these directives. The group should be finished their work either today or tomorrow and we expect to have new directives relatively soon.

Both unions wondered how Sunnybrook could expect in advance that their input would result in a change to the directives. Additionally, both unions query what scientific evidence Sunnybrook had that they considered to be superior to evidence previously relied on by the Provincial Operations Centre scientists.

Our recommendations are the result of the experience of our members in dealing with this crisis, and our combined experience as health and safety specialists for our unions. Again, it must be emphasized that worker health and safety is paramount. When the system puts workers at risk, it also puts patients and the general public at risk.

Occupational Health and Safety Act

This section will discuss the Occupational Health and Safety Act, focusing on the roles of the joint health and safety committees and the Ministry of Labour. During the SARS crisis both ONA and OPSEU provided advice to members about the Occupational Health and Safety Act. The Act sets out the duties of employers and supervisors and the rights and obligations of workers. It also establishes the role of the joint health and safety committee and the powers of the Ministry of Labour. In addition we gave advice to our members about the Regulation for Health Care and Residential Facilities. That regulation governs workers and employers dealing with hazards specific to the healthcare sector. For long periods in many workplaces, and for the entire crisis period in others, it appeared to both unions as if the Occupational Health and Safety Act did not exist — or at the very least, it was as if it did not apply when the workplace hazard was an infectious illness.

The internal responsibility system, although never mentioned in the Occupational Health and Safety Act, is a cornerstone of the health and safety system contemplated by the Act. In theory, all of the parties' rights, duties, and obligations combine to create a system that will allow them to resolve health and safety concerns in the best interests of all. Some believe the internal responsibility system is based on the notion that the workplace parties have equal rights and responsibilities; and that most health and safety problems can be successfully addressed because it is in the interests of the employer and the workers to have a safe and healthy workplace. However, this approach seems to ignore the reality that workers and employer do not have equal power and that it is the employer who controls the workplace. Both ONA and OPSEU have a great deal of experience with workplaces in the healthcare sector where the internal responsibility system simply does not work.

During SARS both unions were aware of many instances where there appeared to be violations of the OHSA and the Health Care Regulation. We want to give you some examples of possible violations.

- Some employers and supervisors failed to provide sufficient, proper, or any personal protective equipment to workers.
- Some supervisors did not appear to understand their responsibilities to ensure that workers' health and safety concerns were addressed.
- Some employers gave little or no instruction to affected healthcare workers, especially those whose concerns were not addressed by the directives.
- Some employers refused to allow joint health and safety committee meetings to address the SARS crisis.
- Both unions received reports that employers had not reported critical injuries or occupational illnesses to the Ministry of Labour, to the joint health and safety committee, and to the trade union.
- Neither union is aware that any employer had introduced and implemented a respiratory protection program prior to the SARS crisis as required by legislation.

- Both unions received reports of employers who had not developed measures and procedures to ensure that the health and safety of pregnant workers was protected. The Health Care Regulation requires them to develop special measures to protect pregnant workers.

The Role of Joint Health and Safety Committees

It is the experience of both ONA and OPSEU that prior to the SARS outbreak the health and safety systems in many hospitals were weak and ineffective. Both unions have received reports about the following kinds of problems:

- Joint health and safety committees met infrequently or not at all
- Health and safety issues were rarely resolved by the committees
- Workplace inspections did not take place
- Legislated training was not up-to-date and
- Workplace injuries and illnesses were not reported to either the joint health and safety committee or the Ministry of Labour as required by the OHSA.

When the SARS crisis occurred, members reported that their employers took the position that there was no role for the joint health and safety committee. ONA and OPSEU quickly took the position that these committees should meet on an emergency basis to address SARS-related health and safety concerns. Although we regularly gave this advice to local union leaders, very few of them were successful in getting the committees to meet. Even when the joint health and safety committees did meet, these meetings were often ineffective. Additionally, both unions raised the issue of the lack of joint health and safety committee involvement at the OHA teleconferences. Still very few committees met regularly.

The following are examples of these problems.

1. On March 26th an ONA labor relations officer for the Scarborough Hospital reported that the union had requested the employer to cooperate and hold emergency meetings of the joint health and safety committee. OPSEU's local president was making the same request. It wasn't until April 1st that the Scarborough Hospital finally agreed to hold a committee meeting. However, the ONA labor relations officer reports that the first full joint health and safety committee meeting did not actually take place until April 16th. The hospital was meeting daily with union leaders but did

not want to involve the committee. When regular committee meetings finally began in April, OPSEU members reported that a number of issues were dealt with successfully.

2. At North York General, workers reported numerous health and safety concerns that indicated the internal responsibility system was not working. The ONA bargaining unit president called the Ministry of Labour for assistance with various unresolved health and safety issues and was told that these were internal matters and not a violation of the Act. No help was forthcoming from the ministry.

3. At the Toronto Rehabilitation Institute, ONA received a report that despite requests to meet, the employer refused to acknowledge the need to have a joint health and safety committee meeting.

4. On June 10th, 2003, after a suspected outbreak of SARS originating in the Lakeridge Dialysis Unit, requests by OPSEU members for a joint health and safety committee meeting were denied. The employer said it did not think a committee meeting was necessary, although it did agree to meet with local union presidents. When the union advised the employer it would consult with the Ministry of Labour about this issue, the employer relented and agreed to allow the committee to meet.

5. Toronto Hospital Corporation (part of the University Health Network), North York General, St. Michael's Hospital, Sunnybrook, and Women's College Health Sciences Centre: All these institutions had to be ordered by the Ministry of Labour to consult with the joint health and safety committee on the employer's fit-testing compliance plan.

6. At the University Health Network (UHN) ONA learned that meetings with the joint health and safety committee were problematic as the employer did not even have an employer cochair. ONA learned in April that UHN had canceled meetings of the committee.

7. At Princess Margaret Hospital, where there was a recommended moratorium on meetings, both cochairs agreed to cancel the April meeting of the joint health and safety committee.

Both unions believe that if the hospital sector's health and safety system had been functioning properly, with safety-conscious and responsive employers, supervisors who were competent as required under the OHSA, and active joint health and safety committees made up of well-trained members, a number of problems could have been avoided and perhaps fewer workers would have become ill with SARS. It is our position that as soon as the SARS crisis was recognized, all employers should have acted aggressively to ensure that training, appropriate equipment, and supervision were in place. Joint committees should have been holding emergency meetings to discuss existing infection control measures to protect workers, and to discuss and consider the directives coming from the Provincial Operations Centre. It would have been useful for the joint committees to meet collaboratively with those in charge of infection control to ensure that the directives were being interpreted in a manner that was appropriate for existing conditions in their own facility.

Effective joint health and safety committees would have been able to quickly assess where the risks of exposure to SARS were greatest and would have worked to ensure that workers understood the directives and could implement them. Effective committees would have known, or could have assessed, the existing knowledge base in different groups of staff, taking into consideration previous training, education, and languages spoken in order to ensure that the measures in the directives were being communicated appropriately and adequately to staff in every department.

Effective joint health and safety committees could have increased their inspection frequency and participated in ensuring that all workers were properly using personal protective equipment, and properly applying safe measures and procedures in their units.

In most cases, this ideal scenario did not take place. Much of the time of both unions was spent offering basic education to members and joint health and safety committee members about their rights under the Occupational Health and Safety Act and the employers' obligation to protect workers' health and safety. Both unions were just trying to get the joint health and safety committees to meet.

The Role of the Ministry of Labour

Both unions also had many issues about the ministry's involvement during the SARS crisis. It is the role of the Ministry of Labour to enforce the OHSA and its regulations. It appeared to both unions that there was a deliberate attempt on the part of the ministry to curtail the enforcement activities of its inspectors from the very beginning of the crisis. OPSEU received a draft protocol dated March 26th that we believe was finalized on April 2nd for all ministry district and regional offices. It prohibited any ministry staff from attending at any SARS-affected worksite, even in the case of a work refusal.

This memo, which we understand to have been operative throughout the crisis, instructs ministry staff who receive a formal worker complaints under the OHSA to refer all such complaints to the district manager. The memo stated that in unusual circumstances, the district manager is to contact the Regional Director. The protocol advises that lawyers at Legal

Services branches and ministry physicians will be available to provide assistance to the district manager.

In bold print the protocol states: "The district manager will handle all SARS complaints personally and over the phone. He or she will not attend the SARS worksite personally and will not send another Ministry employee to the workplace." The protocol says workers should be advised of the internal responsibility system or if technical advice is required, the worker should call Tele-Health Ontario. It also stated that work refusals are to be dealt with in a similar manner. ONA and OPSEU found this approach by the ministry to be one of the most frustrating and possibly dangerous aspects of the SARS crisis.

Events and Issues that Should Have Triggered Ministry of Labour Enforcement Activities

Even before the crisis hit, there were serious problems with enforcement of the OHSA in the healthcare sector. In January 2003, months before the SARS crisis, one author (LMcC) and other Ontario Federation of Labour Health and Safety Committee members met with the director of the Workplace Insurance Health and Safety Policy Branch in the Ministry of Labour and a number of his colleagues to discuss various outstanding health and safety issues. Ministry enforcement was discussed, as was the need for inspectors to have a heightened responsibility to respond when a worker's right to refuse unsafe work is limited, as it is in healthcare facilities. Acknowledging problems with lack of enforcement, the ministry agreed to arrange a meeting of the regional directors and labor representatives to discuss issues around enforcement. This meeting was held at the end of May during SARS II.

1. The fact that a large number of healthcare workers became ill with SARS as a result of workplace exposures should have led the ministry to investigate. Both unions believe that if that many industrial workers suddenly developed a life-threatening work related illness, the ministry would have launched immediate investigations. The illnesses were constantly discussed in the media, as were reports of shortages of equipment, including respirators.
2. The requirement for fit testing of the N95 respirators in the March directives and then from May forward, should have led the ministry to inquire whether fit testing was being done. The ministry was or should have been aware that hospitals may have had no previous experience with this procedure, despite requirements that had existed in the health care regulations since 1993.

3. The Ministry of Labour was involved with the production of the directives, which should have led to more active scrutiny of their implementation where health and safety were affected.

4. There were repeated requests on the part of OPSEU and ONA staff for the ministry to become involved as both unions learned that there were breaches of the directives and contraventions of the act and the regulation. Calls from healthcare workers to the Ministry about unresolved health and safety concerns should have prompted the ministry to enforce its powers under Section 54 of the OHSA. Additionally, one of the authors (LMcC) reported to the OHA teleconference meetings attended by ministry officials that critical injuries were not being reported to the ministry as required by the OHSA. These failures to report should have prompted an immediate Ministry of Labour investigation.

Chronology of Events Involving the Ministry of Labour

It is important to highlight a chronology of events that affected ONA and OPSEU members. It demonstrated to union representatives the Ministry of Labour's lack of involvement throughout the crisis.

During the first round of SARS, which emerged in mid-March at the Grace site of the Scarborough Hospital, approximately 64 employees (paramedics, clerical staff, nurses, and doctors) were diagnosed with SARS as a result of workplace exposures. It was documented in various media, popular and scientific, that healthcare workers were contracting SARS.

1. On March 31st a senior ministry representative spoke with one of the authors (LMcC) about various health and safety issues. The representative was unable to answer questions without first running them by the command center because he had not seen the directives. We questioned why a key ministry official had not yet seen the directives. He was then sent the directives by this author because he did not know when he would be receiving them through ministry channels.

2. On April 1st one of the authors (LMcC) wrote on behalf of ONA to the Ministry of Labour, requesting guidance and clarification on a number of SARS-related issues, among them the specific health and safety needs of pregnant workers. In its response, the ministry cited the section of the Health Care Regulation relating to reproductive hazards, but offered no guidance on how the specific risks related to SARS were to be dealt with. Early in May, the same representative advised me by telephone that the ministry would not be issuing any special guidance for pregnant workers. This was later confirmed at an Ontario Hospital Association (OHA) teleconference, where attendees were advised that neither the Labour nor

Health ministries would issue a directive on the issue of pregnant workers. As a result worker anxiety and mistrust could only increase.

In that same letter to the Ministry of Labour, ONA also inquired about several other health and safety issues. We asked whether the ministry would employ a heightened response to our members' unresolved health and safety concerns and complaints, given their limited right to refuse unsafe work under the OHSA. On April 15th the ministry replied, stating that they were responding to concerns, complaints, and work refusals. However, the ministry did not respond specifically to the query on the possibility of a "heightened response."

3. On April 11th the bargaining unit president at North York General Hospital reported to ONA that the ministry of Labour was advising workers that SARS was not a critical injury under the OHSA. ONA vigorously opposed this interpretation. It was ONA's position that the ministry's refusal to recognize SARS as a critical injury under the Act diminished the employer's responsibility to immediately investigate with a view to preventing a recurrence. It also removed a fundamental right under the Act for worker members of the joint health and safety committee to investigate and prevent further injuries.

It was the position of both unions that suspect cases of SARS were an occupational illness under the OHSA. It was also our position that probable cases of SARS must be considered as critical injuries under the Act. Employers have an obligation to report critical injuries immediately, both to the Ministry of Labour and to the joint health and safety committee. These reports must also be produced, in writing, within 48 hours. These reports are intended to trigger employer, joint health and safety committee, and Ministry of Labour investigations with a view to preventing a recurrence. Employers also have an obligation to report all occupational illnesses within four days.

The Ministry of Labour had an obligation under its own policy to investigate critical injuries to ensure that employers were taking all precautions reasonable to protect workers. Although the Act is silent on the ministry's obligation to investigate occupational illnesses, the ministry's own policy indicates that an inspector shall respond to all reports of occupational illness or disease.

Both unions have been informed that the Ministry of Labour is investigating the two SARS related fatalities; however, to date, neither union has any knowledge of the ministry initiating any form of critical injury or occupational illness investigation into what factors contributed to so many workers contracting SARS.

At the joint OHA/Ministry of Health/Ministry of Labour teleconference meetings, ONA repeatedly asked the Ministry of Labour for its position on SARS as a critical injury. Several ministry representatives on various dates promised a response, yet none fulfilled that commitment.

4. On May 1st a Ministry of Labour representative finally informed ONA that the ministry was taking the position that SARS was not a critical injury. It continued to be ONA's position that the ministry's interpretation of "critical injury" was further endangering workers. One author (LMcC) asked that ONA's position be taken back for further consideration. She agreed.

Later that same day the ministry's representative contacted LMcC. The only question that had come back from her superiors was, "Why do you care?" It was explained that among many reasons, the definition of an injury as critical, triggers investigations, which then should lead to better prevention.

Shortly thereafter, the Ministry of Labour's Provincial Physician advised ONA that the ministry had accepted that "probable SARS" was in fact a critical injury and that it would be calling all healthcare employers to advise them of their reporting obligations under the Act.

5. Between April 15th and 21st nine healthcare workers at Sunnybrook Hospital were diagnosed with SARS following exposure to a SARS patient during a complex and prolonged medical intervention. These exposures and subsequent illnesses were well documented in the popular media and shortly afterwards in scientific journals. In one scientific journal article, the authors speculate on the various reasons that there were so many exposures and illnesses among health care workers. All were related to lack of training on how to minimize exposures during high-risk proce-dures. One worker was documented as wearing a beard while he had his respirator on. No one had advised him to shave it. There had been no fit testing of respirators. To date, neither union is aware of any Ministry of Labour investigation into events at Sunnybrook that contributed to this volume of occupational illnesses.

6. On May 22nd and 23rd news of a new SARS outbreak (SARS II) emerged at St. John's Rehabilitation Hospital and North York General Hospital (NYGH). It quickly emerged that a number of patients who had SARS had been transferred to other Toronto hospitals. In addition, it was discovered that a large number of healthcare workers had contracted SARS during the time that the initial outbreak appeared to be waning. This news was in the media by the time union representatives met with Ministry of Labour managers on May 27th. To the knowledge of both unions, the ministry took no action as news of this situation was revealed.

ONA also received verbal reports that healthcare workers at North York General had been reporting the unusual patient illnesses to their supervisor. Workers reported they were cautioned that they were overre-acting and no action was necessary. This indicated to both unions that the infection control system and the internal responsibility system were inadequate to protect workers, as workers reported that their complaints about a hazard to their own health were discounted. A total of 42 workers from North York General were diagnosed with SARS by early June.

7. On May 27th a number of union representatives met with the director of the Ministry of Labour's Occupational Health and Safety Branch Operations Division and the ministry's regional directors. Union representatives raised a number of enforcement issues at this meeting. OPSEU and ONA specifically raised the SARS issues again and advised the ministry that they believed it was not fulfilling its role. Both unions pointed to the number of occupational illnesses, contradictions in the operations center directives, confusion regarding personal protective equipment within the hospital sector, lack of fit testing, and lack of training. Both unions strongly advised the ministry that it needed to get involved more proactively and that it should not rely on POC directives and internal hospital infection control practitioners to ensure workers' health and safety during the SARS outbreak. This meeting had no apparent effect.

8. By June 6th ONA had received numerous enquiries from individuals seeking answers to their health and safety questions. During SARS II, it became apparent to OPSEU and ONA that many employers were not responding to healthcare workers' concerns about their health and safety. Calls from workers about masks not fitting and their fears of exposure led three of ONA's representatives to call the Ministry of Labour themselves on June 6th requesting the ministry to go into North York General and St. Michael's Hospital to issue orders at least around fit testing and supervisor competency under the OHSA.

9. On June 6th an RN who is a member of the Ontario Nurses' Association initiated a work refusal because her N95 mask did not fit her properly. For the first time, to OPSEU's and ONA's knowledge, the Ministry of Labour became directly involved in the issue of respirators and fit-testing.

The ministry inspector determined that the work refusal was valid under the OHSA. At the investigation meeting on June 9th the inspector issued orders to the employer with almost immediate compliance dates. The orders required the employer to implement a respirator program for all workers with direct patient care in the SARS unit, the ICU, the emergency department, all employees and patient screeners and cleaning staff who were entering the rooms of SARS patients. At this meeting, it appeared to ONA that the employer's focus was on fit testing and training of nurses and doctors. It was ONA who had to remind the employer that fit testing must include all workers who enter SARS patient rooms. Similar orders were also issued to St. Michael's Hospital.

Shortly after this, the ministry advised ONA that it was going to start targeting all Toronto hospitals regarding the fit testing and training issues, starting with category 3 and 2 facilities. This was almost three months into the outbreak. During all this time it had been reported repeatedly that healthcare workers were one of the groups at highest risk of contracting SARS. Over 100 healthcare workers had contracted SARS as a result of workplace exposures and two nurses and a physician had died. Many more healthcare workers were quarantined as a result of workplace exposures, and countless people's lives were disrupted. The emotional and physical toll has yet to be accounted for.

In this same week ONA reported to the Ministry of Labour that we had received complaints that Mount Sinai was refusing to fit test. ONA requested that the ministry include Mount Sinai in the first round of its investigations. On June 11th, one author (LMcC) was advised by the Ministry of Labour that they would visit Mount Sinai either on June 13th or June 16th. (June 13th was later confirmed.)

10. On June 10th Lisa McCaskell wrote to the director of the Ministry of Labour's Occupational Health and Safety Branch Operations Division to follow up on the numerous health and safety issues that had been raised with the Ministry of Labour at the January and May meetings with the ministry and the Ontario Federation of Labour: to date ONA has not received a response.

11. On June 12th and June 13th Barb Wahl, ONA President, wrote to this same director asking for more ministry resources to facilitate the "proactive investigations." She also wrote regarding the disclosure of information under the OHSA, about ONA members who contracted SARS, and requested the ministry to investigate forthwith any and all critical injuries. To date no response has been received.

12. On June 13th, the Ministry of Labour's Provincial Physician advised ONA that the ministry would not be doing any more proactive investigations. Despite Linda McCaskell's questions, the representative would not disclose who in the ministry had made this decision or what had influenced it. It is the position of both unions that critical decisions like these should be a matter of public record. Although the ministry later resumed some proactive investigations, to our knowledge the ministry never visited Mount Sinai.

13. On June 17th, North York General sent the ministry an updated list advising them of all occupational illnesses. The unions are not aware of any critical injury investigations initiated by the ministry at North York General to date, despite this notice having been received.

14. On June 18th, ONA President Barb Wahl wrote to the then-Premier regarding her concern for member and public safety due to the Ministry of Labour's decision to scale back the proactive inspections, and the Ministry of Health's decision to reduce protection to health care workers in its June 16th directives. Although the Premier did respond, ONA was not satisfied with the response as it did not, in the union's opinion,

adequately explain the Ministry of Labour's actions or the Ministry of Health's rationale.

15. On June 28th, one healthcare worker, registered nurse Nelia Laroza, died from SARS following a workplace exposure at North York General Hospital during the second SARS outbreak. The second outbreak was identified May 23rd, approximately two months after the first outbreak. The Ministry of Labour has initiated an investigation into this fatality but ONA has not seen a fatality report at this time.

While it may be that no one factor will be identified as responsible for this worker's death, both unions must ask what responsibility the Ministry of Labour may have in this case given its reluctance to investigate previous occupational illnesses, complaints from workers, and knowledge of possible violations of the OHSA and the Health Care Regulation. Both unions believe it was ONA's formal complaints in June that finally triggered the issuing of orders in some of the facilities. Both Unions believe that if a similar situation had emerged in an industrial setting that the ministry would have acted swiftly and proactively to ensure that all reasonable precautions were being taken to protect workers from further illnesses.

16. On July 19th a second registered nurse, Tecla Lin, died of SARS. She had been exposed early in the first outbreak when she had volunteered to work on the SARS unit of her hospital, West Park Healthcare Centre. Although little was known about SARS when West Park opened its interim SARS unit, the illness was known to be highly communicable, either by droplet or respiratory transmission. West Park has a state-of-the-art respiratory unit, opened in February 2000, featuring negative pressure isolation rooms for highly infectious clients and specific procedures such as protective respirators for staff. The unit is designed to care for patients with complex and multidrug resistant tuberculosis.

Neither union knows if the SARS unit was housed within that special respiratory unit; however, even if it was not, one would have assumed that West Park would be one of the safest hospitals in the province in which to care for highly infectious respiratory illnesses given their reputation and their expertise. The unions await the fatality report from the Ministry of Labour, which may explain what went wrong.

The authors have presented in this chapter many examples of a health and safety system that failed.

It is clear to both ONA and OPSEU that if the culture of worker health and safety is to ever improve in the healthcare sector, the culture of health and safety must be changed by this government and its ministries. The Ministry of Health and Long Term Care when developing directives must incorporate health and safety law directly into the directives. The government must also ensure that the Ministry of Labour will enforce the OHSA and that they will have the means to do so. The ministry must ensure that all employers in this province are complying with this most superior legislation.

Since the mid-1990s, Ontario's Health Care Regulation has required employers to fit test workers before they don their equipment. Section 54 of the OHSA gives the Ministry of Labour the power to enter into any workplace at any time without warrant or notice. However, we are unaware of any instance (prior to June 6th, 2003) when the ministry exercised that power and ordered employers to fit test and develop respirator protection programs. In this instance alone it appears that it has taken the ministry over ten years to finally realize that many employers in health care were not complying with this section of the Regulation under the OHSA.

We must ask ourselves how many other possible violations are not addressed by the Ministry of Labour when enforcement of the OHSA in the healthcare sector is not made a priority. How many other workers can be spared the trauma that healthcare workers endured and how many lives could be saved if only the health and safety of workers were everyone's top priority.

Chapter 2

Presentation to the Commission to Investigate the Introduction and Spread of Severe Acute Respiratory Syndrome (SARS)

Barb Wahl

CONTENTS

Introduction

ONA is the largest nurses' union in Canada, representing 48,000 frontline registered nurses and allied health professionals, who work in hospitals, long-term care facilities, and in the community throughout Ontario. We also represent public health nurses who play an essential role in monitoring compliance with the public health act and health promotion.

For eight years the health and safety of nurses and the public entrusted to their care, has not been a priority of government. During SARS we learned too many bitter lessons about the inadequacies of Ontario's healthcare system, and the vulnerability of our frontline healthcare workers. We experienced insufficient infection control policies, unsafe practices, too little funding, ineffective communications, dangerous staffing levels, and an critical shortage of personal protective equipment that, had it been available, might have prevented unnecessary exposures and ultimately saved the lives of our nurses.

All of these issues link to government underfunding and to the fact that employers and the government ministries failed to live up to their statutory, legal, and moral obligations to protect the health and safety of healthcare workers, leaving public and workers at risk.

From the beginning of the SARS I outbreak, ONA advocated for a full and impartial public investigation, and insisted that nurses be a part of this investigation. The SARS crisis had a terrible impact on nurses' lives and practice. Their voices must be heard. It is vital to identify what happened, what went wrong, and what measures must be put in place to ensure that the health and safety of nurses, other healthcare workers, and the public at large is properly protected. It is our expectation that the lessons learned from SARS will help create a culture that values the critical importance of workplace health and safety. In order to prepare Ontario for future outbreaks, and to ensure that we have the infrastructure in place to deal with them, it is essential to take a look at the broader systemic issues, such as compliance with and enforcement of the Occupational Health and Safety Act. We must also look at the nursing shortage. The following list pinpoints elements that must be addressed:

■ There were inconsistent messages about how to prevent the spread of SARS.

- ■ Immediate action was not taken when frontline registered nurses signaled the alarm about a possible second outbreak
- ■ The economic, physical, and emotional impact of SARS on nurses and other healthcare professionals.

Through the private interview process you will be hearing in greater detail from ONA members and staff about the technical aspects of SARS and detailed occupational health and safety implications.

Hundreds of ONA members were quarantined both at home and at work. They were segregated from their families, friends, and coworkers — and they were ostracized in their communities. We heard from well over 1,000 nurses directly impacted by SARS. ONA members developed SARS after caring for SARS patients. Recently a number of these nurses told the author that they continue to suffer severe emotional and physical repercussions of a disease that we still don't know that much about. They fear their health will never be what it was. Tragically, two nurses died after contracting the infection while caring for SARS patients. Nelia Laroza was an orthopedic nurse at North York General Hospital, site of the second outbreak. She was Canada's first healthcare worker to die from SARS when she succumbed in June 2003. Fifteen other nurses from her unit were also infected with SARS, as was her son Kenneth. Tecla Lin, who had more than thirty-five years of nursing experience, was among the first nurses to volunteer to work on a SARS unit at West Park Lodge, where fourteen infected healthcare workers were transferred. The loss of these two courageous nurses is a tragedy that must not be repeated.

Although SARS may be contained for the moment, it has not been eliminated. Today, frontline nurses believe, based on what they have seen in their workplaces, that Ontario is no more prepared for the outbreak of an infectious disease like SARS than it was in March or June 2003. In fact in some ways we are worse off.

We're still not clear on what protocols are needed to prevent the spread of SARS and to protect frontline nurses when they care for infected patients. There is still tension between those who wish to be proactive, and advocate that we should err on the side of safety, and those who are content to react once the illness reappears. We think that being proactive is crucial. We saw that a reactive approach cost lives.

This government failed to make protection of nurses and other health-care workers and the public a priority. Even before SARS, we knew the provincial Ministry of Labour lacked the commitment to exercise its powers under the Occupational Health and Safety Act. Their job, to conduct random, thorough inspections of healthcare facilities was simply not done.

We need to picture ourselves as nurses working on a SARS unit with no clear direction about which mask to wear, how to make it fit, how often to change it, and what to do when supplies run out. Imagine the fear such professionals experience when intubating a highly infectious patient while wearing an inadequate mask, instead of the Stryker suit or other full protection they know they should have. Think of being told, like some nurses, that they should save their masks in a plastic bag from one shift to the next!

Our nurses tell us that throughout the SARS outbreaks, there was a steady stream of contradictory, confusing, inconsistent, and incorrect information about the means of transmission, infection controls, effective protective gear, and protective protocols healthcare workers needed to follow. This served to heighten nurses' fears and concerns for their own health and safety, and that of their families. And yet, despite how vulnerable and fearful they felt, they stayed at their patients' bedsides! One part-time nurse told me she was off for a few days, and was then asked to work on a unit other than her own. No one told her eleven nurses had called in sick. No one told her she needed to wear a mask. She got SARS. Today, even though nurses are reporting for work, they still feel unsure, unsafe, stressed, and exhausted.

Ontario's lack of preparedness was made worse by:

- too few nurses and too many patients
- insufficient supplies
- insufficient funding
- the lack of a standardized infection control system and containment strategy for hospitals
- inadequate implementation of Ministry of Health and Long Term Care directives for infection control
- inadequate or nonexistent disaster planning or early warning systems; no centralized decision-making authority
- downloading of responsibility for, and lack of, public health services

Supplies and Funding

One of the most disturbing aspects of the entire SARS experience was the discovery that not only was there not enough protective equipment for nurses working on SARS units, but some of the equipment actually failed to protect workers.

Our nurses experienced the following:

- Unapproved, improper respirators were provided.
- There was little or no training regarding the use of respirators and other personal protective equipment.

- There was no fit testing of respirators, despite this being the law since 1993. A mask that doesn't fit is no mask at all.
- Supervisors/managers didn't know enough about the dangers to our nurses' health and safety, or about the Occupational Health and Safety Act and its regulations, and withheld information.
- Critical injuries and illnesses were not reported to the Ministry of Labour as required by law (OHSA).
- The Joint Health and Safety Committees at most hospitals were not given the opportunity to investigate problems and revise procedures necessary to prevent further exposure and illness.

ONA repeatedly requested information from the Ministries of Health and Long Term Care and Labour, but they refused to tell us how many of our members sustained occupational illnesses or how many of those illnesses were critical injuries as defined under the Occupational Health and Safety Act. This, in spite of the fact that the law clearly requires them to do so. Our nurses' fear accelerated because they did not know how many of their fellow nurses were ill.

Because our members have limited rights to refuse work under the Occupational Health and Safety Act, it was our expectation that the Ministry of Labour would have a heightened responsibility to respond to their concerns, but that did not occur. Instead in some cases nurses were forced to move to SARS units, while replacement nurses were brought in from elsewhere to do non-SARS work.

What Are the Answers?

As a start, we must ensure that every healthcare facility has provisions in place to protect the health and safety of all workers. We must have fully functional joint health and safety committees and internal responsibility systems at all facilities. There must be quality assurance controls and inspections so employers are in full compliance with the Occupational Health and Safety Act, and Health Care Regulation.

Only the Safest Equipment

The very safest equipment available is the only option. Full body protection, respirators, gloves, caps, gowns, and eye splash protectors must be effective and provided in sufficient quantities. Safety equipment — particularly appropriate respirators — must fit and nurses must be fully trained in how to use them.

The Nursing Shortage

SARS highlighted the serious nursing shortage and understaffing in Ontario healthcare facilities. It showed that the system is extremely vulnerable when part of our workforce falls ill or is put into quarantine. Government's cuts to healthcare budgets and restructuring of hospitals in the mid-1990s led to massive layoffs and job displacement. Many nurses were forced out of full-time work into part-time jobs.

Only half of ONA members now work full-time. Many of our members hold two or three different jobs to make up full-time hours, working without benefits or disability income protection. During the SARS crisis, the fact that so many nurses worked in more than one facility, reduced the number of nurses available, as one facility after another quarantined its staff.

Fewer Nurses — More Patients Per Nurse

Fewer nurses means more patients for each nurse, and not enough care. It means an overwhelming workload, which is driving thousands of nurses out of nursing altogether. According to the College of Nurses of Ontario if we compare the number of nurses we had in 1994 to the number we have now, the province is currently short 8000 nurses. Other statistics peg that number at 11,000, if we factor in our population growth.

These shortages will grow rather than decrease in the next few years. The Canadian Nurses' Association predicts a national shortfall of 78,000 registered nurses by 2011. A Canadian Institute of Health Information (CIHI) study projects that Ontario will lose almost 10,000 nurses to retirement by 2006.

Working conditions for nurses have resulted in increased on-the-job injuries and illness. The Canadian Labour and Business Centre calculates Canadian registered nurses work almost a quarter-million hours of over-time every week, the equivalent of 7,000 full-time jobs over a year.

During any given week, more than 13,000 registered nurses — 7.4 percent of all registered nurses — are absent from work due to injury, illness, burnout, or disability. That rate of absenteeism is 80 percent higher than the Canadian average. The Centre estimates that overtime, absentee wages, and replacement for registered nurses costs between $962 million and $1.5 billion annually.

ONA believes the answer to the shortage of nurses lies in creating more permanent full-time nursing jobs, paying community and long-term care nurses the same wages as hospital nurses, so that nurses are attracted to these sectors — any of which could be hit by SARS — and immediately doing everything possible to improve workloads and working conditions for nurses. The job picture for nursing will not improve by itself. Ontario must resolve the nursing shortage to ensure we have enough nurses for even normal operations and especially for high demand situations, such as a SARS outbreak.

It has angered our nurses that the Ontario government hired temporary or agency nurses at three times the salary, while our members were sent to SARS units with no extra pay. Hiring agency nurses is more costly. It poses a serious risk to patients and other professionals on the floor, because temporary staff are often not adequately trained for the work they are asked to do. Our members tell me agency nurses often require the help of regular staff to do their jobs.

Some of our nurses who were quarantined or who lost shifts because of SARS containment protocols are still waiting for the compensation for lost wages that the government promised. Nurses who were ill, and the families of those who died, have not received any extra compensation. Those who had SARS told me recently they want the Ontario government to make them an apology, with an assurance of safety should they ever return to work along with equitable remuneration!

No Disaster Planning or Early Warning System

The inadequate way in which SARS cases, or possible SARS cases, were identified has to be extremely disturbing to all of us. At one facility, nurses identified a cluster of patients with SARS-like symptoms and reported the matter to management and the medical staff. Nurses' concerns were dismissed and nothing was done for several days. This led to the second major SARS outbreak. Unfortunately it was similar at other hospitals.

What this tells us is that we are ignoring the signs and symptoms of patients, as reported by the very nurses at their bedsides! The voice of nurses is not being heard! Nurses need whistleblower protection, so that if necessary they can go elsewhere with this type of information. They need respect and recognition as professionals and essential members of the healthcare team. Nurses are tired of being shunted aside and disregarded. They see they are not included in decisions and as a result they, and their patients, are not safe. These are the major reasons why they are leaving the profession.

Who's In Charge?

On matters of public health, public safety, and infection control, all levels of government and stakeholders must work cooperatively, sharing planning, information, and resources, and developing an integrated smoothly functioning infrastructure. There must be a central body that provides leadership, with ultimate authority and accountability.

Throughout the SARS outbreak there were major inconsistencies in how facilities were handling the issues of public visits, screening, follow-up of patients requiring quarantine or having symptoms of SARS, and follow-up after a quarantine period. During the SARS crisis, we witnessed a bureaucratic jumble with no clear decision maker on issues of infection control and the identification and isolation of potential cases.

The Ontario government is responsible for planning and putting in place infection control measures. It is accountable for compliance with health and safety legislation. SARS showed us that the government instead left those responsibilities to individual healthcare administrators or managers.

Public Health Services

A public health system with an extensive public health education system is critical for community protection and safety. The Ontario government downloaded responsibility for public health to municipalities, without providing them with adequate revenue sources. Some programs have been discontinued, others added and made mandatory, without sufficient increases in revenue. As a result, public health, once a vital component of our healthcare system, no longer provides the services that are needed to keep the public informed, to follow up on investigations, to ensure that people are getting the education they need.

ONA members have told us about people in voluntary quarantine during the SARS outbreak who never received a call back from Toronto Public Health to assess their situation. They didn't know if they were healthy and could go back to work or if they should stay home.

Conclusion

SARS exposed the weaknesses in our system. We must correct them and strengthen our infrastructure. It is simply not enough to run newspaper ads proclaiming that healthcare workers are "heroes" — even though we appreciate and agree with the sentiment. If nurses don't get the respect and protection they deserve in their workplaces, all of these accolades mean nothing and only add to their cynicism and frustration. If healthcare facilities in Ontario do not change the way they regard and practice health and safety, more lives will be lost.

PERSONAL PROTECTIVE EQUIPMENT

Chapter 3

Respirators and Other Personal Protective Equipment for Health Care Workers

John H. Lange and Giuseppe Mastrangelo

CONTENTS

Some Emerging Diseases

There are a number of emerging infectious diseases that can be discussed. In some cases, these plagues have been generated by human beings.[1] These man-made plagues have also been part of military operations resulting in thousands of deaths. Today, biological agents have reemerged as weapons of bioterrorism as well as natural occurring events.[2] When the concept of an emerging disease is discussed, it does not include only those yet to be discovered but those that have changed (e.g., influenza —H5N1), [3] as well as some that are reappearing (e.g., antibiotic resistant tuberculosis), especially in the Western world. With increases in worldwide travel, the potential for a newly emerging disease to spread becomes of increasing concern. An individual can now travel to any location in about 24 hours or less, so spread of a new infectious agent, or an old one, can occur more rapidly than detection or response. Thus, there is now a trend toward global public health, especially in view of emerging infectious disease. These concerns relating to the travel have even resulted in health checks for those boarding airplanes, for the purpose of preventing the spread of a particular disease agent. Such actions are reminiscent of the days when quarantine was commonly used to prevent dissemination of a disease.

The occurrence of Severe Acute Respiratory Syndrome (SARS) in July 2003[4] raised the public's awareness of emerging diseases. Undoubtedly SARS will be followed by other emerging diseases. There have been a number of potential diseases (e.g., monkeypox) that recently arose but did not become worldwide epidemics. One of the viruses of greatest concern today, however, is influenza. It is known that three to four influenza pandemics occur each century, with the most recent in 1968. Based on the historical occurrences of these pandemics, the next one will likely occur in the next year or so. It will most likely involve the avian flu virus H5N1 which appears to be subject to a high mutation rate and is the most likely candidate for the next major outbreak. Four emerging diseases (SARS, monkeypox, tuberculosis, and influenza) are discussed, but are only a few of those that pose a real hazard (e.g., viral hemorrhagic fever) to public health and healthcare workers.

Severe Acute Respiratory Syndrome

SARS was a classic emerging infectious disease that quickly spread across the world. This resulted in an awakening of the Western world as to its vulnerability to emerging infectious diseases. The corona virus (CoV) family is the agent responsible for SARS. These viruses were historically noted as agents that were responsible for the common cold, along with other viruses (e.g., rhino viruses). The CoV SARS emerged in November 2003 in the Guangdong Province of southern China, although it appears to have been known as early as July 2003.[5] It is likely that this virus existed in a milder form for many years and mutated to a virulent strain.[6] Such occurrences are not usual and have been best illustrated by influenza outbreaks, most notably those of 1918, 1957, and 1968. SARS made worldwide headlines [7] and demonstrated how fast an infectious agent can spread, even from isolated locations in the world. The last reported case occurred in April 2004, and it was as a result of a laboratory accident.

A high infection and mortality rate became the hallmark of SARS. What also became apparent from SARS is the vulnerability of those in the health care industry to emerging diseases.[8] It was reported that healthcare workers were between 20 and 80% of SARS cases,[9] which makes them one of the groups most vulnerable to emerging diseases, especially since they will see the first case of a disease well before any outbreak is recognized. This is best illustrated by the death of Dr. Carlo Urbani, who first identified the disease in Vietnam and later died from this agent through occupational transmission. Some have suggested that the disease should even be named in his honor. Overall healthcare workers also have higher risks for other hazards, such as injuries, along with the more traditional infectious diseases,[10] making this population particularly vulnerable. Risks from infectious disease become of even greater importance when they are superspreaders, those that can result in multiple secondary cases of the disease from a single source contact (e.g., transmission to large numbers of people), as occurred with SARS.[11] SARS has become a class example of a superspreader.

Initially it was thought that the SARS CoV was not spread by respiratory transmission, but it has later been shown to be transmittable by this route.[12] Thus, as with many respiratory viruses, SARS can be spread from person to person by droplet along with other airborne routes (e.g., nebulization).[13] This demonstrates that precautions for all routes of transmission must be considered for emerging infectious diseases.

Many of the cases of SARS in healthcare workers have been attributed to the lack or poor use of respiratory and personal protective equipment (PPE).[14,] Studies on SARS reported an infection rate ranging from about 2 to 25% even when precautionary measures were taken (Table 3.1).

It has been known for some time, even before the SARS event, that bioaerosols can be generated through a number of mechanisms. These mechanisms include: exhalation droplets (coughing, sneezing, shouting, or talking), medical procedures (e.g., nebulization), fomite transmission, and body wastes (e.g., feces).[23] The higher the exhalation velocity the smaller the particles and the larger the number being formed, and

Table 3.1 Studies on SARS CoV Infection in Healthcare Workers Using N95 Respiratory Protection as Well as Other PPE[a,b]

Location	Infection Rate[^]	Comments
Singapore[15]	~4%	HCW had the highest % infection rate of any group. Recommended using powered-air purifying respirator (PAPR) along with other PPE. Besides use of an N95, other types of PPE were not identified. HCW in low-risk areas were given surgical masks.
Hong Kong[16]	~4%	All infected workers used a N95 respirator, but did not constantly use other PPE.[a] Eye shields were used. In high-risk areas, some PPE was used more than once by the HCW. Study reports that surgical masks are not protective. Personnel in low-risk areas contracted SARS.
Singapore[17] (Hsu et al., 2003)	5%	Contact cases resulted from an index patient. After initiating infection control[b] no further transmission occurred from index patient.
Hong Kong [18] (Li et al., 2003)	~2%	Strict infection control required.[a] Fit testing was not mentioned. Spot checks were conducted to ensure compliance. Type of mask used was not described.
Canada[19] (Scales et al., 2003)	17%	Contact cases[b] resulted from an index patient. A higher rate (30%) of infection for HCW from the index patient was reported for those using a surgical mask.
Hong Kong[20] (Tsui et al., 2003)	~25%	Study reports[c] that "precautions could not prevent all HCW from contracting SARS". Eye shields were used instead of goggles. percentage age represents total HCW infection rate.
Hong Kong[21] (Seto et al., 2003)	~5%	No HCW was infected that used all PPE. Thirteen that omitted using one type of PPE became infected. Study reported that N95 respirators and surgical masks to be effective in prevention, but not paper masks.

Note: All studies reported initiating infection control, including hand washing. Studies reported are for populations and case reports on a single or limited number of individuals.

[a] PPE included N95 respirator, goggles or eye protection, gowns, and gloves.

[b] PPE included N95 respirator (unless noted), gowns, and gloves.

[c] % healthcare workers who were reported to be infected by occupational exposure.

From Lange, 2005a, with permission from the Chinese Medical Association.[22]

generally this will result in the particles remaining suspended in the air for a longer period of time.[24] Even large particles can dry out and thereby be reduced in size, as well as resuspension of particles that have been deposited on surfaces.

The mortality rate from this virus was about 10%, with many of these deaths occurring in healthcare personnel.[25] Some have suggested that if proper PPE, including respirators, were employed by healthcare workers the number of cases in this group would have been diminished.[26] This is supported by the suggestion that the attack rate for SARS is higher for healthcare personnel than for inpatients.[27] An increased attack rate for healthcare workers is suggested as resulting from their close proximity to patients. In many cases the cause of SARS in healthcare workers can be directly traced to the aerolization of virus through nebulizers or similar equipment with cessation of use resulting in a reduction of cases.[28] Aerosolization appears to be a major hazard for many organisms that can be transmitted from person to person (Koley, 2003).[29]

Other forms of person to person transmission not involving neubulizers could be responsible, and appear to have also occurred.[30] However, with any new disease, there is a learning curve and other practical considerations that must first be identified before preventative measures can be fully realized. This was certainly the case for SARS CoV, and these events provide us with a valuable lesson in handling future disease events of this nature.

Infectious dose is a big question for any emerging infectious disease. For SARS the infectious dose has not yet been estimated or clearly evaluated. Studies have reported that some infectious disease agents have an infectious dose of one organism (e.g., smallpox, tuberculosis).[31] However, others require more than one organism (e.g., *Yersinia pestis*) for initiating an infection (pneumonic plague, also called black or bubonic plague). Using a published one-hit type model,[32] it was estimated that there is a 7.7% risk of infection for one hour of exposure to an organism such as tuberculosis. Since infection rates for SARS have been reported to be 2 to 25%, and in some cases even higher using hypothetical infection risk data, applying a one-hit model for a single infectious dose, it can be suggested that the SARS CoV falls in the category of having one infective dose (single viral particle) causing disease. [33]

Monkeypox

Monkeypox belongs in the group of viruses (orthopox viruses) that includes smallpox (Variola) and cowpox. This virus was first discovered in laboratory monkeys in 1958 and later was observed in central and western Africa, but has been also reported to exist in other animals including squirrels, rats, mice, and rabbits.[34] The first case in humans was reported in 1970. In June 2003, this disease was first identified in the United States and was associated with those keeping pet prairie dogs.[35] Occurrence of this disease in the United States included a family cluster with illnesses ranging from a minor febrile rash to severe neurological symptoms.[36] Concern quickly arose that a potential outbreak from this virus may occur, especially since cases appeared in multiple states.[37] The clinical features of monkeypox are similar to those of smallpox except there is swelling of the lymph nodes in cases of monkeypox.[38]

Limited information exists on person to person transmission of monkeypox, although it is believed that this can occur through respiratory droplet.[39] However, it has been suggested that person to person transmission is limited in nature and may only occur for a few generations, although epidemiological studies in Africa suggest a higher rate may occur.[40] Fleischer et al.[41] suggested that risk to healthcare personnel from this virus is limited, although others have raised concerns.[42] The U.S. Centers for Disease Control and Prevention (CDC) suggest that vaccination against this disease can be achieved through inoculation by the smallpox vaccination.[43] Although this vaccine is not 100% effective, it appears to be sufficient to provide herd immunity against this virus.

Antibiotic Resistant Tuberculosis

Tuberculosis (TB), a disease of the lower respiratory system, is generally spread by airborne transmission. This disease is caused by bacteria of the genus *Mycobacterium*, with the most common species being *Mycobacterium tuberculosis*. There are other species in this genus that can cause TB as well (e.g., *Mycobacterium bovis*). Commonly this disease spreads through coughing and sneezing resulting in person to person transmission. TB can also be spread by other mechanisms (e.g., milk), although these are generally less common today. WHO has estimated that about 2 million people die each year from TB and it has been estimated that about one-third of the human population is infected, with a higher percentage of men being infected than women. TB has become the forgotten plague.[44]

One of the biggest issues with TB today is the occurrence of antibiotic resistant strains, especially multidrug resistant (MDR) strains.[45] Historically there were two drugs, isoniazid and rifampin, commonly used in the treatment of TB.[46] Drug susceptible TB can be cured in about six months while MDR forms can take two years and require treatment with drugs having more side effects. The majority of MDR forms have occurred as a result of improper treatment regimes and in patients who do not undergo a complete course of treatment.[47] Prison populations have become a reservoir for TB. The rate of TB in prisons can be fifty times greater than the surrounding population and can result in a high percentage (>50%) of these being MRD cases.[48] The MRD cases are often caused by inadequate and incomplete treatment due to cost and transfer of personnel, missed cases, as well as other reasons. Employees in prisons and healthcare workers are at great risk of infection. In one study, one-third of new cases in New York state prison employees were a result of occupational exposure.[49] In Russia, prisons, which have a population of about 1 million, account for more than half of the new cases of TB (the Russian population is around 150 million).[50] The occurrence and rise in the incidence of MDR strains in such places as Russia and other countries are a serious risk to global health.[51] TB outbreaks are also occurring in other populations, especially where crowding exists (e.g., homeless shelters).[52]

It has been estimated that over 4% of new TB cases are MDR today in Eastern Europe, Latin America, Africa, and Asia. The rise of TB and MDR strains has become an important international problem and is of global importance [53] and is not restricted to the nondeveloped countries[54]

Influenza (Avian Flu)

Influenza is probably the most important emerging disease that exists today. Every year several new strains of the flu emerge and result in millions of cases and an untold number of deaths. Historically this virus or group of viruses has resulted in major pandemics which occur several times each century. There are three groups or types of influenza viruses, A, B, and C. Influenza type A can infect humans, birds, horses, pigs, seals, whales, and other animals and is the cause of worldwide pandemics.[55] Type B is found only in humans and causes epidemics but not pandemics. Type C causes mild illness in people but cannot cause epidemics or pandemics.[56] It has been suggested that the next pandemic of influenza will result from the Asian flu or avian flu (influenza A/H5N1).[57] These viruses have a high mutation rate as a result of the lack of proofreading mechanisms for its DNA that would normally allow for repairs. This results in antigenic drift of the virus, often going from a low pathogenicity to one that is much greater. The reservoir for this specific virus appears to be birds (e.g., chickens). Influenza can survive in water up to 4 days at a temperature of 72°F and 30 days at 32°F. Thus, infected particles can exist for a long period of time in the environment.

Birds can allow spread of the disease as well as transfer it to and from humans. Transfer between birds and humans allows a mixing of genetic material, which can result in a new strain that is highly pathogenic.

Currently, there have been few reports of person-to-person transmission of these new viruses, but due to the high mutagenicity of A/H5N1, it is only a matter of time until a strain emerges that can be effectively transferred among people. The occurrence of a novel strain that can be transmitted from person to person along with other routes (e.g., fomite) would potential signify the start of a pandemic. This can be illustrated in the communicable transfer of influenza and cold viruses (e.g., rhinal virus).

There are now suspected cases of avian flu being transmitted from person to person.[58] In Vietnam thirteen cases and nine deaths were reported from this virus suggesting it is highly pathogenic.[59] More recently there has been a reported case in Cambodia, which may also be a result of person to person transmission, suggesting that this virus is spreading.[60] However, in other outbreaks human to human transmission does not appear to have occurred and no healthcare workers have been infected.[61] Other outbreaks have also been reported (e.g., South Korea) with and without human cases. There remains the question as to whether human to human transmission is occurring at this time; [62] however, as mentioned, it is not a question of whether it will occur but when.

As with other infectious diseases, it is healthcare workers who will be most affected by an influenza outbreak. Thus, occupational protection is paramount in this population.

Protection Against Emerging Infectious Diseases

In general, there are four ways for preventing the spread of and infection by emerging diseases. For healthcare workers three are of importance: vaccination, barrier mechanisms, and use of PPE. A fourth method exists, isolation, but this is not practical as a preventive measure for healthcare workers. No vaccine is available for some diseases (e.g., SARS). This leaves barrier methods or PPE as the measures for preventing transmission to healthcare personnel or other potentially exposed persons. Barrier methods, which are not discussed in this chapter, include physical barriers such as tents over the patient and increased air flow. These barrier methods would be considered forms of engineering control. Increasing the number of air changes has been shown to be effective in preventing transmission of SARS [63] and may be useful for other infectious diseases. Separation of patients from each other and isolation procedures for preventing spread of an infectious disease to healthcare workers is of equal importance, as is the use of PPE. Use of various barrier methods has been shown successful in Vietnam, where respirators were not initially available, and such methods alone mitigated the spread of the diseases.[64] When considering PPE it must be remembered that its use is the form of protection of last resort: engineering controls are the preferred and the best methods for prevention.

Many of these factors for protecting healthcare personnel have been overlooked or circumvented due for the most part to inadequate preparation and financial factors. This can also be said to be a result of bad science in making educated choices regarding protection.[65] In illustrating this, the recommendations exist in numerous governmental and even peer-reviewed publications on the dangers of using inadequate respirators.[66] Numerous reports in the literature suggest what are in truth inappropriate methods of respiratory protection,[67] and represent a lack of critical thinking by some commentators that has influenced selection practices regarding PPE. Certainly, these problems exist throughout the healthcare community on a daily basis. As has been seen with SARS, it is highly likely that such errors and mistakes will occur during the next outbreak and the issues and concerns raised in this chapter will reappear.[68]

At least one report[69] suggested that eye protection and N95 respirators be worn by patients. The general problem with N95 and related respirators, which are disposable respirators, is maintenance of the face seal.[70] Since respirators remove hazards as they are breathed in by the user (by air entering the respirator through a filter and not leaving), respirators worn by a patient would have little impact on the release of agents.[71] It would also add physiological stress[72] to those using the device, most likely increasing cardiopulmonary stress to the individual and worsening their situation with no benefit to those nearby. This suggestion of the patient using a respirator is a good illustration of the lack of understanding in the healthcare community on the use and application of PPE. In many ways it is also an attempt to reduce associated financial costs and training requirements. With little doubt, the cost of training and and equipment is high, but this is dwarfed by the cost of treating the illnesses that arise from not using effective PPE. Such differences in cost for prevention as opposed to later treatment have become well known in healthcare [73] and are one of the reasons for emphasizing preventative activities, but apparently not for those working in this field as related to emerging diseases. Preventative measures have been well established in occupational environments other than health care.

Although not related to PPE directly, but worth mentioning, is the issue of quarantine, used historically to prevent the spread of communicable diseases. Such use in developed countries has not been seen for decades, but such use and associated issues were raised during the SARS event.[74] In today's society it has been questioned as to whether enforcement of quarantine measures would be effective. In Singapore the police were instructed not to arrest people with potential SARS for fear of driving those with the disease underground,[75] which could result in a large number of hidden cases and continuation of the epidemic. However, in Toronto, police were used to quarantine people in hospitals and serve notices, track down infected persons, and enforce isolation procedures.[76] One of the issues with quarantine measures and forced isolation is related to PPE. In most, if not all cases there was not a sufficient quantity of PPE, including respirators, to provide protection to those in quarantine as well as the personnel treating and assisting in these measures. When there is a shortage of such materials there can be difficulty in asking those affected to take drastic measures and place themselves at risk of contracting a contagious disease. The SARS event demonstrates the logistical problems and ethical issues associated with quarantine and involuntary requirements of various personnel. These issues are most critical for those who may potentially be infected and there is inadequate preparation in using other measures for preventing spread of the disease. With the existence of modern PPE, which can be highly effective in preventing spread of a disease, use of other measures such as quarantine could be considered as a secondary option.

Personal Protective Equipment

PPE can be described as any equipment that can be used by an individual for protection against a hazard. For infectious disease, this equipment mostly consists of gloves, suits of some type, and respirators. If the respirator is not a full face mask, eye protection may also be necessary. The most common type of PPE for hands is surgical gloves. They are certainly in common use in hospitals and readily available. When used with a surgical or related type of gown, they can be pulled over the sleeve providing full body protection. A half-mask respirator provides no protection for the eyes. This will require use of eye protection, such as glasses, goggles, or a face shield. In some cases it may be prudent to use glasses/goggles and a face shield. Others types of PPE that may be used will include shoe covers or some type of boot. In general, shoe protection is not for drop hazards but to prevent biological fluids from leaking into the shoe or onto the skin of a healthcare worker. Covering the head and hair may also be required in some cases. Regardless, any healthcare worker who comes in contact with a potentially infected patient must shower and wash thoroughly upon completion of work or in some cases after contact with the patient. As for most infectious diseases, such as SARS, complete showering does not appear to be necessary; however, for other diseases, such as a viral hemorrhagic fever (e.g., Ebola virus), it may be a prudent practice. One of the problems for many of the emerging infectious diseases, including those that can be considered "well known" (e.g., Marburg virus) is that little is understood about the organism's life history and etiology. In the case of the Marburg virus, its reservoir hosts and related information are mostly unknown, with the virus surfacing from time to time and then disappearing.[77] However, the importance of the hazards posed by these agents cannot be understated, especially since some, like Marburg virus, can have a 92% fatality rate.[78] It appears that these organisms are amplified in healthcare settings resulting in a high fatality rate among healthcare personnel, as was seen on a global scale with SARS. The occurrence and potential of such diseases have been known and recognized by modern health care, but with little public concern until recently. These hazards from emerging infectious disease agents prompt a new perspective on practices for PPE.

Hand washing, although not included as PPE, is of equal importance. This activity is often neglected, but it is an important part of disease transmission prevention. It is often considered that many do not know how to properly wash their hands, with few references providing any practical guidance. It is suggested that this should consist of vigorous washing with soap and water for a time period of 15 seconds. Disinfection of the hands requires washing (scrubbing) for at least 15 seconds with soap and water. This must including washing between fingers, washing wrists, and turning off the faucet with the towel used for drying.[79] Koley suggested hand washing can also be accomplished using alcohol-based solutions.[80]

Respirators

Respirator protection and its use are not a simple selection and application process where one can be picked off a shelf and used. For any respiratory usage there must be an appropriate selection process, training, and fit testing. There are also administrative requirements in implementing a program. Each of these has its own importance and applicability to the use and implementation of a respirator program. There may be differing requirements, depending on the individual country, as to how a program will operate as dictated by regulatory requirements governing use of respirators. Most follow the United States Occupational Safety and Health Administration (OSHA) requirements regarding respirators. These requirements are described in the Code of Federal Regulations (CFR 29 CFR 1910.13), as an example for industry. The construction and maritime industries have their own CFR codes, but the requirements are the same or very similar. In addition to these requirements, OSHA has also published regulations for other types of PPE, such as foot protection. For the most part these requirements are not directly applicable to healthcare workers, but they should be aware of the existence of such regulations and requirements. In some specific OSHA standards there are detailed requirements for respirators and some of this information can be used to assist the HCW in selecting an appropriate respirator. As has been previously discussed, the focus of this chapter is respirator protection for healthcare workers to protect them from emerging infectious diseases. Protection against these organisms will involve preventing physical penetration of the organism through the respirator. Thus, this will limit the types of filters which can be selected.

As was observed for SARS, there appeared to be cases or events of the disease in healthcare workers which resulted from inadequate use of respirators or selection of an improper respirator, along with inadequate engineering controls.[81] Certainly there has been considerable debate as to the appropriate respirator for this virus.[82] Much of this confusion arose as a result of a lack of information on the spread and dissemination of the disease, even among healthcare workers. Since it has been recently realized that this virus can be easily transmitted via an aerosol route, selection of the appropriate respirator and its correct use is paramount. During the occurrence of a new emerging infectious disease, it is unlikely that a great deal of information will be available at the time of such event, about the life history of the organism, or at least the specific strain or variety, although lessons from SARS would suggest that airborne transmission is a likely mechanism of concern and should be a focus of preventative measures. However, it should also be noted that SARS was effectively controlled using masks and special barriers,[83] demonstrating that other forms of control are necessary besides respirators. This also reinforced the historical concept of engineering controls as the first choice of protection, with respirators being the least desirable.

For the most recent emerging infectious disease, SARS, N95 respirators were suggested to protect healthcare workers.[84] Others have suggested that N95 respirators are not adequate for protection against these infectious agents.[85] Seto et al. [86] suggested that surgical masks were protective against SARS while paper masks were not, while Derrick and Gomersall[87] reported that even double surgical masks are ineffective against ambient particles, that is, desiccated or airborne "droplets" containing viral particles.[88] N95 respirators were originally suggested for use against the inhalation of tuberculosis and became the standard for protection against this disease since 1994.[89] These, N95, are considered disposable respirators.

The CDC[90] as well as other public agencies [91] through their publications recommended that the N95 respirator be used in protecting HCW against the SARS virus.[92] Most agencies throughout the world follow the recommendations issued from the CDC. As mentioned, the use of N95 respirators was supported in a number of other publications.[93] This recommendation and supporting publications influenced the healthcare community and led to the increased use of N95 respirators. The design of these respirators appears to derive from guidelines related to protection against tuberculosis.[94] However, as noted,[95] characteristics of a virus in regard to its properties and penetration in regard to the type of respirator is much different than for a bacterium, such as tuberculosis. It appears that little thought was given as to the characteristics of the organism when making recommendations for respirator protection against the SARS CoV. When conducting any form of respirator selection, it is fundamental to include in this process the agent in question. Unfortunately, in the rush to provide protection for healthcare personnel during this crisis, such a process was not given much consideration. In many ways, this may have been a result of inadequate information on the organism's transmission. However, logic would dictate that if a respirator is being selected, this is based on a concern that an airborne route is one mechanism of transmission. When going through the selection process, care must be taken not to follow a recommendation without independent evaluation, as was done when implementing the N95 for SARS.

Respiratory protection is mostly designed to protect the respiratory tract, the primary location being the lungs. In many cases this is true for the vast majority of occupational hazards. However, infectious agents can be taken up and affect the entire respiratory system from the nasopharyngeal area to the alveoli. In general the larger the particle the more likely it will be deposited in the upper part of the airway. So particles greater than 30 μm will more likely be deposited in the upper nasopharyngeal region, about 5 to 30 μm in the lower nasopharyngeal, 1 to 5 μm in the tracheobronchial, and less than 1 μm in the pulmonary or alveolar region.[96] Since viruses are in the general range of 20 to 400 nm (0.02–0.4 μm) they will be more predominantly deposited in the pulmonary region. However, bacteria have a much wider size range (about 0.4 to 10 μm), therefore deposition can vary to a greater degree from the lower nasopharyngeal to the pulmonary region. The velocity also changes dramatically from region to region within the lung, with the higher air velocity occurring in the upper regions and the lowest in the alveolar. The size of the infectious agent will, in part, dictate the type of filter required for a respirator. For example, *Bacillus subtilis. Mycobacterium bovis* and *Cladosporium cladosporioides* have sizes of 0.8, 0.9 and 2.1 μm, respectively.[97]

Types of Respirators

Each type of respirator has its own applicability, and generally they fall into one of three groups: paper masks, air-purifying respirators (APR), and supplied air respirators (SAR). SARs are also sometimes called type

C respirators. It should be noted that a N95 respirator is a half-mask, but is classified as a disposable respirator. Most of the APRs historically used in most occupational settings were not disposable. APRs cannot be used in atmospheres that are deficient in oxygen, although this is generally not a concern for those in healthcare settings. Surgical masks have also been used and reported as a class of respirator.[98] These devices are not respirators but rather barrier devices, which have limited ability to prevent deposition of infectious agents into the respiratory system. One study reported that surgical masks allow 42% of droplets nuclei to penetrate.[99] In another report by Grinshpun the filtration efficiency for surgical masks was reported to be 20%.[100] Regardless of which evaluation or estimate is used, this represents a large amount of penetration and poor filtration efficiency by surgical masks. It is likely that a paper mask would have a similar failure in providing protection for the user against airborne droplets containing infectious disease agents. Respirator types, excluding paper masks, fall into three general categories, which are quarter, half-, and full-face masks. As a general rule, quarter masks are not commonly used, at least in the United States, and are not recommended for protection of the respiratory system. Half-masks provide protection for the respiratory system and include the mouth and nose, but do not provide any protection for the eyes. Full face-masks on the other hand are protective against exposure relating to the mouth, nose, and eyes. Historically, the N95 respirator was suggested for protection of healthcare workers.[101]

For infectious disease, the mechanism for preventing transmission of the organisms from the environment into the respiratory tract is mechanical filtration. Particles or infectious agents are trapped in filters by sedimentation, impaction, interception, or diffusion. This mechanical filtration can involve either simple straining or depth filtration. Simple straining involves capturing particles because they are larger than pores in the filter. For depth filtration the particles are able to penetrate the filter and become adhered at some depth within the filter itself. Both of these involve a direct physical barrier against penetration by the particle, which in this case is an organism, bacteria or virus, and act in a similar manner as a particle. The other factor that is important for particles is electrostatic interactions. Electrostatic forces are important for small particles, which include the viruses and the smaller bacteria. The larger bacteria, which would include tuberculosis, would be captured, most likely due to its size, and would fall into a true physical capture, which can be evaluated as straining and depth for a filter. The stronger the charge on the filter and particle the higher the attraction will be, and the chance of collection will increase. Charged filters will slowly lose their charge over time resulting in their ability to only eliminate particles by simple mechanic filtration. The advantage of electrostatic filtration is that these forces are active over some distance.

There are no filters that have a 100% efficiency in capturing particles and the chances of capturing a single particle or in the case of an organism depends on a number of factors. Capture efficiency will depend on the shape and size of the organism, electrical charge, and type of filter being employed. Based on a report by Grinshpun, respirator performance depends on the size of the microbe and on its aspect ratio.[102] It was also reported that there does not appear to be any growth of microbes in the filters used to collect and prevent deposition into the respiratory tract. Historically, filters were designed to protect against dust, mists, and fumes and not bacteria or viruses. The two types of filters discussed with APRs are high-efficiency penetrating air (HEPA) filters and ultra-low penetrating air (ULPA) filters. HEPA filters are the most readily available and are commonly used in many industrial settings. These filters are defined as being 99.97% efficient for monodispersed particles that are 0.3 μm (300 nm) in size or larger. Based on this definition alone, most bacteria and many viruses would be effectively captured using this type of filter. However, if this strict definition is used, some viruses will not be collected, such as the SARS CoV. This virus has a size of around 0.060 to 0.080 μm (60–80 nm). When HEPA filters have been tested using the sodium chloride test aerosol it has been shown that the particle size distribution is much lower than that reported by standard convention (0.3 μm). This test found a count median diameter (CMD) of 0.075 +/- 0.020 μm with a geometric standard deviation of 1.86 (http://www.cdc.gov/niosh/topics/respirators). This value from the sodium chloride test suggests that SARS CoV can be captured, especially if an electrical change is included in this assumption. One of the issues associated with selecting a filter and its size comparison with the infectious agents is the dose or number of infectious agents required to cause the diseases. Since for many of the emerging diseases this number will not be known, it must be considered to be low, with the potential of one infectious particle being able to initiate the process. With this taken into consideration, there is a requirement for a high level of protection and efficiency of filters along with the seal and related factors in preventing exposure or inhalation of the agent. Fortunately, the HEPA filters will mostly be effective and efficient for bacterial and related type organisms, since most are larger than 0.3 um and very few are in the lower range of 0.4 μm. However, the range of viruses is within the standard value set for HEPA (0.3 μm) as well as the count medium diameter determined by the sodium chloride test. This concern has resulted in some suggesting that ULPA be used for viruses, especially those that are less than 0.3 μm such as SARS CoV.[103]

ULPA filters have a standard reported efficiency of 99.999% for monodispersed particles that are 0.12 μm in size or larger. This particle size filtration is closer to that of the smaller viruses, and when electrostatic attraction is included should be relatively efficient in collecting very small infectious particles. What is difficult in using ULPA filters is the higher cost and difficulty in obtaining them. In the event of an outbreak, it may be difficult if not impossible to obtain ULPA for the type and model of respirator that is being used by a specific institution or facility. Thus, in most cases HEPA filters, which are much more readily available, will be employed, especially during the initial stages of an event. However, it is conceivable that a facility can keep in stock a limited supply of ULPA for the type of respirator they employ.

The type of filter selected may be as important as the type of respirator. Since different organisms are of varying size, the filter should be selected on the basis of the kind or class of organism. Since bacteria are large as compared to most viruses HEPA filters would appear to be most appropriate, especially when the sodium chloride test is considered along with electrostatic charge. This is often true for many viruses in addition to SARS. Even for viruses that are small, such as SARS CoV, HEPA filters appear to have been effective in preventing disease occurrence in healthcare workers when the respirator was properly used. If there is no knowledge of the type of virus or its family, then the higher level of filtration may be necessary, which is the ULPA. The only respirator system that will not require some type of filtration is supplied air respirators, which obtain breathable air from an outside source.

As noted earlier, the type of respirator suggested for SARS and emerging infectious diseases has been N95, originally for protection against tuberculosis and other bacteria. During the SARS outbreak some reports suggested that they are effective while others suggest that they are not.[104] These are single use respirators and were not designed to be cleaned or washed. The N class of respirator can also be purchased as N99 or N100. The numerical value refers to the efficiency of the filter. The N100 is actually 99.7% efficient while the others are 95 and 99% efficient. This efficiency is for an airflow rate of 85 liters per minute for penetration of median aerodynamic sized particles of 0.3 μm. The N means it is not resistant to oil. There are also R and P respirators which are resistant to oil and oil proof, respectively.[105]

The level of respirator above the disposal forms (e.g., N95) are half mask respirators. These are also types of APR, light weight, small in size, and easily maintained, and in many cases they can be reused. Many of these respirators are made of soft plastic or elastometric materials that easily mold to the contours of the face. For protection against infectious diseases, these respirators can be used with either HEPA or ULPA filters.

Limitations of these respirators are that they do not cover the eyes and have resistance for breathing. It has been suggested that this additional stress to the pulmonary system can be "significant" and result in confounding health effects.[106] However, when evaluated in the SARS outbreak, these respirators with HEPA filters appeared to be effective in preventing disease in healthcare personnel when some effective form of eye protection was included.

Full-face respirators are similar to the half-mask APRs except they cover the eyes. These respirators suffer from the same limitations as half-masks, except the face piece can fog up or become scratched. However, clearly the advantage is that they provide protection against exposure to the eyes (conjunctiva). Both half-mask and full-face respirators are negative pressure respirators, as is the N95, and can result in face seal leakage. The negative pressure is created inside by the user when drawing in air through the filters. Facial protection (e.g., face shield and eye protection) without a respirator has been suggested to be ineffective when evaluated in dentists for the presence of antibodies against viruses.[107] This suggests the great importance of respirators when considering protection against microorganisms in healthcare settings.

Powered air-purifying respirators (PAPR) are similar to full-face respirators except they are positive pressure. These respirators filter the air, using HEPA or ULPA filters, using a small motor that draws the air. The advantage of this type of respirator is that the user does not experience pulmonary stress in moving air across the filter. However, these respirators can be bulky and the battery limits the amount of time it can be used. From a practical point of view, they would not be highly effective in most healthcare settings and may result in patient insecurity. On the other hand, they can be every effective and provide a high degree of protection for the user and would be considered more economical than attempting to use a supplied air system (SAR). They would also be applicable for those that cannot tolerate the stress of using a negative pressure APR as well as those having respiratory or pulmonary impairments but are required due to limitations of personnel to treat patients. Some have suggested that PAPRs be used for emerging infectious diseases, including SARS.[108] One study reported that healthcare workers preferred a PAPR over N95, in general, and as they were used over time became more acceptable.[109] This suggests that healthcare personnel can adapt to levels of PPE, although its acceptance probably requires training and effort to change perceived concepts.

Supplied air respirators provide breathing air from an outside source. This air must be categorized at a minimum as grade D. This type of system is often used in research facilities that perform experiments on highly dangerous viruses. SARs are not effective for treatment of patients with an emerging infectious disease and would require establishment of special rooms for such use.

Other types of respiratory systems have also been tested (surgical helmets). It was reported that helmet type respiratory systems are not adequate for protection against SARS and probably most other microorganisms.[110] Thus, selection of a respirator or system must be undertaken with care. However, it is interesting to know in the study on helmets that N100 respirators appear to be effective in filtration, which was used as a comparative system. N100 respirators were reported to have a protection factor of 100 using a PortaCount® Plus quantitative test system. This suggests that N100 respirators are effective as a respiratory protective device along with demonstrating the importance of proper fit testing. This quantitative fitting testing system has been reported to count particles in the size range of 0.02 to 1 μm, which is in the range of many viruses and small bacteria.[111]

One of the difficult activities that are involved with emerging infectious diseases is respirator selection.[112] Table 3.2 provides some guidance on selection of a respirator. This table provides, in general, the advantages and disadvantages for different types of respirators. Certainly when making a selection there are a large number of factors to be considered.[113] Some of the selection factors that can be important in regard to emerging infectious diseases include, type of agent, its mechanism of spread, and applicability to methods of medical treatment. Selection of an appropriate respirator during the SARS event became a major issue. Little thought went into what type of protection was needed for the organism causing the outbreak. The N95 was suggested as the respirator of choice because it was commonly used in infection control, however, this was designed for bacteria (tuberculosis), which has different characteristics as an airborne particle in comparison to a virus.

During outbreaks of infectious disease, healthcare workers in hospitals are not the only occupational groups that can be considered to be at high risk. Those that are on the front line of the healthcare system are also at great risk (e.g., prehospital healthcare workers). In many cases ambulance workers, other emergency personnel, and those in ambulatory medical environments are at similar risks and will be the first to see an infected patient. Little information and guidance has been presented in the literature on this population of workers.[114] The problem with this group is that they are often not in a very structured work environment, as compared to a hospital setting, and do not have the benefit of being prepared for such an occurrence. They also are not likely to have received much instruction on infectious disease transmission and prevention, especially in using respirators and other forms of PPE. Table 3.3 provides a description of suggested PPE, including respirator, for this category of healthcare worker.

Table 3.2 Respirator Selection Involving an Emerging Infectious Disease

Type of Respirator	Criteria of Selection
Surgical Mask/ Disposal Masks	Not recommended for emerging infectious diseases. Has no face seal protection and no protection factor can be established through fit testing. One study did report some effectiveness during the SARS outbreak, suggesting at least a lower rate of infection than without use. Paper masks were reported to be ineffective during the SARS outbreak. Use may create a false sense of protection for HCW. These respirators may provide some protection against large droplets.
N95	Historically used for bacteria, especially tuberculosis. Reports suggest that it is not completely effective for the SARS CoV; however, some indicated that it did have effectiveness. Does not provide protection for all mucous membranes (conjunctiva) and is negative pressure, disposable, half-mask type respirator. Has been suggested to have limited to poor face seal, with roughly about a 10% leakage overall. Particularly not effective when medical procedures involving aerosolization are involved. Not considered by some as the best selection for an emerging infectious disease, but considered better than no respirator or the use of surgical masks.
N100	Suffers from the same problems as N95s, but has been suggested to have a higher protection factor. However, generally not recommended for emerging infectious disease for the same reasons discussed under N95.
Half-mask with HEPA filter	Half-masks which are elastomeric respirators that can be fit tested and by regulation are considered to have a protection factor of 10 and can be considered to have a good face seal; although they are negative pressure respirators. Studies have reported a much higher protection factor than that presented by regulatory standards. When combined with HEPA or ULPA are suggested to be effective for most emerging infectious disease agents. Does not provide protection for mucous membranes, conjunctiva. Based on the SARS incident, when combined with other forms of PPE these are effective.

TABLE 3.2   Respirator Selection Involving an Emerging Infectious
Disease *(Continued)*

Type of Respirator	Criteria of Selection
Full-face with HEPA filters	These provide protection for the entire face including the conjunctiva. Suffer from the same limitation as half-masks in that they are negative pressure respirators. Has been suggested to be the best practical protective measure for emerging infectious diseases; however, can cause anxiety and fear with patients first encountering healthcare workers using such equipment. Has a high protection factor, both as reported in testing and by regulatory published values. Use also requires inclusion of other forms of PPE.
PAPR with filters	Protects the entire face as do full-face respirators, except they are with HEPA positive pressure. This is the highest level of practical protection. Limitations due to bulkiness and battery supply. These are battery powered. Suggested for use if aerolization procedures are being used.
SAR or type C	These are air supplied respirators where the air source is remote from the work area. OSHA has established requirements for the air used in these respirator systems (e.g., grade D, no less than 19.5% oxygen). These respirators are some times known as SCBAs (self-contained breath apparatus), which are commonly used in firefighting. For the most part they are impractical for use in a health care setting, although used in research activities involving highly infectious disease agents.

Note: Examples of respirator selection for the emerging diseases mentioned are
as follows: tuberculosis — N95; SARS, monkeypox, and influenza — half-mask
with HEPA filter. This is for initial responses and does not include activities that
would aerosolize biological materials.

It attempts to address some of the practical aspects of the working
environment, while affording the best protection to the worker.

Table 3.3 can also be used for those in an ambulatory setting as well.
This table is primarily a sliding scale for PPE, especially respirators. It
employs a scenario of the presence and occurrence of an outbreak. It
does not take into account the type of organism and the severity of
diseases which it can cause. Most emerging infectious disease that will
be acute in nature and pose a hazard of being an epidemic will likely be
viral in nature. It is also likely that the main route of spread will be by
aerosol droplet, although other routes, such as fecal contamination, are

TABLE **3.3** **General Suggested Guidelines for PPE Regarding Emergency Response Personnel, Including First Responders and Prehospital Healthcare Workers⁺**

Type of event and personnel	Respirator	Other PPE
Outbreak (initial reports) outside regional area, unlikely to be locally at this time		
Nondirect contact Personnel	None	Gloves/hand washing
Direct contact Personnel	Surgical mask	Glove/protective suit Hand washing
Epidemic occurring outside region but no cases in local area, low risk of cases		
Nondirect contact Personnel	Surgical mask	Gloves/hand washing
Direct contact Personnel	N95	Glove/protective suit Hand washing
Epidemic, chance cases may emerge		
Nondirect contact Personnel	Surgical mask	Gloves/protective suit Hand washing
Direct contact Personnel	Half-mask	Glove/protective suit Hand washing/eye protection
Epidemic occurring locally		
Nondirect contact Personnel	N95	Gloves/protective suit Hand washing/eye protection
Direct contact Personnel	Half-mask	Glove/protective suit Hand washing/eye protection

⁺ Nondirect contact personnel include police and fire, who will generally not be directly involved with treatment of the patient. Direct contact personnel are paramedics, ambulance and related mostly nonhospital or ambulatory personnel. Any respirator, including N95 must be fit tested and those using such equipment properly trained and instructed. Surgical masks are recommended only as a barrier device against large droplets and cannot be fit tested; however, as noted in the text, they are not highly effective and will at best prevent large droplets from entering. All respirators must use HEPA filters and the user fit tested..

also possible and should never be excluded. If consideration is given for other diseases related to bioterrorism, then issues of bacteria and related organisms pose a risk as well, as seen for anthrax and plague.[115] For some diseases, such as plague, it appears that the person is not highly contagious as thought by many, especially in the initial stages of disease, including prodromal.[116] The problem with this scenario is that such conditions vary greatly from organism to organism and no universal rule can be established. From observation and study of the SARS incident, the requirement of fit testing and training is paramount and cannot be excluded in this population. Lack of these functions could easily void any benefit that would be provided from PPE.

Fit Testing and Fit Checks

Respirator fit testing is one of the most important functions for ensuring a respirator will properly provide the "selected" level of protection. This term, *respirator fit testing*, is sometimes improperly used to include two different types of activities, fit testing and fit checks. Only respirators that allow a complete seal to the face can be "evaluated" through the use of various fit testing agents. Thus, paper masks or other "types" of respirators, which would include surgical masks, cannot be fit tested. The minimum category of respirator that can be fit tested or have a fit check conducted is a half-mask air purifying respirator. Much of the importance of fit testing and proper fit of a respirator was forgotten during the SARS event.[117] However, the importance of such testing cannot be overemphasized. Based on reports in the literature, it can be said that use of respirators and their appropriate fitting was considered in light of economic factors, rather than those related to preventing the spread of disease. Much of the failure in protection, especially as related to surgical masks and N95 respirators, has been attributed to a lack of appropriate fit testing.[118] Table 3.1 suggests that the rate of failure, although it may not be directly as a result of inadequate or lack of fit testing was 2 to 25%. Even if only 25% of these cases are a result of poor fit or no testing, it represents a large number of preventable cases. This is an important concept and practical application is paramount in providing the best protection. Overall, this aspect has had only limited attention in the literature.

A respirator fit check is performed each time a respirator is put on. It is commonly called a positive/negative pressure fit check. Although these tests are the "least precise" for determining seal to the face of a respirator, this has been suggested to be the best method of determining adequate fit for protection at the time of using a respiratory device. The positive pressure involves the wearer covering the inlet with the palm of the hand or impermeable material, such as a small plastic bag, and gently exhaling. If leakage does not occur and the wearer detects a slight positive pressure inside the mask it is considered that the test was successful. For negative pressure, the same is done except the inhalation values are covered and the wearer creates a negative pressure by slightly inhaling. Most leaks will occur around the nasal area of the face/respirator. The advantage of these tests are that they are quick, rapid, easy to perform, and can be accomplished by the user. If there is any impairment that interferes with the face/respirator seal, then the user cannot employ a respirator. The most common impairment for this seal is facial hair, such as a beard. However, if the facial hair does not interfere with the respirator/facial seal then it would be acceptable (e.g., mustache). Training respirator users to perform fit checks is not difficult, but can be time consuming. In many ways this is the difficult aspect of training for this type of testing, in that there can be considerable cost involved. In many ways, some have taken this as a tradeoff and elected not to initiate training and use as a result of economics taking priority. However, as seen in Canada during the SARS events, this occurred with tragic consequences.

Fit testing is required by many regulatory agencies, including OSHA.[119] The purpose of this testing is to ensure that there is an adequate protection factor (PF) for the respirator when it is in use. Fit testing basically evaluates the seal of the respirator with the person's face (face-fit). Failure to implement this basic activity in a healthcare setting is most likely to be a result of inadequate understanding of respirators, time required to implement, cost associated with respirators and equipment for testing, the need for a trained person to conduct and perform these functions, and maintenance costs associated with keeping such an associated program in place. Costs associated with fit testing are even greater than for fit checks. Simply applied, it's an issue of time and money. Use of a respirator that has an inadequate fit may actually be more dangerous than not wearing a respirator. This is because use of a properly fit respirator may lead to a false sense of protection by the wearer, whereas in reality there is actually little or no protection. Most agencies throughout the world copy or reference the OSHA criteria for fit testing and respirator use. According to OSHA standards for asbestos and lead abatement, a person using a respirator must be fit tested at least once a year. In addition, they are required to be medically approved to use a respirator as well. However, these criteria are derived for exposure to airborne asbestos or lead, both being considered chronic toxicants. Infectious diseases are considered for the most part to be acute hazards. Since the time period for an emerging infectious disease may not be known, annual fit testing would be appropriate with a need of refit testing upon discovery of an outbreak within a facility or location. For global or regional events, it may be prudent to fit test first responder personnel upon discovery of an event or occurrence. Such activity could be incorporated as part of the periodic program in respirator and other PPE use. As in the SARS outbreak, it has been suggested that many of the healthcare workers who used respirators and became infected were not adequately fit tested. Thus, fit testing is important and required for healthcare personnel who will employ these devices during an outbreak of an infectious disease [120] Persons who are fit tested must also be adequately trained and such training directed to the specific hazard (e.g., infectious disease).[121] Commonly, fit testing is incorporated as part of employee training in respirator use.

The first part of any respirator fit test involves selection of a respirator for the individual. Respirators are available in small, medium, and large sizes. The actual size and shape of a specific listing of a respirator may also vary from manufacture to manufacture. Before fit testing is conducted, appropriate selection by the healthcare worker must be undertaken to ensure that the respirator is adequate and comfortable. This is also true for any type of PPE selected. For healthcare workers such activities as this can be performed during orientation, and also as part of continuing education. May regulatory agencies in the United States require a specific number of hours of continuing education, and respiratory protection and its applicability in disease prevention could easily be included as part of this in-service training. This would eliminate some of the economic trade-offs that are required as part of a respiratory program and allow such use to serve two functions, continuing education and prevention.

Fit testing can be classified as either qualitative or quantitative. There are four commonly described qualitative fit testing agents (Table 3.4). Others may be used but they must meet the requirements established by the governing regulatory agency, as in the case of OSHA, are specified by regulatory standards. When fit testing is qualitative a set protection fact is assigned for those successfully completing the test. Each of the test agents listed in Table 3.4 are considered equivalent in determining a respiratory PF, although in practicality, each has its advantages and limitations, which are discussed later.

If the person has a change in his or her facial features or something that interferes with the face seal, after fit testing, retesting would then be required. Examples of conditions that would require retesting are: significant gain or loss of weight (e.g., usually considered +/-20 pounds or 10% of body weight) scarring, plastic (cosmetic) surgery, or dental changes.[123] This condition would be for those that have a significant change for the facepiece-to-face seal. Those using respirators must be trained to recognize conditions that may require refitting. This would be an important issue for healthcare workers and would be critical as part of any training program.

TABLE 3.4   Types of Qualitative Fit Testing Agents

Agent	Detection of Agent
Isoamyl acetate (banana oil)	Odor Threshold
Saccharin	Taste
Irritant Smoke	Irritation to respiratory system
Bitrex™	Taste

Note: According to OSHA standards (e.g., asbestos) the PF for each of the qualitative test agents is 10. A PF of 10 means that there will be a 10-fold concentration outside the mask as compared to inside. However, in many cases, the PF is adjusted for the type of respirator being employed. Regulatory agencies (OSHA) have assigned PFs for various respirators (Table 3..5), although these values do not correspond to that of experimental data. There have been published experimental values for many respirators and masks as related to protection against tuberculosis. These values for surgical masks, N95, elastomeric half-mask (with HEPA filter), and PAPR (with HEPA filter) are 2.5, 17.5, 46.9, and 236, respectively.[122] Most fit testing is conducted for negative pressure respirators (half- and full-face masks), although these tests can also be used for PAPRs and SARs, which are generally positive pressure. PAPRs and most SARs are positive pressure respirators and in many cases due to being positive pressure, fit testing would not be warranted. There are some SARs which are not positive pressure (negative pressure), but they are not commonly used in most industries today. SARs are used in research laboratories where the most dangerous infectious diseases (e.g., smallpox, Marburg) are being studied, so they do have use and applicability in some aspects of emerging infectious diseases. Quantitative fit testing can also be conducted for these respirators as well.

TABLE 3.5   Types of Respirators and Published Protection Factors

Half-mask	10
Full-face mask	50
Powered air-purifying respirator	100
Supplied air respirator	1000+

Each fit testing agent has its advantages and disadvantages. However, if the test agent is employed properly it can be effective in evaluating the fit of a respirator. Historically, the fit test agents commercially available were banana oil, saccharin, and irritant smoke. Most recently Bitrex™, which is denatonium benzoate, has become commercially available for fit testing.[124] This agent results in a bitter taste in the mouth of subjects who do not have an adequate fit.

The major advantage of irritant smoke in fit testing is that those who do not have a proper fit experience an involuntary cough. This cough is a result of the irritant smoking and giving off hydrogen chloride and tin fumes. The reaction with moisture in the air (humidity) causes a white smoke having a pungent odor. This is the result of a reaction of stannic chloride with moisture in the air. The mixture of HCl and tin results in irritation to mucous membranes (respiratory tract and eyes) at relatively low concentrations. However, at higher concentrations, even for short exposure periods, it can result in coughing, chest pain, and choking. The OSHA Permissible Exposure Limit or Level (PEL) for HCl is 5 ppm and NIOSH listed an immediate danger to life and health (IDLH) concentration at 100 ppm. Stannic chloride products on the other hand are considered less toxic than HCl. Today most manufactures do not directly recommend using these agents for fit tests although they continue to be commonly employed. This recommendation by manufacturers is a result of the potential hazards exhibited by HCl, although from a practical point of view the actual hazard is very low and could be considered by some as nonexistent.

Saccharin and banana oil involve the subject detecting exposure by taste and odor, respectively. Both exhibit a low risk of toxicity to those exposed. Exposure limits published by OSHA and NIOSH are 100 ppm (PEL) and 1,000 ppm (IDLH).[125] The disadvantage of these agents is that not everyone can detect them and users may not accurately identify to the tester that they detect the taste or odor.

Bitrex™ is the most recent agent identified for fit testing, and has undergone evaluation against other fit testing methods.[126] This agent has a bitter taste, has been used to denature alcohol, and has been incorporated into other products to prevent accidental poisoning, although its effectiveness for preventing accidental poisoning has been questioned [127] This highly bitter taste is considered to be an advantage over saccharin and banana oil, while exhibiting a low level of toxicity to the user. Material safety data sheets for this chemical do identify that short-term exposure can result in asthma and contact uricaria, although this is apparently based on a single case event.[128]

A study comparing Bitrex™ with a quantitative method (PortaCount® testing), using a N95 respirator suggested a higher fail rate with the quantitative agent as compared to the qualitative agent.[129] Overall, such testing suggests the applicability of using quantitative over qualitative testing for highly infectious agents.

All fit agents required that those being fit tested be able to detect the agent. For the odor and taste testing agents, this is accomplished before the test is conducted. When using an irritant smoke detection is performed at the end of the test itself. The OSHA standards, as for asbestos, provide a detailed protocol for testing with these agents (irritant smoke, saccharin, and banana oil).

Quantitative fit testing provides the actual amount of leakage for the respirator face seal. Historically there were two methods for quantitative: sodium chloride and dioctylpthalate (DOP). Other tests that have been employed include ethylene gas, freon, methylene blue, and paraffin oil. Currently there are commercial quantitative fit testing devices available (e.g., PortaCount® Plus). In many cases these commercially available devices have replaced the older and more difficult system for quantitative respirator fit testing. To conduct a quantitative test, it is necessary to measure the test chemicals' concentration inside and outside the respirator. This measure is performed by instrumentation. The advantage of this type of testing is that it does not rely on the subjectivity of the person being tested, although the instruments for evaluation are expensive and require a trained operator. There are commercially available instruments that can perform these tests and are now computer based.

Regardless of the type of fit testing conducted, it must be performed, according to OSHA, or the applicable regulatory standard, once a year. It is also necessary that personnel use the respirator in a proper manner. Ideally the wear factor, which is a measure of the percentage of time a respirator is actually worn, will be 100%, especially in situations where the organism has an infectivity rate of about one "organism."

Most if not all publications on respirator use, especially related to the SARS incident, note that fit testing and related practices (e.g., applicable training) are of importance in ensuring protection. In many of these cases they did not recommend use of a proper respirator or suggest an inadequate respirator or its inappropriate use that would not provide protection.[130] The improper use of a respirator would negate any benefit gained through fit testing and fit checks. One study of workers (asbestos) who received annual or periodic training on respirator use found that there was a less than effective use of these protective devices.[131] Thus, for those who have little training it is likely that the "efficiency" will even be much less, although, their exposure potential could be of much greater concern and of greater risk in that these emerging diseases are acute hazards. In many ways, these factors contribute to selected trade-offs in establishing an effective respiratory program against economics. However, one case of infection in a healthcare worker would greatly outweigh any savings associated with not implementing an effective PPE program.

Respirator Program

It is not uncommon that respirators are not properly used [132] Much of the improper use is a result of failure to train personnel and a lack of understanding by management.[133] As with any device, appropriate training, quality assurance, and supervision are required. OSHA regulations require a written program governing the selection and use of respirators. These criteria and requirements can be found in 29 CFR 1920.134. The Written Standard of Operating Procedures needs to contain necessary information so that an effective and applicable respirator program can be administered. As with any program, it must be specific to the requirements of the users and the hazards that they may face. This would be more critical for those involving agents that have a high acute fatality rate.

Training may be one of the most important aspects of a respirator program. Most healthcare workers do not receive training on PPE as part of their formal educational process. During the SARS incidents, it was indicated in the literature that a lack of training had a direct responsibility for an increased rate of disease among healthcare workers.[134] Many of the healthcare workers did not receive any training on proper use of respirators at the time of an outbreak in their facility; although, as a result of the new concerns for bioterrorism, many are beginning to receive this type of training. Any training should include information on how to use respirators, fit testing (including checks), inspection, cleaning/maintenance, and selection. The frequency of training appears to be important in that most training lacks long-term effectiveness.[135] However, as noted with the SARS event, most healthcare personnel have not even received the most basic training. Peer feedback intervention does appear to be most effective and that, when combined with periodic training, may provide a high level of prevention.[136] There may also be a benefit to providing information on emerging infectious disease, in general. Understanding of the agent can aid in providing a logic in how selection is conducted for the hazard and the limitations of PPE in such circumstances. Thus, the availability of respirators by themselves is not effective without an effective program. It should also be noted that some of the groups that are least effective in using and applying PPE are the highest trained professionals. This alone may be one of the greatest hindrances in establishing an effective program and applicability of training. Time is often critical to these people and they do not consider training on PPE to be of great importance. Problems such as this are one of the many limitations that need to be evaluated and researched in developing effectiveness of

PPE for healthcare personnel. Since disease does not select by occupation, training and application must be across the spectrum of those who work in health care.

Summary

Respirators are considered a secondary form of protection with engineering control the primary protection. However, for emerging infectious diseases engineering controls are difficult and in some cases impossible to establish, especially for a large number of cases and at the initial outbreak. The inability to effectively employ engineering controls makes respirators and related PPE a primary form of protection for the worker. Selection of an appropriate respirator and PPE is critical and has been shown to aid in reducing the number of occupational cases of a disease. Appropriate and effective use of PPE has been demonstrated during the outbreak of SARS. Care must be taken that the basis of selection is not for economic reasons but rather as a preventative measure. Many of the emerging diseases, as well as historical infectious agents, are easily spread by airborne and related routes (e.g., fomite). Even when respirators and PPE are employed, the lack of proper training and support can greatly reduce effectiveness. Communication on the applicability and proper use of respirators, along with fit testing and fit checks are imperative to a good program and protection of healthcare personnel. It must be remembered that healthcare workers refers not only to those in hospitals and ambulatory settings, but can include nursing homes. As an example, a nursing home could serve as a reservoir for a disease along with the healthcare workers employed there. The SARS event, which elucidated the importance of emerging infectious disease and the potential global implications, will not be the last emerging disease. Additional precautions are need in preparation for the next and inevitable event.

Chapter 4

Airborne Pathogens: Selection of Respiratory Protection

Mark Nicas, Ph.D., M.P.H., C.I.H.

CONTENTS

Introduction

Human respiratory tract infection by a variety of pathogens can occur via inhalation of airborne organisms, although for most pathogens it is thought that infection via droplet transmission or hand-to-mucous-membrane contact is more important. In general, the proportion of respiratory infections that occur via alternate exposure pathways has not been determined via experiment or epidemiology. For example, firm experimental evidence that respiratory illness due to rhinoviruses is transmitted via direct hand contact with nasal membranes and the conjunctiva of the eyes has led to a common belief that most all common colds are transmitted by this exposure route. However, an intervention study in which mothers periodically applied a virucidal iodine solution to their fingers upon the appearance of cold symptoms in a sick family member found that infection incidence among the mothers was decreased by 66% relative to a control group of mothers who applied a nonvirucidal placebo solution. [1] Although the iodine solution may not have killed all virus particles on the treatment mothers' fingers, it is possible that inhalation transmission was responsible for the remaining secondary infections in these women.

Despite general uncertainty about the predominant routes of exposure, the default assumption among infection control professionals seems to be that droplet transmission and hand-to-mucous-membrane contact are far more important than inhalation transmission. This assumption was evident in the initial recommendations for protecting health care workers who attended patients with severe acute respiratory syndrome (SARS). In fact, one group of hospital epidemiologists published a study in which they argued that SARS was not transmitted via inhalation.[2] Subsequent studies presented strong circumstantial evidence for airborne transmission of the SARS corona virus,[3] although it must be noted that the proportion of SARS cases that occurred due to different exposure pathways is still unknown. It should go without saying that when the medical profession is confronted with a new human pathogen, simply assuming that inhalation transmission does not occur can lead to serious adverse consequences. At the same time, the occurrence of inhalation transmission does not preclude the possibility of infection by droplet transmission and hand-to-mucous-membrane contact.

The purpose here is not to attempt an apportionment of respiratory infection incidence by different routes of exposure. Instead, this chapter describes a risk-based approach to selecting respiratory protection given that one knows or assumes that inhalation transmission can occur. It must be recognized at the outset that the air-purifying respirators available to healthcare workers will not entirely eliminate airborne pathogen exposure. Different types of respirators will reduce pathogen exposure to different degrees. Therefore, one must consider what degree of exposure reduction is necessary. The general procedure is to consider the pathogen's dose–response characteristics, estimate airborne exposure intensity, specify an acceptable or target risk of infection, and choose a respirator that will reduce exposure intensity to meet the target risk level. Although the overall procedure is described, this chapter does not offer respirator recommendations for specific pathogens, because assigning an acceptable risk value is a complex sociopolitical process that should properly be undertaken by public health agencies.

The Infectious Dose Model

The infectious inhalation dose model used in this chapter is probabilistic or "nonthreshold." It is assumed that a single organism can infect the host with a probability denoted α, such that there is no threshold number of organisms (other than one) required for infection. [4] Let d denote the integer number of organisms deposited in the appropriate region(s) of the respiratory tract. The risk (probability) of infection conditioned on the values d and α, denoted $R_{d,\alpha}$, is given by:

$$R_{d,\alpha} = 1 - (1 - \alpha)^d \qquad (4.1)$$

The logic of the equation is as follows. The probability that a single deposited organism will not infect the host is $1 - \alpha$. The probability that D organisms will not infect the host is the product of the independent probabilities that each organism will not infect the host, or $(1 - \alpha)^d$. The risk of infection is the complement of the probability of not being infected.

If the value of α is the same across human hosts, the infectious dose 50% value, denoted ID_{50}, can be loosely described as that deposited dose which produces a 50% likelihood of infection in any host. A probabilistic model can account for variability in susceptibility to infection by treating the α value as variable across hosts, as will be explained.

A deterministic or "threshold" model assumes that if the host receives some threshold number (or greater) of organisms, infection is certain to occur, whereas if the host receives fewer than the threshold number of organisms, infection is certain not to occur. Variability in susceptibility to infection is reflected by interindividual variability in the threshold dose value, and the ID_{50} is that deposited dose which will infect 50% of the population with certainty.

In general, the available data on airborne pathogens are too sparse to permit distinguishing between the probabilistic and deterministic frameworks. For example, in an analysis of experimental data for inhalation anthrax, it was found that a deterministic model provided a better fit to guinea pig mortality data than did several probabilistic models, but that a probabilistic model provided a slightly better fit to monkey mortality data than did a deterministic model. [5] This chapter uses the probabilistic model because it is consistent with observed dose-infection response data for a variety of organisms, [6] and it tends to be more health conservative (tends to produce a higher risk estimate) than a deterministic model.

A traditional way to account for variability in host susceptibility to infection is treat the α parameter in Equation 4.1 as a standard beta random variable. However, if it is known or suspected that α varies across hosts, it is generally acceptable to use the population average value of α, denoted μ_α, in the risk calculation.[7] For a given individual, assigning the population average value of α may greatly underestimate or overestimate the person's infection risk. On the other hand, the individual's true value for α will not be known a priori, so it is proper to view a given individual as randomly drawn from the population and to assign that individual the expected α value. Given an integer dose d, infection risk is:

$$R_d = 1 - (1 - \mu_\alpha)^d \qquad (4.2)$$

Equation 4.2 requires a specified dose, but in the risk estimation framework the dose received will be uncertain and is appropriately treated as a random variable. This chapter assumes that the random dose D is adequately modeled as a Poisson variable with mean μ_D. The unconditional risk of infection R is found by summing R_D across all possible dose values D = d weighted by the corresponding probabilities that D = d. The resulting equation is:

$$R = 1 - e^{-\mu_D \times \mu_\alpha} \qquad (4.3)$$

where the quantity e is the base of the natural logarithm.

Equation 4.3 involves just two quantities — the expected dose μ_D and the average host susceptibility value μ_α. Estimating the μ_D value will be described subsequently. The μ_α value is approximately related to the ID_{50} as follows:

$$\mu_\alpha = \frac{0.693}{ID_{50}} \qquad (4.4)$$

In Equation 4.4, the ID_{50} is properly interpreted as that expected dose of pathogens (μ_D) which corresponds to a 50% chance of being infected, in which case the ID_{50} value need not be an integer and can be less than one. However, the ID_{50} value cannot be less than 0.693 because μ_α cannot exceed one. Equation 4.4 indicates that as the ID_{50} value decreases (the pathogen has higher infectivity), the μ_α value increases (each pathogen has a greater likelihood of infecting a host). Estimates of the ID_{50} value

Table 4.1 The Estimated Average Infectivity Parameter μ_α and the Corresponding ID_{50} Value for Select Respiratory Tract Pathogens for which Inhalation Transmission Is Known to Occur

Pathogen	μ_α	ID_{50}	Source
Mycobacterium tuberculosis	1	NA	8
Coxiella burnetii	0.9	0.8	9
Francisella tularensis	0.7	1	10
Coccidioides immitis spores	0.4	1.7	11
Variola major virus	0.1	6.9	12
Bacillus anthracis spores	1.5×10^{-5}	4.6×10^3	13

are available for some pathogens. Table 4.1 lists μ_α and ID_{50} values for several respiratory tract pathogens.

Figure 4.1 illustrates the intuitively sensible relationship between the unconditional infection risk R, the expected dose, and the ID_{50} value (the average host susceptibility). Expected dose values of 0.1, 2.0 and 10 are considered across a range of ID_{50} values from 0.7 (μ_α = 0.99) to 70 (μ_α =.01). For a fixed expected dose, R increases as the ID_{50} value decreases. For a fixed ID_{50} value, R increases as the expected dose μ_D increases.

Respiratory Protection

The effect of wearing a respirator is to reduce the expected dose received (decrease the μ_D value by some fraction) and thereby reduce infection risk. Let μ_P denote the average or expected penetration through the respirator system (a fraction between 0 and 1) of the airborne pathogen. Different types of respirators have different associated μ_P values. The new unconditional risk is given by:

$$R = 1 - e^{-\mu_P \times \mu_D \times \mu_\alpha} \tag{4.5}$$

By setting μ_P = 1 (that is, no respirator use), Equation 4.5 is the same as Equation 4.3.

Face Seal Leakage versus Filter Penetration

For most air-purifying respirators tested and certified by the National Institute for Occupational Safety and Health (NIOSH), the extent of pen-

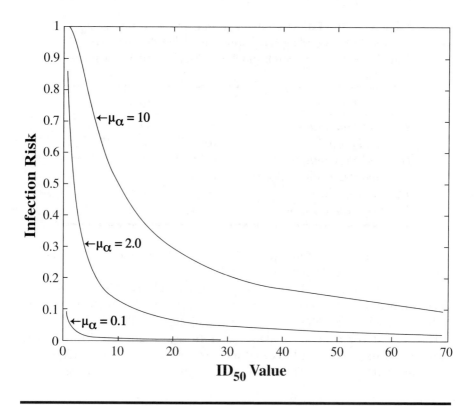

Figure 4.1 The unconditional infection risk computed by Equation 4.3 given alternative expected dose values of 0.1, 2.0 and 10, across a range of ID50 values from 0.7 (μ_α = 0.99) to 70 (μ_α =.01).

etration is primarily determined by inward leakage of unfiltered contaminated air around the respiratory inlet covering (for example, inward leakage around the face seal perimeter of a tightly fitting face piece) and not by contaminant penetration through the respirator's filter elements.

To explain, consider the type 95 filtering-face-piece respirator (N95 FF), which is the NIOSH-certified respirator most commonly used by healthcare workers in the United States. According to NIOSH regulations, the N95 filter may allow up to 5% of test particles with aerodynamic diameters on the order of 0.3 μm to penetrate through the filter. However, that same filter would permit a lower percentage of larger 3-μm aerodynamic diameter particles to penetrate through the filter, and those larger particles are expected to carry substantially greater numbers of pathogens due to their greater volumes. For example, a 3-μm diameter particle has 1,000-times greater volume than a 0.3-μm diameter particle. Therefore, assume that the N95 filter permits 0.5% of the pathogens in the air volume inspired through the filter elements to penetrate into the respirator. Next, a N95 FF could readily permit 5% of the inspired air volume to enter the respirator through small leaks around the face seal perimeter; none of the pathogens in this air leakage would be removed in the process. The overall pathogen penetration into the respirator is a volume-weighted average of the penetration through the face seal leaks and through the filter. Given the values offered (5% air entry through face seal leaks with 100% pathogen penetration, and 95% air entry through the filter with 0.5% pathogen penetration), the overall penetration value is 5.5%, of which the great majority (90%) is due to face seal leakage:

$$\text{Overall \% Penetration} = \underbrace{(.05) \times (100\%)}_{\text{Faceseal Penetration}} + \underbrace{(0.95) \times (0.5\%)}_{\text{Filter Penetration}} = 5.5\%$$

The Assigned Protection Factor and Average Penetration

The degree of exposure reduction due to respirator use has traditionally been summarized by the assigned protection factor (APF), which is the inverse of the assumed overall penetration (a fraction between 0 and 1) into the respirator. For the N95 FF class, NIOSH recommends APF = 10, which signifies that the assumed overall penetration is $1 \div 10 = 0.1$ (or 10%).[14] However, a recent analysis of data collected from various workplace studies in which half-mask respirator penetration was measured while subjects performed their normal job duties indicates that APF = 5 is more appropriate for the N95 FF and other halfmask respirators.[15] In addition, the μ_p parameter in Equation 4.5 is better estimated by $0.4 \times (1 \div \text{APF})$. The reason involves the variability in respirator penetration from wearing to wearing, and the manner in which an APF value is statistically derived.[16]

In brief, respirator penetration values are thought to vary lognormally across different wearing periods, and the APF is usually equated with the inverse of the 95th percentile of the penetration values. However, the computation of unconditional infection risk properly involves the average penetration value, which is less than the 95th percentile penetration value. If one assumes that the lognormal distribution of respirator penetration values has a geometric standard deviation value of 2, the mean penetration value is 0.4 times the 95th percentile penetration value.

Table 4.2 lists suggested APF values, along with the corresponding μ_p values equal to $0.4 \times (1 \div \text{APF})$, for several types of respirators that might be used by healthcare workers. Figure 4.2 depicts the reduction in infection risk afforded by three types of respirators across a range of pathogen ID$_{50}$ values from 0.7 ($\mu_\alpha = 0.99$) to 70 ($\mu_\alpha = .01$), given that the expected dose

Table 4.2 Suggested Assigned Protection Factors and the Corresponding μ_P Value for Different Classes of Respirator Devices that Could Be Used in the Healthcare Environment

Class of Respirator	Assigned Protection Factor	Mean Penetration Value, μ_P	Source
Half-mask filtering-face piece filter types N95, N99, and N100	5	.08	17
Half-mask elastomeric face piece filter types N95, N99, and N100	5	.08	18
Full elastomeric facepiece filter type N100	50	.008	19
Hooded powered air-purifying with high efficiency respirator filter used in the pharmaceutical industry	1000	.0004	20

without respiratory protection is $\mu_D = 10$. The three respirator types are: (1) a N100 half-mask filtering face piece, APF = 5; (2) a full face piece equipped with N100 filters, APF = 50; and (3) a hooded powered air purifying respirator equipped with high efficiency filters, APF = 1000. Clearly, the higher the respirator's APF value (the lower the μ_P value), the lower the risk of infection. Whether any of the three respirators in the Figure 4.2 scenario is deemed "adequate" depends on the acceptable risk of infection for the pathogen involved.

Estimating Exposure Intensity

Of the three input factors in Equation 4.5 (μ_D, μ_P, μ_α), the most uncertain value is the expected exposure intensity μ_D. For a given type of exposure scenario (for example, healthcare worker entry into a room to attend a patient), μ_D can vary across pathogens, across patients, across time for the same patient, across different procedures performed on the patient, and across ventilation conditions. However, it is possible to make a first pass exposure estimate if one is willing to make assumptions or has some measurement data available. Exposure estimates involving coughing by a patient and a laboratory accident are discussed below.

A Coughing Patient

Cough Particles

A cough emits several hundred particles of saliva fluid that span a wide range of sizes. Table 4.3 list the average number of particles in different

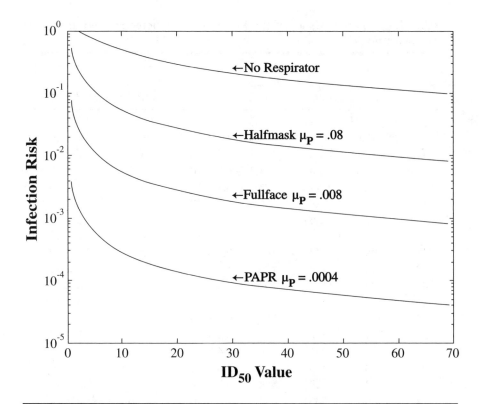

Figure 4.2 The unconditional infection risk computed by Equation 4.5 for an expected dose of 10 pathogens across a range of ID50 values from 0.7 ($\mu_\alpha = 0.99$) to 70 ($\mu_\alpha = .01$), given the alternative use of: (1) no respirator; (2) a half-mask N100 filtering-facepiece respirator; (3) a full face piece respirator with N100 filters; and (4) a hooded powered air-purifying respirator with high efficiency filters (equivalent to N100 filters).

diameter ranges emitted per cough. The original data were collected by Louden and Roberts,[21] and were used by this author and colleagues to examine the risk of airborne infection due to pathogens carried by respirable particles.[22] The term *respirable* refers to particles with aerodynamic diameters < 10 μm, which can reach and deposit in the alveolar region of the lungs. In contrast, "inspirable" particles have aerodynamic diameters in the range 10 μm to 100 μm; these larger particles do not reach the alveolar region, but can be inspired and deposit in the thoracic and head airways regions of the respiratory tract.

A brief explanation of Table 4.3 is as follows. The observed (reported) particle diameter ranges are listed in columns 3 and 4. Because emitted respiratory particles rapidly lose water by evaporation, the observed diameters are assumed to be approximately one-half the original diame-

Table 4.3 Characteristics of the Respiratory Particles Emitted in the Average Cough

$d_{0,min}$ μm	$d_{0,max}$ μm	$d_{eq,min}$ μm	$d_{eq,max}$ μm	$\bar{v}_0$ cm³	Mean Number Particles	Number of Pathogens per Particle[a]
2	5.81	1	2.9	3.8×10^{-5}	121	3.8×10^{-5}
5.8	11.6	2.9	5.8	3.8×10^{-10}	100	3.8×10^{-4}
11.6	17.4	5.8	8.7	1.7×10^{-9}	6.2	1.7×10^{-3}
17.4	22.4	8.7	11.2	4.2×10^{-9}	3.3	4.2×10^{-3}
22.4	52	11.2	26	3.1×10^{-8}	18	3.1×10^{-2}
52	112	26	56	3.2×10^{-7}	64	3.2×10^{-1}
112	170	56	85	1.5×10^{-6}	58	1.5×10^{0}
170	228	85	114	4.2×10^{-6}	31	4.2×10^{0}
228	288	114	144	9.1×10^{-6}	20	9.1×10^{0}
288	346	144	173	1.7×10^{-5}	12	1.7×10^{1}
346	406	173	203	2.8×10^{-5}	5.3	2.8×10^{1}
406	464	203	232	4.3×10^{-5}	4.3	4.3×10^{1}
464	524	232	262	6.3×10^{-5}	3.5	6.3×10^{1}
524	582	262	291	8.9×10^{-5}	2.7	8.8×10^{1}
582	700	291	350	1.4×10^{-4}	5.0	1.4×10^{2}
700	878	350	439	2.6×10^{-4}	0.50	2.6×10^{2}
878	1172	439	586	5.8×10^{-4}	5.0	5.8×10^{2}
1172	1468	586	734	1.2×10^{-3}	1.8	1.2×10^{3}
1468	1762	734	881	2.2×10^{-3}	1.3	2.2×10^{3}
1762	2058	881	1029	3.7×10^{-3}	0.33	3.7×10^{3}
2058	2352	1029	1176	5.6×10^{-3}	0.67	5.6×10^{3}
2352	2942	1176	1471	9.8×10^{-3}	1.7	9.8×10^{3}
2942	3532	1471	1766[b]	1.8×10^{-2}	0.67	1.8×10^{4}

[a] Expected number of pathogens per particle with volume = $\bar{v}_0$ standardized to $C_F = 1 \times 10^6$ mL^{-1}.

[b] The reported range was > 1471 μm. To assign an expected number of pathogens per particle, the upper range limit was set at 1491 μm + 295 μm = 1766 μm. The 295 μm increment equals the span of the preceding diameter range.

ters.[23] Therefore, columns 1 and 2 list the original particle diameter values, denoted d_0. The observed particle diameters listed in columns 3 and 4 are denoted d_{eq}, because it is assumed that the observed particle sizes were at equilibrium (water loss due evaporation equaled water gain due to condensation). The quantity $\bar{v}_0$ in column 5 is the mean initial volume (cm³) of a particle in a given size range, based on the assumption that particle diameters were uniformly distributed across the size range. Column

6 lists the average number of particles emitted per cough in each size range; these are based on data collected for a total of 90 coughs from three subjects. Column 7 lists the expected number of pathogens carried by a single particle in each size range, based on the assumption that the viable pathogen concentration in the saliva aerosolized by coughing, denoted C_{Sal}, equals 1×10^6 mL^{-1}. The value in column 7 is the product of the $\bar{v}_0$ value in column 5 and 1×10^6 mL^{-1}.

The Pathogen Emission Rate

The emission rate into air of respirable and inspirable pathogens, denoted G (# hr^{-1}), is modeled as the product of the patient's coughing rate, denoted E (# hr^{-1}), the respirable or inspirable particle fluid volume per cough, denoted V_F (mL), and the viable pathogen concentration in saliva C_{Sal} (# mL^{-1}):

$$G = E \times V_F \times C_{Sal} \tag{4.6}$$

Based on the Table 4.3 data, it is estimated that the respirable and inspirable particle fluid volumes emitted per cough are, respectively, 6×10^{-8} mL and 2×10^{-4} mL. However, the health status of the subjects involved in the Louden and Roberts study was not reported. [24] The author suspects that the subjects did not have respiratory tract infections, and that the cited particle fluid volumes per cough underestimate the true values for patients with respiratory tract infections and more productive coughs.

A study by Louden and Brown on a series of pulmonary tuberculosis (TB) and pneumonia patients reported the coughing rates summarized in Table 4.4.[25] It is evident that patients can vary by an order of magnitude in their coughing rate. For most pathogens, there are no available quantitative measurements of viable organism concentrations in saliva or other respiratory fluids. However, for *Mycobacterium tuberculosis* (*M. tb*), a study by Yeager et al., measured viable *M. tb* bacilli concentrations in the saliva and sputum of 22 pulmonary tuberculosis patients.[26] For *M. tb* in saliva, the range was 1×10^2 mL^{-1} to 6×10^5 mL^{-1}, with an average of 7×10^4 mL^{-1}. For *M. tb* in sputum, the range was 6.6×10^4 mL^{-1} to 3.4×10^7 mL^{-1}, with an average of 8.4×10^6 mL^{-1}.

Given the differences in cough rate, the orders-of-magnitude range in pathogen concentrations in saliva and other respiratory fluids (as indicated by the *M. tb* bacilli data), and likely differences in the particle fluid volume per cough among respiratory disease patients, a wide range of pathogen emission rates is to be expected. It is reasonable to infer that individuals termed "superspreaders" (as used in the recent SARS literature) or "dangerous disseminators" (as used in the older TB literature) are those

**Table 4.4 The Coughing Rate Among 96
Pulmonary Tuberculosis Patients and 48
Pneumonia Patients**

Pulmonary TB	Coughs hr⁻¹ (%)	Pneumonia (%)
< 1.5	34/96 (35%)	3/48 (6.3%)
1.5 to 3	15/96 (16%)	4/48 (8.3%)
3 to 6	12/96 (13%)	4/48 (8.3%)
6 to 12	18/98 (18%)	8/48 (17%)
12 to 24	9/98 (9.2%)	15/48 (31%)
24 to 48	6/98 (6.1%)	13/48 (27%)
> 48	2/98 (2.0%)	1/48 (2.1%)

infrequently encountered persons with high values for the factors E, C_{Sal}, and/or V_P, such that their pathogen emission rates are very much higher than average.

Pathogen Removal Pathways

Particle-associated pathogens emitted into room air are removed by exhaust ventilation, by particle settling due to gravity, and by pathogen death due to environmental stress. Each loss mechanism can be quantified by a fractional removal rate with the unit of inverse time, for example, hr^{-1}. If room air were perfectly mixed, the fractional removal rate due to exhaust ventilation denoted λ_{vent} (hr^{-1}) would equal the room supply air rate denoted Q ($m^3\ hr^{-1}$) divided by the room volume denoted V_{room} (m^3). However, room air is seldom if ever perfectly mixed, and if the HCW were in close proximity to the patient (the pathogen emission source), λ_{vent} would be closer to $0.5 \times (Q \div V_{room})$.

The fractional removal rate due to particle settling denoted λ_{settle} (hr^{-1}) is the particle terminal settling velocity denoted V_{TS} ($m\ hr^{-1}$) divided by the height of the room denoted H (m).[27] In turn, a particle's settling velocity depends on its aerodynamic diameter. Because particles of different sizes are emitted in a cough, in theory one needs to separately consider all pertinent particle sizes. However, if the concern is primarily with respirable pathogens, a representa-tive particle size can be taken as an aerodynamic diameter of 4.5 μm, for which V_{TS} = 2.2 m hr^{-1}. If the room height were 8 feet, then H = 2.44 m, and λ_{settle} = (2.2 m hr^{-1}) ÷ (2.44 m) = 0.90 hr^{-1}.

The fractional removal rate due to pathogen die off denoted λ_{dieoff} (hr^{-1}), varies with the pathogen and environmental conditions. In general, there is little information available on the die off rates of airborne pathogens. Airborne *M. tb* bacilli exhibit a fractional die off rate of 0.12 hr^{-1}.

The Airborne Concentration and Inhaled Dose

Given values for G, λ_{vent}, λ_{settle}, λ_{dieoff} and V_{room}, the estimated steady state concentration of respirable pathogens in room air, denoted C_{SS} (# m^{-1}), is:

$$C_{SS} = \frac{G}{\left(\lambda_{vent} + \lambda_{settle} + \lambda_{dieoff}\right) \times V_{room}} \tag{4.7}$$

The expected number of pathogens that deposit in the respiratory tract is the product of the airborne concentration C_{SS}, the healthcare worker's breathing rate denoted B (m^3 hr^{-1}), the healthcare worker's duration of exposure denoted T (hr), and the fraction of inspired particles that deposit in the respiratory tract denoted f_{dep}. Therefore, the expected dose is given by:

$$\mu_D = C_{SS} \times B \times T \times f_{dep} \tag{4.8}$$

A Hypothetical Example — Consider a scenario in which a healthcare worker attends a patient with a viral respiratory tract infection for a total of T = 4 hours. The patient has 20 productive coughs per hour, or E = 20 hr^{-1}. Perhaps based on polymerase chain reaction analysis of saliva, it is estimated that $C_{Sal} = 1.0 \times 10^7$ mL^{-1}. Due to the productive coughing, the respirable particle fluid volume is $V_F = 3.0 \times 10^{-7}$ mL. For these input factors, the respirable pathogen emission rate is G = 60 hr^{-1}. Assume that the patient room is 15 ft × 15 ft × 8 ft such that $V_{room} = 50$ m^3, and receives 6 nominal air changes per hour of dilution supply air such that Q = 300 m^3 hr^{-1}. For these V_{room} and Q values, $\lambda_{vent} = 0.5 \times (300$ m^3 hr$^{-1}) \div (50$ m$^3) = 3.0$ hr^{-1}. If the focus is on respirable pathogens, $\lambda_{settle} = 0.90$ hr^{-1}, as previously computed. Assume the airborne pathogen die off rate is $\lambda_{dieoff} = 0.69$ hr^{-1}, which signifies a 1-hour pathogen half life in air. Based on Equation 6, $C_{SS} = 0.26$ m^{-3} for this set of input factors. Next, assume that the healthcare worker's breathing rate is 1.2 m^3 hr^{-1}, as estimated for an adult performing light work.[28] About 20% of 4.5-μm particles in an inhaled air volume will deposit in the alveolar region, although 90% overall will deposit in the respiratory tract. If the target site for infection is the alveolar region, then $f_{dep} = 0.2$, and based on Equation 4.7, $\mu_D = 0.25$ pathogens. If the target site for infection is the entire respiratory tract, then $f_{dep} = 0.9$ and $\mu_D = 1.12$ pathogens.

Based on Equation 4.3, if the virus has infectivity parameter μ_α =.069 (ID_{50} = 10), and if the alveolar region is the target site with μ_D = 0.25, the healthcare worker's infection risk is.017 (1.7%). In the alternative, if the entire respiratory tract is the target site with μ_D = 1.12, the healthcare worker's infection risk is.074 (7.4%). Wearing a half-mask filtering face piece respirator with APF = 5 (μ_P =.08) would reduce infection risk to 1.4 $\times$ 10^{-3} (0.14%) if the alveolar region is the target site, and to 6.2 $\times$ 10^{-3} (0.62%) if the entire respiratory tract is the target site. Wearing a hooded powered air-purifying respirator with APF = 1000 (μ_P = 4.0 $\times$ 10^{-4}) would reduce these respective infection risks to 6.9 $\times$ 10^{-6} (.00069%) and 3.9 $\times$ 10^{-5} (.0039%). The choice of respirator device depends on the level of infection risk one is willing to accept. In the author's view, the powered air-purifying respirator should be used.

A Laboratory Accident

The Airborne Concentration and Inhaled Dose

— An accident such as dropping a culture tube can release a very large number of pathogen-containing particles into the air. If the pathogens rapidly disperse throughout room air, the initial airborne concentration, denoted C_0 (# m^{-3}), is the product of the volume of material aerosolized denoted V_M (mL), and the pathogen concentration in the material denoted C_M (# mL^{-1}), divided by the room volume V_{room} (m^3):

$$C0 = \frac{V_M \times C_M}{V_{room}} \tag{4.9}$$

Similar to the previous coughing patient scenario, pathogens will be removed from room air by exhaust ventilation, by particle settling due to gravity, and by death due to environmental stress. Assuming there is no further release of pathogens into room air, the pathogen concentration over time, C(t), will decrease in a manner approximated by the expression:

$$C(t) = C_0 \times e^{-\left(\lambda_{vent} + \lambda_{settle} + \lambda_{dieoff}\right) \times t} \tag{4.10}$$

If an individual spends T hours in the room (where T would likely be less than one hour) immediately subsequent to the release, the average exposure over the T-hr interval, denoted $C_{average}$, is given by:

$$C_{average} = \frac{C_0}{T \times \left(\lambda_{vent} + \lambda_{settle} + \lambda_{dieoff} \right)} \times \left[1 - e^{-\left(\lambda_{vent} + \lambda_{settle} + \lambda_{dieoff} \right) \times T} \right] \quad (4.11)$$

In turn, the expected dose is given by:

$$\mu_D = C_{average} \times B \times T \times f_{dep} \quad (4.12)$$

A Hypothetical Example — Consider a scenario in which 5 mL of a liquid culture containing viable pathogens at $C_M = 1 \times 10^8$ mL^{-1} drops onto the floor of a laboratory, and 0.01% of the fluid is aerosolized into particles with aerodynamic diameters of 10 μm. An individual will spend 15 minutes in the laboratory cleaning up the spill in a manner that will presumably not cause additional pathogens to become airborne. Assume the room has volume 100 m^3 and receives 10 nominal air changes per hour such that $Q = 1{,}000$ m^3 hr^{-1}. Based on Equation 4.8, the initial airborne pathogen concentration is $C_0 = 5 \times 10^2$ m^{-3}. Even though the laboratory air is not perfectly mixed, the lack of an ongoing point source of emission means that $\lambda_{vent} = Q \div V_{room} = 10$ hr^{-1}. For a 10-μm aerodynamic diameter particle, $V_{TS} = 10.8$ hr^{-1}. If the room height is $H = 3$ m, $\lambda_{settle} = V_{TS} \div H = 3.6$ hr^{-1}. Assume the airborne pathogen death rate is $\lambda_{dieoff} = 0.69$ hr^{-1}, which signifies a 1-hour pathogen half life in air.

Based on Equation 4.10, if $C_0 = 5 \times 10^2$ m^{-3} and $T = 0.25$ hr, $C_{average} = 136$ m^{-3}. For a 10-μm aerodynamic diameter particle, f_{dep} is approximately 0.8, where deposition is primarily in the head airways region. Assume that the breathing rate is $B = 1.2$ m^3 hr^{-1}. Based on Equation 4.11, the expected dose is $\mu_D = 33$ pathogens. If $\mu_\alpha = 0.069$ (ID$_{50} = 10$) and the target site for infection is the entire respiratory tract, infection risk is 0.9 (90%).

As one might suspect, this high risk level requires a highly protective respirator. Wearing a hooded powered air-purifying respirator with APF = 1000 ($\mu_p = 4.0 \times 10^{-4}$) would reduce infection risk to 1.3×10^{-2} (1.3%), which is still substantial. Either a more protective supplied-air respirator needs to be used, or entry into the room should be delayed until the airborne pathogen concentration declines to a lower level. According to the latter strategy, one can use Equation 4.9 to find a suitable waiting time T_{wait}; the predicted concentration $C(T_{wait})$ would become the new starting concentration value in Equation 10. For example, if $C_0 = 5 \times 10^2$ m^{-3} and $T_{wait} = 0.5$ hr, the airborne pathogen concentration $C(0.5$ hr$) = 0.39$ m^{-3} (which is a 99.98% reduction from the initial concentration). If an individual entered the room at this point and was present for 15 minutes ($T = 0.25$ hr), the new $C_{average} = 0.11$ m^{-3}, the new $\mu_D = 0.026$ pathogens, and the new infection risk without respirator use is 1.8×10^{-3}.

For this exposure, even use of a N95 FF (μ_p = 0.08) might be deemed adequate because infection risk would be reduced to 1.4×10^{-4} (0.014%). On the other hand, if there is potential for aerosolization of more pathogens during clean up activities, a more protective respirator should be considered.

Conclusions

The risk-based approach to selecting respirators as outlined here requires numerous model inputs, and some values such as the pathogen concentration in a patient's respiratory fluid will likely not be known at the time respirator selection must be made, if ever. However, what this chapter has termed model "inputs" or "parameters" would likely be termed "risk factors" by infection control professionals. And if there is uncertainty about the model input values on which a respirator decision is to be quantitatively based, there is equal uncertainty about the risk factor values on which a decision might be qualitatively based. An advantage of the quantitative approach is that one can develop a numerical sense of the magnitude of uncertainty in infection risk. The risk-based approach also identifies information needs and makes the respirator selection process transparent. Others may dispute the input values that are used, but the selection process is clear to everyone, and the acceptable infection risk value is specified.

That said, the problem remains that key model input values may be unknown, especially if the transmissible disease is due to a newly emerged pathogen. The author has no method for precisely forecasting risk in the face of such uncertainty, but suggests the following health conservative approach – wear one of the hooded powered air-purifying respirators (PAPRs) with high efficiency filters for which the Occupational Safety and Health Administration administratively set APF = 1000.[29] The PAPR will permit far less penetration than a N95 FF, and if pathogen airborne exposure intensity and infectivity turn out to be high, the healthcare worker stands a much better chance of not being infected than if he or she wore a N95 FF. If it is subsequently determined that pathogen airborne exposure intensity and infectivity are quite low, it is easy to step down from the PAPR to a filtering face piece respirator. In the author's view, the reverse strategy of using a N95 FF until proven inadequate by infected and perhaps dying healthcare workers, who also may serve as secondary disease vectors, is not acceptable.

THE MULTIFACETED SYSTEMS INVOLVED IN PROTECTING AGAINST OCCUPATIONALLY ACQUIRED RESPIRATORY INFECTIOUS DISEASES: PUBLIC HEALTH AND OCCUPATIONAL SAFETY, A SHARED RESPONSIBILITY

III

Chapter 5

Knowledge Gaps and Research Priorities for Effective Protection Against Occupationally Acquired Respiratory Infectious Diseases: A Canadian Perspective

Dr. Annalee Yassi and Dr. Elizabeth Bryce

CONTENTS

Executive Summary

On March 12, 2003, the World Health Organization (WHO) announced a global outbreak of an atypical pneumonia that was quickly named Severe Acute Respiratory Syndrome (SARS) and shortly thereafter determined to be caused by a novel coronavirus. The virus spread internationally along travel routes and caused the well-documented nosocomial outbreaks in the Greater Toronto Area, China, Hong Kong, Vietnam, and Singapore. Contact, droplet and airborne precautions were reportedly instituted in affected hospitals; however, they were apparently incomplete, intermittently applied, or only partially effective. The Canadian outbreak resulted in 438 cases, 51% of these were healthcare workers[HCWs]) with three related deaths.

The objective of this report is to summarize our findings from an analysis of the key domains, as pertinent to improving the effectiveness of facial protective equipment (FPE) in preventing occupational-associated respiratory disease transmission in healthcare workers.

Summary of Evidence Available from the Scientific Literature

SARS was a disease largely spread by respiratory droplets.

Epidemiology and Transmission

The lack of spread within the community and the recent information on relatively low R_0 values for SARS coronavirus (SARS CoV) indicate that SARS is less contagious than influenza and other similar respiratory infections. It is important to emphasize that the consistent application of basic infection control precautions terminated outbreaks in Vietnam, China, and Singapore. Large outbreaks occurred early in the emergence of the disease when the causative agent was not recognized and infection control procedures not in place. The literature makes it fairly clear that failure to implement appropriate barrier precautions was responsible for most nosocomial transmission. As such, attention to understanding why there was a failure to implement appropriate precautions, and how best to promote compliance in future, is an important topic for study.

Although largely spread by the droplet route, there is indirect evidence that the generation of aerosols and the lack of control of aerosols at source was an important factor in hospital dissemination. The relative lack of transmission within the community also suggests that sneezing and coughing may not generate highly infectious aerosols in contrast to hospital-based mechanical procedures. The relative role of aerosol transmission in disease scenarios traditionally thought to be spread by the droplet route is unknown, as is our understanding of the role of mucosal contamination and autoinoculation in acquisition of infection.

As patients with SARS did not appear to transmit disease unless they had symptoms, recognizing the disease in patients presenting to a hospital was probably one of the most important factors in limiting spread. Once the disease was recognized, all the outbreaks in 2003 were able to be contained, using a variety of different infection control strategies. The development of new laboratory tests for the SARS CoV provides optimism that identifying SARS patients will become easier in the future. This is an area of important research that is already ongoing, and will lead to greater protection of healthcare workers against SARS. However, specific clinical diagnosis of disease can never be relied upon to protect against emerging diseases.

Risk Assessment

In hospitals, the risk of disease transmission appeared to vary widely, but several factors were quickly identified as being important determinants of risk. Patients were only able to transmit disease if they were symptomatic and the patients with the most severe illness seemed to pose a greater risk. Working in close proximity to a patient resulted in a higher risk of disease transmission to HCWs. Added to the individual risk of the source patient were the risks associated with the hospital environment in terms of whether the patient wore a mask in the hospital, was nursed in isolation, and the state of the hospital ventilation system. Further, whether the patient underwent aerosol generating procedures also influenced the risk of disease acquisition for an individual healthcare worker. Therefore, for HCWs who do not work in an area of a hospital where patients who acutely ill and who may require one of the above procedures, the risk of acquiring SARS is also quite low.

Risk Management

The occupational health literature has extensively documented that controlling hazards at the source is the most effective means of protecting workers.

Controlling Aerosols at Source

The only consistent form of source control applied during the SARS outbreaks was having patients wear a surgical mask, a simple and likely effective method of limiting SARS CoV exposures, but which was not formally evaluated for its effectiveness. Many other potential forms of source control exist, such as the installation of filters on the exhaust port of nebulizer masks, and fitting anaesthesia machines, pulmonary function machines, ventilators, and manual ventilation units with filters The effectiveness of these measures remain to be studied.

Isolation and Ventilation

The extent to which isolation of SARS patients within an institution is useful in reducing risk of transmission is not known, but this practice could be defended on general infection control grounds — as it is wise to minimize the number of potential exposures. The available evidence also suggests that procedures likely to generate high concentrations of aerosols should be performed only in designated areas where a higher level of protective measures can be employed.

Inadequate hospital ventilation systems in the general patient area were identified as an important determinant of "superspreading" of SARS in one hospital in Hong Kong, likely in combination with aerosol-generating procedures. This observation is similar to that of a recent study of nosocomial-transmitted tuberculosis in Canadian HCWs that also found ventilation systems outside of isolation rooms was an important determinant of infection. While there has been much interest in the importance of having SARS patients nursed in negative pressure rooms, more research is needed to identify if there is any added benefit of negative pressure rooms beyond that of isolation and adequate ventilation throughout the hospital.

Environmental Decontamination

Studies have shown that SARS CoV is easily killed with standard disinfectants. It is also known that SARS can survive for several days on surfaces, and for longer periods in stool, especially stool from patients with diarrhea. Recommendations regarding surface decontamination and hand washing thus appear to be well-grounded for SARS, in that the virus appears to be better able to survive outside the human body than most other common respiratory viruses. The practical importance of these findings and the role that fomite transmission of SARS plays in spreading the disease in hospitals is not known.

Personal Protective Equipment

While there is an extensive literature on the performance of personal protective equipment (PPE), especially respirators with regard to particle penetration of some bioaerosols, how this performance translates into protecting healthcare workers from infectious diseases in not clear. Two observational studies have shown that using any mask regularly is more protective than not using a mask regularly. N95 masks have been shown to reduce exposures to airborne particles to a greater extent than surgical masks. However, it is still unclear whether N95 masks offer significantly better protection from acquiring disease than surgical masks. Small studies have shown that wearing gowns, gloves, goggles, and caps were protective in univariate analyses, but not in final models. It is not clear if the lack of these effects is due only to small sample sizes and confounding effects or to true limited effectiveness. It is also not clear how some HCWs contracted SARS while working with what should have been adequate PPE during aerosol-generating procedures. It will be important to study whether the failures to protect HCWs in these circumstances were due to failure in efficacy of controls, or in the effectiveness in their use. Failures in efficacy would imply that better PPE (i.e., N95 masks, PAPRs) may be needed to adequately protect HCWs from SARS in these circumstances. However failure in effectiveness in the use of PPE would imply that less complicated infection control guidelines, which focus on the key protective factors, combined with the appropriate safety climate and incentives for compliance may ultimately be more successful in reducing infections. Further we have found that there is relatively little information on how important the transocular route is for disease transmission and how existing eye protection reduces this risk to HCWs.

Fit Testing

Review of the scientific literature prior to the advent of SARS provides clear evidence that fit-tested N95 masks provide an extra degree of protection to exposure to organisms *transmitted by the airborne route,* primarily tuberculosis. It is equally as clear that any leak in the seal negates the additional benefit this type of respirator provides. Thus it is important that HCWs know how to verify that there are leaks around their masks. Fit testing minimizes the chance of leakage. However, the relative importance of fit testing as opposed to fit checking is unclear. The information from a study by Huff using a nebulized solution containing Tc^{-99m} suggests that fit testing does have a valuable role to play in reducing the risk of exposure to aerosolized droplets.

The educational value of the fit-testing exercise cannot be dissected from the actual fit-testing benefit, nor should it be. The limited studies demonstrating the importance of an HCW conducting a fit check each and every time to ensure a good seal, suggests that fit testing annually is less important than ongoing assessment of the ability of HCWs to achieve an effective seal through fit checking. As noted above, with respect to N95 versus surgical masks, fit testing reduces exposure to infectious particles, but whether it reduces the risk of infection is unknown. Whether fit testing is needed in a given institution should be based on an assessment of the potential risks of infectious exposures to airborne organisms in the facility.

Adherence to Infection Control Guidelines

Current research suggests that individual factors are less important than organizational and environmental factors in affecting the level of compliance with use of PPE, and specifically facial protection. The literature also indicates that the theoretical or laboratory derived protectiveness of different types of PPE needs to be carefully evaluated with field studies, as compliance in the workplace is usually much less than in idealized research settings. The available evidence supports the view that users as well as infection control and occupational health experts need to be consulted *before* required workplace practices are established and PPE is selected. Once the PPE and work practice requirements are set, workers do need to be trained, but the available evidence indicates that knowledge deficit is not a major barrier to compliance. Noncompliant staff generally know they are noncompliant. This suggests that a focus on training content or methods to increase knowledge may not yield much change in compliance.

Even in circumstances where the key factors in protecting HCWs are known, the challenge of changing workplace behavior will remain. A number of interventions such as educational outreach visits, posted reminders, interactive educational meetings, and other multifaceted approaches have been shown to be very successful in changing the behavior of physicians around the use of clinical practice guidelines. However, research on knowledge translation in the workplace setting pertaining to infection control guidelines is lacking.

Feedback to workers on their adherence to precautions has been identified as an important factor in facilitating compliance with infection control practices. However, the type of feedback that is most effective in achieving compliance is not known and the optimal timing of feedback and the optimum feedback frequency are also not known. Time and equipment to permit worker adherence to infection control guidelines must be available.

Most of the reviewed studies were observational in nature. Many of the research questions raised here need to be investigated as controlled intervention studies in "real-world" situations.

Summary of Key Factors Identified by Healthcare Workers in Focus Groups

The healthcare workers who participated in focus groups spent the greatest amount of time discussing organizational factors.

Organizational Factors

Foremost among their concerns were the lack of consistency with safety instructions and the frequently changing directives which were commonplace during the SARS outbreaks. This was a source of much anxiety for healthcare workers both in British Columbia and Ontario. Coupled with this was the diversity of views on the role of regulatory agencies, such as the Ministry of Labour and the Workers Compensation Board. Many workers saw the measures imposed as being somewhat draconian, while others saw some measures, such as the requirement for fit testing as long overdue.

Workplace attitudes toward safety were also seen as important. Paramount to these were the attitudes and actions of management and the perceived importance of occupational health and safety, both of which were important determinants of the safety climate within hospitals.

Healthcare workers also expressed support for the development of evidence-based and practical infection control policies that includes representation from frontline workers. Ensuring adequate resources for infection control was also seen as a priority. In order to improve worker adherence to infection control guidelines, focus group participants felt that better enforcement of infection control guidelines was needed, but should not rely on nurses needing to "police" other professionals. Participants also saw the need for more accommodation of worker concerns and infection control guidelines for patients and visitors.

Safety training, in terms of infection control training was also discussed at length. Focus group members expressed their views that repeated training was needed and that better tracking methods in order to monitor who has been trained and who requires training should be developed. Workers felt that the appropriateness of the "train-the-trainer" model needs to be evaluated in terms of the existing time constraints on frontline workers. It was also felt that hospitals need to develop specific policies to address issues for part-time staff, physicians, residents, and students.

Communication about safety within healthcare organizations was seen as having a key role in protecting HCWs, especially during the SARS outbreaks. Face-to-face "town-hall" meetings were seen as necessary in order to build worker confidence in hospital infection control policies during SARS. A variety of communication media were seen to be more effective than any single strategy and workers identified a need for communication strategies to be adapted for the large, multicentered organizations which have developed in recent years. Similarly, recent organizational changes have resulted in fewer frontline managers, formerly responsible for much of the communication with other HCWs. Communication between employees, units, and especially between occupational health and infection control was seen as being important in creating safe workplaces.

Focus group participants discussed fit testing at length, but the value of it was not universally accepted, as different institutions used different methods and workers often saw these inconsistencies as sources of concern for the whole process. The participants also identified the need to address the increased amount of worker fatigue which existed when HCWs work with full PPE. They also felt that the effect of casualization and outsourcing of the workforce needed to be evaluated in terms of their effect on worker health and safety.

Environmental Factors

Environmental factors were the least discussed issues in the focus groups. The topics that were discussed included the role of isolation rooms for patients with suspected communicable diseases, the availability of ante-rooms for HCWs to change into PPE, and the use and availability of negative pressure rooms. Participants also discussed the importance of environmental decontamination, primarily hand washing, and the well-documented problems with the availability of specific PPE during SARS, especially with respect to N95 masks and face shields or goggles.

Individual Factors

Knowledge of infection control procedures and the rationale behind them was seen as being important, but not sufficient to ensure proper infection control procedures. Attitudes such as professionalism and belief in effectiveness of infection control guidelines, as modified by past experiences were identified as having important influences on worker adherence to procedures. The additional burden on HCWs that wearing full personal protective equipment imposed was also seen as being a key determinant. The increased time constraints, increased workload, and discomfort associated with wearing PPE were felt to be important barriers to worker adherence to recommendations. The peer environment, especially the compliance of other occupational groups (including physicians), and the feedback from peers, were also identified as important factors, which could exert a positive or negative influence on individual worker actions. Attitudes of family members, in particular the fear that family members expressed toward contracting SARS, also influenced the actions of HCWs on the job.

Priorities for Further Research

Taking into account the evidence from the literature review, the priorities identified through the focus group analysis and a proposed framework for assigning research priorities, the following areas for further research were identified:

1. Improving workplace health and safety through organizational factors: How best to bring about meaningful knowledge translation.
    a. How can the safety climate of healthcare institutions be improved? What approaches best facilitate an organizational culture that promotes safety?
    b. What are the best mechanisms to provide communication to frontline workers in order to ensure appropriate infection control practices?
    c. What are the best mechanisms to provide feedback to frontline HCWs in order to ensure infection control measures are practical and feasible while still enhancing safety?
    d. What are the best ways to train HCWs on appropriate use of personal protection equipment?
    e. How have changes to the healthcare workforce in terms of increased casualization and increased outsourcing of services affected workplace health and safety?
    f. What key components of an occupational health program are needed to improve or maintain worker health and safety in healthcare facilities?
2. Epidemiology and transmission of SARS:
    a. How do respiratory droplets produced by aerosolizing procedures differ from those produced by more "natural" methods such as coughing or sneezing, in terms of their size, their spread, and their infectivity? This question is key because it addresses the issue of the hierarchy of precautionary measures.
    b. Do infectious organisms survive on barrier equipment and clothing and for how long? This has implications for environmental decontamination, reuse of barriers versus the use of disposables, and the potential importance of autoinoculation through contaminated PPE.
    c. How able are respiratory tract pathogens to cause disease through the transocular route?
3. Risk reduction through engineering controls and personal protective equipment:
    a. What is the relative effect of engineering controls to maximize particle fallout or decrease viability of organisms (e.g., temperature, air exchange, relative humidity)? There may be simple yet effective measures to decrease these aerosols that could have significant impact on reducing the risk of exposure.

b. What design criteria are required to minimize generation and dispersal of infectious aerosols in medical equipment such as anaesthesia machines and ventilators? This question addresses the relative effectiveness of decreasing aerosols at source.

c. What is the added benefit of nursing high risk patients in a negative pressure atmosphere over physical isolation and adequate ventilation throughout hospitals? There has been a great emphasis on hospitals improving access to this technology, yet evidence to support their use is lacking.

d. What is the effectiveness of facial protection against bioaerosols? (In conjunction with question 2c, above, answers to this question will clarify the relative importance of full facial protection, versus eye protection, versus nose and mouth protection.)

e. What is the relative importance of fit testing versus fit checking of respirators? The reason for selecting this as a priority is less an issue of burden of disease but more an issue of stakeholder interest, the implications for where resources are expended and the potential extrapolation of this knowledge to other airborne illnesses.

Introduction

The recent events regarding SARS, particularly the morbidity and mortality in Canadian HCWs, focused attention on the adequacy of facial protection in preventing airborne and droplet-spread transmission of infectious agents. Facial protection traditionally consisted of a mask, and in some circumstances, protective eyewear. During the SARS outbreak, widely divergent opinions emerged, on the adequacy of facial protection, ranging from the view that N95 masks* (originally used in industrial applications and advocated for airborne diseases such as tuberculosis) were unnecessary for agents mainly spread via droplets, to the belief that a higher level of protection, for example, powered air purifying respirators

(PAPRs) was required under certain circumstances. The "science" behind respirator selection and use was also a contentious issue as the need for fit testing was questioned and there was confusion regarding the approval criteria for N95 masks. Similarly, there were conflicting views regarding protective eyewear, and expert opinion varied as to the need for safety glasses versus splash goggles or face shields. Clearly, there was

---

* The terms *mask* and *respirator* are used interchangeably in this report, which reflects their usage in health care. However, the authors recognize that the two words have very different meanings in the occupational health and occupational hygiene fields, as described later in the report.

a need to evaluate the adequacy of facial protection to ensure that HCWs are protected in future outbreaks, not only for SARS, but also against a variety of new and emerging respiratory pathogens.

In light of these observations, The Change Foundation issued a request for applications in September 2003 for a grant to undertake a review of the relevant literature on facial protection that would also address the concerns of front-line healthcare workers. A research team in Vancouver was assembled and wrote a proposal which was accepted in October 2003. The project was conducted over the period from November 1, 2003 to March 31, 2004.

This report was written by a unique interdisciplinary collaboration of researchers based in Vancouver, BC, with a strong track record of relevant research in this subject matter. The team included experts in the health of healthcare workers — with researchers from both occupational medicine and occupational hygiene; nationally and internationally renowned infection control experts; and specialists in public health and epidemiology. We also had clinicians and a representative of frontline care providers.

The objective of this document is to summarize our findings in an analysis of the key domains identified by The Change Foundation as pertinent to improving the effectiveness of facial protective equipment (FPE). These include: (1) a review of the scientific literature dealing with bioaerosols, filtration, and how this influences the design and performance of FPE; (2) a review of the scientific literature of the organizational, environmental, and individual factors that influence the effectiveness of occupational health and safety in general, and infection control procedures; in particular (3) an analysis of these factors as identified through a series of 15 focus group discussions involving frontline healthcare workers; and (4) the identification of a framework for assigning priorities for further research and a list of identified priorities derived from the gaps identified in the literature review and the priorities of frontline HCWs.

It was not our goal to define what specific policies are needed to protect workers from infectious diseases such as SARS, but to identify what is already known about SARS and other respiratory tract nosocomial infections with regard to worker safety and to identify areas where further research should be directed.

Literature Review

Methodology

This literature review was directed at understanding what scientific knowledge already exists with respect to the efficacy of facial protective equipment in preventing the transmission of respiratory infections, and the effectiveness of protective measures when used in the real world. The following describes the methodology used for this section of the project.

The research team developed a list of key words to be used in searching several databases for articles published since 1988 that relate to infection control practices, occupational health and safety issues, organizational behavior, and other issues of importance in protecting workers against respiratory infections in healthcare settings. Literature searches were conducted using Medline, EMBASE, the Cumulated Index of Nursing and Allied Health Literature (CINAHL), Web of Science, and OSHROM. Citations were divided into two broad categories, (1) the applied and basic science of bioaerosols and how various types of protective equipment perform in preventing the transmission of respiratory tract pathogens; and (2) the organizational, environmental, and individual factors that influence the effectiveness of infection control procedures in general, and the use of facial protective equipment in healthcare settings. We have retained these two categories for purposes of discussion here.

These initial searches produced lists of 462 citations and 379 citations, respectively. The research proposal expected that the committee would design a data abstraction form, collect data from each article, and summarize the data using a weighting formula based on the number of studies and the study design used. This methodology, which is similar to that of the Cochrane Reviews for clinical trials, was found to be unworkable in practice for this project given the time-frame involved. The topic areas were too broad, the study designs too varied, and the numbers of citations were too many, to be summarized in this manner.

Instead, a series of research topics were then developed by the research team for each of the two broad categories "basic science and efficacy" and "factors influencing effectiveness." The titles we found that related to these categories were next reviewed to eliminate citations which did not directly relate to the objectives of the study. This resulted in the literature review list being shortened to include 316 and 267 citations. The research topics were then divided between the research committee members to write summaries, using articles on these lists as reference materials. Secondary reference materials, derived from these primary references, were also added to the source reference list. The drafts from each group were merged, then the compiled version reviewed by the team as a whole, and the summary of the evidence, the gaps in the evidence, and the recommendations for further research were then determined with consensus from the research team.

Literature Review of the Basic Science and Efficacy of Facial Protection

This section is divided into two parts. The first part discusses the science of airborne particles and the evolution of respirators and summarizes the advantages and disadvantages of respirators in the healthcare setting. The second part summarizes current scientific knowledge on performance data of respiratory protective devices and addresses issues surrounding the application of the science to the healthcare setting.

The Science of Airborne Transmission and Use of Respirators

During every breath the respiratory system takes in a mixture of solid particles, liquid droplets, vapors, and gases.

What Is an Aerosol? Where Do Airborne Droplets or Droplet Nuclei Fit In?

Collectively, these suspended particles and their carrier gases are known as aerosols. Aerosols made up of solid particles are called "dusts" or "fumes," while aerosols made up of liquid particles are called "fogs," "mists," or "sprays." Droplets are ejected from the respiratory tract during coughing, shouting, sneezing, talking, and normal breathing. The size and number of droplets produced is dependant on which of these methods generated the particles. These droplets may contain contagious material such as bacteria or viruses, including the SARS coronavirus.

The Infectious Agent as a Particle

A number of scientific studies have shown that a NIOSH-certified respirator such as the N95 effectively filters aerosols, containing microbes such as *Mycobacterium tuberculosis*. Brousseau et al.,[1], Qian et al., [2]and Lee et al.[3] demonstrated that biological particles including those contained in droplet nuclei, will be deposited in airways and filters in the same manner as nonbiological particles, and that the most important characteristics of these particles are aerodynamic diameter and shape. The biological state does not appear to influence the way in which particles are collected and retained by a filter.[4] All particles, whether they are liquids, solids, or microorganisms, can be filtered by a particulate filter. The efficiency of the filtration is dependant on particle size, shape, and electrostatic and hygroscopic interactions.

How Long Do Respiratory Droplets Remain Airborne and Where Are They Deposited? — Typically, a person breathes between 10 to 20 m³ (10,000 to 20,000 liters) of air daily. Where airborne particles are deposited within the airways is primarily a function of particle size [5, 6, 7, 8, 9]. Larger droplets (generally greater than about the 50 to 100 μm size range), settle more quickly than smaller particles, and exposure to these is typically the result of direct contact with the skin surface including mucous membranes of the eyes, nose, and mouth or onto inanimate surfaces in the immediate vicinity of the infectious patient. These larger droplets are normally not inhaled into the lungs because they are trapped by cilia and mucus in the nose and mouth. However, they can be deposited in the pharynx if the HCW is in close proximity to the infectious patient.

Small particles and droplets less than 10 μm in size are likely to remain in the air long enough to be swept around by air currents and may be inhaled by a susceptible host within the same room. Therefore, when working in close proximity to a patient, one can be exposed to respiratory droplets following a cough, sneeze, or a high velocity exhalation, or during endotracheal intubation, bronchoscopy, or similar invasive procedures.

Droplet nuclei are at the lower end of the spectrum for droplet diameter and can travel considerable distances in the air and may be readily inhaled into the lung. Droplet nuclei are typically smaller than 5 μm, and exhibit a settling velocity in still air of about 1 m per hour.

Inhaled particles greater than about 3 μm will deposit in the upper respiratory tract and particles less than 2 μm will be deposited in the alveolar regions. Particles near 0.3 μm will have the least deposition (about 14%), while either larger or smaller particles will deposit with much higher efficiency, often approaching 100% deposition. From the perspective of infectious disease spread by the airborne route, particles deemed "inhalable" fall in the size range of 0.1 to 10 μm in diameter. The effects of the high relative humidity in the respiratory tract can result in the relative size of particles increasing in aerodynamic diameter which, in turn, can affect the site of deposition in the respiratory tract.[10] For inhaled infectious particles, the location of receptors in the respiratory tract for particular pathogens also influences their ability to cause disease.[11]

Table 5.1 and Figure 5.1 summarize the size of respiratory droplets and how it relates to the time they remain aloft and their potential to transmit disease.[12]

Size of Particles Produced by the Human Respiratory Tract — It should be noted that although there may be nearly 2 million particles extruded from a sneeze compared to fewer than 100,000 from a cough, more infective droplets may be released in a cough because of the deeper origin of particles in the lungs.[14] A further complicating matter is the effect

TABLE 5.1 Behavior of Infectious Aerosols in Still Air and Route of Exposure

Diameter in μm	Time to Fall 3 m	Route of Exposure
100	10 sec	Direct contact with skin or mucous membranes
40	1 min	Direct contact
20	4 min	Direct contact
10	17 min	Direct contact Some deposition in mouth or nose
6–10	Several hours	Deposition in nasal passages
0.06–6	Many hours	Deposition into lungs

Note: Small particles (< 6 μm) do not settle out at an appreciable rate, but spread so that as distance (r) from the source increases, the relative concentration of particles in air decreases in proportion to r^3.[13] This equation does not consider the effects of droplet evaporation or convective disturbances. Thus, as the distance from the source doubles, the aerosol concentration declines eightfold.

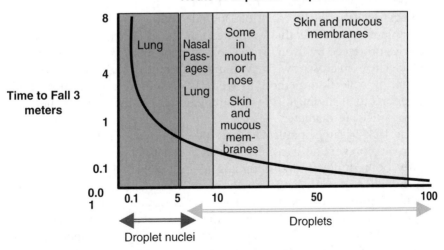

Figure 5.1

of relative humidity on the infectious droplets. The size of aerosolized droplets ejected by a patient is likely to be reduced very quickly in air of low relative humidity (e.g., below about 20–40%) and high temperature (above 20°–25°C). While a droplet of pure water will evaporate fully if relative humidity is less than 100%, a droplet that contains soluble material, such as sodium chloride, will reach an equilibrium state based on the mass of the sodium chloride contained in the droplet and the relative humidity of ambient air. Since respiratory secretions contain an isotonic concentration of sodium chloride, it cannot be assumed that smaller respiratory droplets, potentially containing microorganisms, will fully evaporate in ambient air. However, if the relative humidity is low enough (less than approximately 40%), then even a particle containing soluble material will evaporate completely, leaving behind a residue particle consisting of the dried solute and any other solid matter that was contained in the original droplet, possibly including microorganisms. If these biological agents are not damaged by the drying process, they can potentially infect a susceptible host.

Duguid [15] in a study conducted in 1946, reported on the size distribution of aerosols produced from the nose and mouth during various activities (Table 5.2) collected under experimental conditions. According to the results, the size of the expelled particulate determines the fate of the particle in air. This study also reported on the composite size distribution of particulate captured on slit samplers as seen in Tables 5.2 and 5.3.

The lack of very small particulate was likely an artifact of the methodology available at the time of Duguid's study.[15] A more recent study by Papineni and Rosenthal[16] reported on the production of respiratory particles produced by five normal subjects using an optical particle counter and electron microscopy. They found substantial variability person to person, by collection method, as well as by method of exhalation. Table 5.4 reports their findings in two size fractions, less than and greater than 1 μm particulate diameter.

The lack of larger particulate was also an artifact of the methodology employed, preventing a comparison between the two studies, although clearly both studies found coughing to be associated with the greatest dissemination of particulate. Papineni and Rosenthal examined the exhaled breath of three subjects by electron microscopy and found 36% of the exhaled particulate was < 1 μm in diameter, 64% was > 1 μm in diameter. [17]

In a study published in 2004, Fennelly[18] developed a specially constructed chamber to become the first study ever to report the size ranges of infectious particulate disseminated by patients with active tuberculosis. Although it has been long known that *M. tuberculosis* is disseminated through droplet nuclei, the organism had never before been cultured from the respiratory exhalations of patients in a clinical setting. Of particular

Table 5.2 The Percentage Size Distribution of the Larger Droplets as a Function of Expiratory Activity

Diameter in μm	Sneezes	Coughs with Mouth Closed	Coughs with Mouth Open	Speaking Loudly
0–5	0 (0%)	0 (0%)	0 (0%)	0 (0%)
5–10	36 (1.2%)	24 (0.8%)	8 (2.7%)	20 (0.7%)
10–15	94 (3.1%)	119 (3.9%)	39 (1.3%)	84 (2.8%)
15–20	267 (8.9%)	337 (11.2%)	127 (4.2%)	200 (6.7%)
20–25	312 (10.4%)	346 (11.5%)	189 (6.3%)	224 (7.5%)
25–50	807 (26.9%)	767 (25.6%)	577 (19.2%)	597 (19.9%)
50–75	593 (19.8%)	468 (15.6%)	593 19.8%)	531 (17.7%)
75–100	260 (8.7%)	285 (9.5%)	341 (11.4%)	352 (11.7%)
100–125	144 (4.8%)	160 (5.3%)	231 (7.7%)	260 (8.7%)
125–150	105 (3.5%)	125 (4.2%)	202 (6.7%)	214 (7.1%)
150–200	115 (3.8%)	115 (3.8%)	253 (8.4%)	179 (5.9%)
200–250	82 (2.7%)	96 (3.2%)	165 (5.5%)	99 (3.3%)
250–500	118 (3.9%)	113 (3.8%)	213 (7.1%)	197 (6.6%)
500–1000	59 (1.9%)	40 (1.3%)	52 (1.7%)	41 (1.4%)
1000–2000	8 (0.3%)	5 (0.2%)	10 (0.3%)	2 (0.07%)
Total	3000 (100%)	3000 (100%)	3000 (100%)	3000 (100%)

interest is the wide variability between patients of the size ranges of the infectious particulate. This study used an Andersen multiple stage impactor to determine the size ranges of the infectious particulate. During sputum-induction procedures, the mode infectious particulate size was 1.1 to 2.2 μm (49% of total) while 90% of the sample recovered was between 0.65 and 3.3 μm. In a patient coughing naturally (not induced) the mode size was 2.1 to 3.3 μm, and 100% of particulate was larger than 1.1 μm aerodynamic diameter.

Principles of Filtration as Applied to Respirator Particulate Filters

Since the mid-1950s, filtration of aerosol particles by fibrous filters has been extensively studied and the relationships between particle size and filtration efficiency as well as mechanisms of filtration firmly established. Aerosol particles, whether solid or liquid attach firmly to their contact surface and fibrous filters are designed to maximize the chance that these particles adhere to the filter material while allowing gases to continue through the filter. Five basic mechanisms dictate how a particle is captured by the filter material: inertial impaction, interception, diffusion caused by

TABLE 5.3   Composite Size-Distribution Table for the Droplets Expelled
During Sneezing, Coughing and Speaking

Droplet Diameter in μm	One Sneeze	One Cough with Mouth Closed	Counting Loudly "1 to 100"
< 1	Remain airborne		
1–2	26,000 (2.6%)	50 (10%)	1 (0.4%)
2–4	160,000 (16%)	290 (5.8%)	13 (5.2%)
4–8	350,000 (35%)	970 (19.4%)	52 (20.8%)
8–16	280,000 (28)	1,600 (32.5%)	78 (31.2%)
16–24	97,000 (9.7%)	870 (17.4%)	40 (16)
24–32	37,000 (3.7%)	420 (8.4%)	24 (9.6%)
32–40	17,000 (1.7%)	240 (4.8%)	12 (4.8%)
40–50	9,000 (0.9%)	110 (2.2%)	6 (2.4%)
50–75	10,000 (10%)	140 (2.8%)	7 (2.8%)
75–100	4,500 (0.45%)	85 (1.7%)	5 (2%)
	Fall at once to ground		
100–125	2,500 (0.25%)	48 (0.96%)	4 (1.6%)
125–150	1,800 (0.18%)	38 (0.76%)	3 (1.2%)
150–200	2,000 (0.2%)	35 (0.7%	2 (0.8%)
200–250	1,400 (0.14%)	29 (0.58%)	1 (0.4%)
250–500	2,100 (0.21%)	34 (0.68%)	3 (1.2%)
500–1000	1,000 (0.1%)	12 (0.24%)	1 (0.4%)
1000–2000	140 (0.014%)	2 (0.04%)	0 (0%)
Approximate Total	1,000,000 (100%)	5,000 (100%)	250 (100%)

TABLE 5.4   Mean Droplet Concentration (per liter of air) in Exhaled
Breath for Five Subjects

Droplet Diameter (μm)	Coughing	Mouth Breathing	Nose Breathing	Talking
	Mean (SD)	Mean (SD)	Mean (SD)	Mean (SD)
< 1	83 (63)	12.5 (10.7)	4.7 (4.1)	19.2 (9.5)
> 1	13.4 (13.2)	1.9 (2.3)	0.7 (0.67)	3.3 (1.2)

Brownian motion, gravitational settling, and electrostatic attraction. Mechanical filters rely upon the first four methods for particle capture. Electrostatic filters use electrostatically charged filter fibers or electrets to increase the particle capture and often can use a much looser weave of filter fibers as a result. This loose weave has a much lower resistance per unit area of filter medium and is typically not pleated.[19,20,21]

The most important parameter for characterizing how a particle will deposit is particle size. An increase in particle size will cause increased filtration by the interception and inertial impaction mechanisms whereas a decrease in particle size (below 0.3 μm) will enhance collection by Brownian diffusion. As a consequence, there is an intermediate particle size region where two or more mechanisms are simultaneously operating yet none is dominating. This is the region where the potential for particle penetration through the filter is at the maximum and the efficiency of the filter a minimum.[22] The fibrous filters found in most respirators have minimum filter efficiency in the vicinity of 0.3 μm. The 0.3 μm particle is referred to as the most penetrating particle size (MPPS) and is the basis for respirator testing (worst-case testing) and certification pursuant to International Standard Organization EN149:2001, NIOSH 42 CFR Part 84, and Australian Standard AS1716.

What Is a Respirator?

A respirator is a personal protective device that is worn on the face, covers at least the nose and mouth, and is used to reduce the wearer's risk of inhaling hazardous airborne gases, vapors and particulate matter or aerosols. Note that the term *mask*, as in surgical mask, is used to refer to a device that is worn by a person to minimize the spread of airborne contaminants from that person's respiratory tract and to protect other persons from exposure. As such, surgical masks are therefore not recognized by regulators as an approved design for respiratory protection, even though they may offer some degree of protection.

Aerosols containing bacteria, viruses, fungi, and other biological material (bioaerosols) are filtered in a similar manner as nonbiological particulate material. Brousseau et al.[23] affirmed that the most important parameters for aerosol filtration, whether biological or nonbiological, are the physical characteristics of the aerosol such as aerodynamic diameter and shape.

The main types of respirators are classified as follows:[24, 25]

1. Air-purifying respirators — remove contaminants from the air
   a. particulate respirators — filter out aerosols;
   b. chemical cartridge/canister respirators — filter out chemical vapors and gases

2. Air-supplying respirators— provides the wearer with a source of air other than the surrounding air
   a. airline respirators — supplied by breathable air via a hose from a remote source; and
   b. self-contained breathing apparatus (SCBA) — uses its own compressed air supply.

The discussion here will be confined to the first group, particulate respirators, which can be further divided into:

1. Disposable filtering face piece respirators (fabric type with 2 straps), where the entire respirator is discarded when it becomes unsuitable for further use due to excessive breathing resistance, unhygienic condition, or physical damage;
2. Reusable or elastomeric respirators, either half-face or full-face, where the face piece can be cleaned and reused but the filter cartridges are discarded and replaced when they become unsuitable for further use; and
3. Powered air-purifying respirators (PAPRs), where a battery-powered blower moves the air through the filters to the face.

Assigned Protection Factors (APFs)

The level of protection afforded by a particular class of respirators is based on its assigned protection factor (APF). An APF is a measure of the anticipated level of workplace respiratory protection that would be provided by a properly functioning respirator or class of respirators to properly fitted and trained users.[26,27,28] The APF is a special application of the general protection factor (PF) concept. The PF is the ratio of the amount of contaminant to which a person would be exposed without a respirator, to the amount of contaminant to which a person is exposed with a respirator. This is determined by comparing the amount of contaminant inside the facepiece, C_i to the amount of contaminant outside the respirator, C_o, such that

$$PF = C_o/C_i$$

Since Ci is equal/greater than C_o the protection factor is always equal to or greater than unity.

APFs are used as a regulatory requirement used in establishing what type of respirator to use in a given situation. It is calculated by multiplying the APF by the eight-hour exposure limit for a particular contaminant to which the worker may be exposed. For example a respirator with an APF

TABLE 5.5 **Assigned Protection Factors**[29]

| | Respirator Style | | | |
Respirator Class	Half Face-Piece	Full Face-Piece	Helmet/ Hood	Loose-Fitting Face-Piece/Visor
Air purifying		100	—	—
Powered air purifying	50	1000	1000	25
Supplied air (continuous flow)	50	1000	—	—
Self-contained breathing apparatus (SCBA)	—	10,000	—	—

of 10 will allow the worker to work in an atmosphere up to 1000 ppm where the eight-hour exposure limit for the contaminant is 100 ppm (i.e., $10 \times 100 = 1000$).

This regulatory mechanism has not been applied to bioaerosols, since no exposure limits have been established for any disease-causing micro-organisms. Nevertheless, it is assumed that the higher the APF, the greater the level of protection for the worker. Thus a device with an assigned protection factor of 10 allows for penetration through the filter medium up to 10%; an APF of 25 allows penetration up to 2.5% and an APF of 1,000 allows penetration up to 0.1%. The actual risk of disease transmission associated with these APFs is unknown and likely varies markedly depending on the organism of interest and the clinical situation. From a practical perspective, a properly fit-tested particulate face piece respirator or elastomeric half-face piece respirator commonly provides protection factors from several dozen to several hundredfold levels of protection when assessed by quantitative fit testing techniques. The APFs for different types of respirators are shown in Table 5.5.

Fit Testing — Assessing Respirator Face Seal Leakage

All facial seal dependent respirators — those with elastomeric perimeters that are specifically designed to form a seal with the skin of the face — are required to be fit tested in order to check for evidence of leakage at the facial seal. This is a requirement for all North American, United Kingdom, European, New Zealand, and Australian jurisdictions when a worker is required to wear such a device for protection against airborne contaminants in overexposure conditions. Overexposure conditions exist when a worker is working in an environment where the eight-hour occupational exposure limit (OEL) for a particular contaminant could be exceeded.

The primary role of fit testing is to ensure that the wearer has selected a respirator brand, model and size that properly seals with his or her face.[30] Fit tests are designed so that the filter penetration of the test substance is negligible and that any entry of the test agent is solely the result of any existing leaks along the facial seal. Fit tests are also useful for training wearers in proper donning procedures including how to conduct a fit check (negative or positive pressure tests). A fit check should be carried out every time the wearer dons the device.

Determining face piece fit involves qualitative or quantitative fit testing. Qualitative fit testing relies on the wearer's subjective response to taste, odor, or irritation. Quantitative fit testing involves methods that measure pressure differentials or particulate concentrations inside versus outside the face piece. The various fit test methods, both qualitative and quantitative, are described in the 2002 edition of CSA Standard Z94.4.[31]

A 1998 study on N95 performance,[32] has shown that if no fit testing was conducted, one could experience considerable leakage. The average exposure experienced by the 25-person panel in this study was measured at 33% of ambient level — which is below the performance requirements for the N95, set at equal or less than 10% leakage. When the panel was fit tested, the average exposure was reduced to 3% of ambient. Another study by Coffey et al.[33] demonstrated that fit testing screens out poorly fitting respirators. For example, in an initial screening of the various brands available on the market, the employer would not be aware that the brand originally chosen would provide relatively poor fit when compared to another brand. Researchers observed large variability for filtration effectiveness among the 21 models tested, and that some models were far more effective than others. Without fit testing, the 95th percentile penetration ranged from 6 to 88% among the 25 subjects.

Coffey et al.[34] using a panel of 25 subjects (men and women) chosen to represent face lengths from 93.5 mm to 133.5 mm and lip lengths of 34.5 mm to 61.5 mm, examined eighteen different brands of N95 filtering face piece respirators. This study is representative of the wider range of face sizes that would be found in the healthcare field (e.g., encompassing almost 95% of the U.S. working population). The respirators were evaluated both qualitatively and quantitatively without fit testing in order to judge how the different brands of respirators would function "off the shelf." Without fit testing, the 5th percentile simulated workplace performance (SWPF) values ranged from 1.3 (virtually no protection) to 48. Only three of the eighteen respirators had a 5th percentile SWPF greater than 10 (the nominal protection factor expected of a N95 mask). There remained a large variation between models in the percentage of people passing the various fit-test methods (Bitrex®, saccharin, PortaCount®, generated aerosol). One model of the eighteen had a high pass rate for all methods.

The respirators returned different results for different test agents. Passing the Bitrex® fit-test method resulted in 12 of the 18 models providing adequate protection. Passing the PortaCount Plus® fit test resulted in 12 of 13 models providing adequate protection, while six respirators were unable to pass with any subject.

Lee et al. [35] conducted quantitative fit tests with respect to TB exposure on a number of different brands of respirators. Fit test pass rates increased significantly when a well-fitting brand was chosen for the test subjects. Initial screening of the various brands indicated great variability in filter penetrations. They selected two brands because: (1) their medium/regular models fit the greatest proportion of subjects; (2) they provided the highest fit factors; and (3) the greatest proportion of employees rated them as comfortable to wear. The latter is an important consideration for an effective respirator program. Among 1860 individuals who were fit tested, 99.6% were successfully fit tested with one or the other brand.

Qualitative fit testing involves exposing the subject to a substance which can be either smelled, tasted, or is irritating to the upper respiratory tract. Assessing fit on a particulate filter respirator has traditionally been based on the saccharin test. Several years ago Bitrex® (denatorium benzoate) was introduced as an alternate substance to saccharin. McKay & Davies[36] assessed the relative effectiveness of the two test agents and found Bitrex® more effective. All study subjects correctly detected Bitrex® in an induced leak test (sensitivity 100%). Nine of the twenty-six subjects were unable to detect saccharin in the presence of the induced leak. The authors claim that Bitrex® is a better test agent for qualitative fit tests and helps to minimize false negative fit tests.

An Overview of Respirator Performance and Certification Process

Filter media typically consist of fibers made from fiberglass, cellulose, or more commonly today, plastic polymers such as polypropylene. Particles can be captured by a number of mechanical methods including: interception, inertial compaction, sedimentational or gravitational settling, Brownian diffusion, and by a nonmechanical method — electrostatic attraction.[37,38,39,40]

Designing a respirator involves balancing filtration efficiency versus worker comfort. Filtration effectiveness increases with filter thickness and density when the primary method of filtration is based on mechanical methods, as is the case for filter material used up until the mid-1990s.[41,42,43] A thicker, denser filter will cause an increase in the effort required to inhale or exhale through the filter material thereby reducing worker comfort because of increased breathing resistance. This limitation imposed by the filter design and material of the day was lifted recently with the introduction of plastic polymers microfibers as the building material for the filter.

In June 1995, the National Institute for Occupational Safety and Health (NIOSH), the agency responsible for certification of respiratory protective devices in the United States (and recognized by Canadian and other jurisdictions and agencies) issued new regulations for certifying nonpowered particulate respirators under federal statute — the Code of Federal Regulations, specifically 42 CFR Part 84. The new regulations replaced the older 30 CFR Part 11 regulations in force at the time.

The impetus for change was in response to the recognition in the mid-1980s that workers in healthcare and correctional facilities were exposed to airborne TB without adequate respiratory protection. Specifically, the older 30 CFR Part 11 dust/mist/fume type particulate respirators were not found effective as filtration devices for airborne biological agents. Furthermore, the traditional use of surgical masks in the healthcare setting was seriously challenged by a number of organizations including the Centers for Disease Control and Prevention,[44] NIOSH,[45] and the Occupational Safety and Health Administration (OSHA). The CDC revised its guidelines for respiratory protection in HCWs in 1994 to include the recommendation that respirators used to protect healthcare workers from TB have a minimum of 95% efficiency for 1 μm when tested at 50 L/min airflow.[46] At the time only high efficiency particulate-(HEPA) rated respirators could meet these criteria. Dust/mist/fume-type respirators had not been tested at the time in accordance with the new CDC criteria. Tests conducted at a later date on the filtration effectiveness of 30 CPR Part 11 versus 42 CFR Part 84,[47] clearly indicate most 30 CFR Part 11 dust/mist devices failed to meet the new test criteria, particularly at the higher flow rates (85 L/min).

Under 42 CFR Part 84, a new filter classification system was created that distinguishes nine classes of filters based on three filtration efficiencies and three series of filter degradation resistance. The three efficiency levels are 95, 99, and 100% (99.97 % actual) tested at the NIOSH-prescribed test flow rate of 85 L/min, a flow rate considered a moderate workload for human subjects. A "95," "99," and "100" rated respirator is allowed particle penetration of 5, 1, or 0.03%, respectively. The test particulate used was in the size range that is considered the most penetrating particle size (MPPS) — generally considered as particle in the 0.1 to 0.3 μm range.[48] The 0.3 μm particle forms the basis for testing filters.

Filtration efficiency depends on particle size. An increase in particle size will cause increased filtration by the interception and inertial impaction mechanisms, whereas a decrease in particle size will enhance collection by Brownian diffusion. As a consequence, there is an intermediate particle size region where two or more mechanisms are simultaneously operating yet none is dominating. This is the region where the potential for particle penetration through the filter is at the maximum and the efficiency of the filter a minimum. For fibrous filters, such as found in most respirators, the minimum filter efficiency is generally known to occur in the vicinity of 0.3 μm. This is the basis of the widely used dioctyl phthalate (DOP) or sodium chloride tests for high efficiency particulate filters (HEPA) and 42 CFR Part 84 particulate filter devices (95/99/100 series), which make use of mono-disperse 0.3 μm diameter DOP or NaCl particles for testing the filter.

Chen and Huang have shown that if a polypropylene filter is electrically neutralized, the filter efficiency is reduced by a factor of 36 to 68%.[50,51] Other studies have shown that decreasing the electrostatic charge on the filters by using an isopropanol wash, penetration of N95 respirators increased from an average of 2% to as high as 43.5%.[52] The authors also demonstrated that penetration of N99 respirators went from an average of 0.23% to as high as 53.3%; penetration of P100 respirators went from an average of 0.001% to as high as 3.92%. These studies reinforce the fact that such respirators rely heavily upon electrostatic attraction, and if exposed to industrial aerosols, such as oily mists or certain other chemicals, the efficiency of these respirators can fall dramatically. That is the reason that "N" designated respirators cannot be used in work environments where one could be exposed to oil mists. In that case, a "P" type respirator must be selected, as noted below. This is of little consequence to the health care setting. "N" type devices are suitable for most health care applications.

Temperature and relative humidity have historically been shown to have an effect on respirator efficiencies.[27] However, recent testing of newer electrostatic respirators suggests that the effects of relative humidity on filtration efficiency are no longer very significant, most likely due to technological advancements in the filter media.[53] Polypropylene, the basis for most 42 CFR Part 84 respirator filter media, is a highly hydrophobic material — the fibers do not absorb water.

With respect to TB exposure, NIOSH has approved all filter media of respirators certified as 42 CFR Part 84 compliant for use against TB exposure, since the filters are more efficient at the 1 μm size than at the most penetrating particle size (0.3 μm) size. Since individual viruses are smaller than the most penetrating particle size, they will be effectively filtered by all 42 CFR Part 84 compliant respirators. Polypropylene filter media have also proved to be highly effective in filtering particles in the size range typically associated with viruses and fungal spores. Of greatest concern are viruses carried on droplets near the most penetrating particle size, as they have a higher probability of penetrating a respirator than an individual virus.

However, most viruses which cause respiratory and gastrointestinal disease in humans, must be contained in large droplets (>5 μm) in order to survive outside the body and transmit disease from person-to-person. This includes such common respiratory pathogens as influenza, respiratory syncytial virus (RSV) parainfluenza viruses, the common coronaviruses and others. The notable exceptions are measles, varicella zoster virus (chickenpox), and smallpox which apparently can survive in small diameter droplets or droplet nuclei and can be transmitted by air over long distances.[54]

Typically disposable particulate respirators are constructed from a filter material in the shape of a formed cup or loosely in the shape commonly called the "duckbill." Approved half–face-piece devices are designed to sit on the bony framework of the face — over the jawbone and cheekbones. Approved half-face-piece respirators are designed to form a secure seal where the device meets the skin of the face in accordance with performance criteria established by NIOSH. Approved devices are supplied with two straps; typically, one is designed to be placed over the back/top of the head, the other around the neck. This is to ensure the device is pulled both up and down over the jawbone and cheekbones to facilitate the seal with the face.

The skin-to-respirator seal is important since the space within the respirator is under negative pressure during inhalation. As a consequence, air-purifying respirators are classified as "negative-pressure" devices unlike respirators that are supplied by either ambient-pressure air (PAPRs) or high pressure air (airline or SCBA). The latter are classified as "positive-pressure" devices. Accordingly, positive pressure devices provide the wearer with a higher level of protection than negative pressure devices when the respirator is not able to form a seal with the face. Loose fitting PAPRs are not positive pressure according to the respirator classification system.

Half face piece respirators — both filtering face-piece type such as the N95s as well as elastomeric devices — are assigned protection factors of 10 (see Table 5.5).

Performance of Surgical Masks and Air-Purifying Particulate Filter Respirators

Surgical masks were developed to prevent the wearer's exhaled secretions from contaminating the operative field.[55] However, these devices also have been used for decades, in the healthcare industry and by the general public, as protective devices to prevent exposure to various respiratory pathogens. Surgical masks are constructed of a filter material and cut basically in the shape of a rectangle. The device is placed over the nose and mouth and held in place by straps placed behind the ears or around the head but more usually around the back of the head and neck. The device fits fairly loosely and a tight seal is not feasible where the outside edge of the mask meets the skin of the face. Most users in the healthcare industry tend to wear surgical masks rather loosely; considerable gaps are usually observed at the peripheral edges of the surgical mask along the cheeks, around the bridge of the nose, and along the bottom edge of the mask below the chin.

Standard surgical masks are considered a Class II device by the U.S. federal Food and Drug Administration (FDA) which require premarket sales approval. This means that to obtain approval as an item for sale, the manufacturer must demonstrate to the satisfaction of the FDA that the new device is substantially equivalent to similar masks currently on the market.[56] There is no specific requirement to prove that the existing masks are effective and there is no standard test or set of data required supporting the assertion of equivalence. Nor does the FDA conduct or sponsor testing of surgical masks.

Concerns surrounding health care worker exposure to TB gave greater prominence to the use of surgical masks as protective devices for health-care workers.[57] Moreover, several studies conducted in the early 1990s showed that air leakage occurs both around and through surgical masks. Chen et al.[58] demonstrated that surgical masks are highly variable when challenge tested with 1 µm particles, with results ranging from 5 to 100% penetration. In another study, Chen and Willeke[59] observed 40 to 60% penetration for one model of surgical mask over the particle size range of 0.3 to 1.0 µm and 80 to 85% penetration for the other brand tested over the particle size range of 0.3 to 2.0 µm range. Weber et al.[60] assessed eight brands of surgical masks and found penetration ranging from 20 to 100% for particles in the 0.1 to 4.0 µm aerodynamic diameter range. These and related studies led the CDC in 1990 to recommend the use of NIOSH-approved respirators as superior protective devices against TB aerosols.

Wake et al.[61] conducted a filter penetration study on a wide variety of devices available in the United Kingdom. Single strap dust masks (non-UK, non-NIOSH approved) typically sold in hardware stores, proved highly ineffective when challenged with microbiological aerosols of *Bacillus subtilis* subsp. *globigii, Micrococcus luteus,* and *Pseudomonas alcaligenes* allowing penetration up to 100%. Surgical masks allowed penetration up to 83% of the bioaerosol. Surgical masks made with polypropylene fibers, offered better protection, ranging from 0.9 to 25% penetration. Dust/mist and dust/mist/fume (approved by the UK and equivalent to 30 CFR Part 11 filters) allowed penetration from less than 0.01 to 0.9%. Filtering face piece (N95 equivalent — FFP3 approved) proved the most effective in filtering the bioaerosols, allowed penetration from 0.02 to 0.4%.

Another study by Brosseau et al.[62] found filter penetration highest and most variable for the surgical masks when compared to NIOSH-approved respirators. Geometric mean penetration of the filter material was about 22% for surgical masks versus that of 0.02% geometric mean penetration for respirator-type HEPA filters when challenged by both nonbiological (0.55 μm latex spheres) and biological test particulates (*Mycobacterium abscessus* and *Pseudomonas fluorescens* aerosols). *M. abscessus* is in the range of the most penetrating particle size —0.3 μm aerodynamic diameter.

With respect to testing the efficacy of surgical masks, a number of manufacturers routinely conduct biological testing on their products such as the Viral Filtration Efficiency (VFE) or Bacterial Filtration Efficiency (BFE) tests. The BFE and VFE tests typically aerosolize solutions of bacteria or viruses into 3.0 μm particles, which are far easier to filter than if 0.3 μm droplets were used. Occasionally, the investigator may run the droplets through a drying chamber, so that droplets evaporate and only individual viruses or bacteria are challenging the filter. There is no requirement for manufacturers to run these types of tests, but they are still very commonly done; the filtration efficiencies reported from BFE and VFE tests are very high (nearly always >99.999%), so they make the devices appear far more effective than they may actually be. This is an issue of concern for anesthesia and respiratory breathing system filters, pulmonary function filters, and heat–moisture exchanging filters, as there is no requirement for NaCl or DOP challenges to determine filtering capabilities. Manufacturers typically report results of BFE and VFE tests (typically >99.999% efficiency), and these devices are considered "bacterial" or "viral" filters. However, there are now breathing system filters and pulmonary function filters that claim to be >99.999% efficient at removing bacteria or viruses, but which may show 70% or less efficiency when challenged with NaCl or DOP tests at 0.3 μm.[63] These filters are routinely used to filter microorganisms at the source, such as on anesthesia machines, pulmonary function machines, ventilators, and manual ventilation unit. The lack of meaningful standards for these devices, along with the use of BFE and VFE test data, has created an environment in which healthcare workers think they are far more protected than they actually are.

However, even with the use of highly efficient, modern filter media, exhaled air may escape or enter unfiltered around the edges of the mask.[64, 65] Surgical masks cannot be fit tested. To illustrate the ineffectiveness of facial seal of the surgical mask, Tuomi[66] conducted particle penetrations studies on several brands of surgical masks. One test involved normal positioning of the surgical mask on the test head/breathing machine; the other test involved tape-sealing the edges of the surgical mask to the test head. The overall filtration efficiency of the nontaped versus taped mask measured 33% and 67%, respectively across most of the particle size range (0.2 to 10.0 μm) with a greater difference noted for the larger particle sizes (above 2 μm).

Powered Air-purifying Respirators

A powered air purifying respirator or PAPR is basically an air-purifying respirator in which a blower pulls ambient air through air-purifying filters (housed in cassettes or canisters), and then supplies purified air to the face piece.[67,68] This is accomplished by the addition of a battery-operated blower. Certain models of PAPRs do not provide a seal with the face.

PAPRs can be fitted to the following face pieces:

- tight-fitting or face-seal dependent
  - half face-piece type
  - full face-piece type
- non-tight-fitting or non-face-seal dependent
  - loose fitting helmet/hood
  - loose fitting face piece/visor
- full-body suit

The PAPR used predominantly used in the healthcare industry is the loose fitting face-piece/visor type which carries an assigned protection factor (APF) of 25. Facial seal dependent or tight fitting PAPRs provide a higher level of protection than their loose fitting counterparts and are assigned a protection factor (APF) of 1,000. Tight fitting PAPRs also allow fit testing. Loose fitting PAPRs cannot be fit tested.

All types of NIOSH-certified PAPRs meet the CDC requirements for protection against tuberculosis when fitted with a HEPA filter. At this time, there are no certified 42 CFR Part 84 filters, including filters rated at 95, 99 or 99.97% efficiencies, available for PAPRs. Only 30 CFR Part 11 NIOSH-certified HEPA filters are currently approved for use with PAPRs. HEPA filters are highly effective and equivalent to an N100 filter.

Loose fitting PAPRs provide a viable alternative in the health care industry where a worker, who is required to wear respirator, cannot achieve a proper fit as determined by a failed fit test, or is fully bearded. Note that a loose-fitting PAPR provides a higher level of protection than a tight-fitting half face piece respirator (filtering face piece type, or elastomeric face piece fitted with particulate filters) — refer to Table 5.6 for a comparison of protection factors for the various devices available.

PAPR-Like Devices

A study by Derrick and Gomersall[69] found the Stryker® and the Stackhouse FreedomAire® powered-air supplying surgical helmets offer very little protection against airborne 0.02 to 1.0 μm diameter particles. It should be noted, however, that these devices are not sold as "respirators" and

Table 5.6 Advantages and Disadvantages of Various Types of Respirators as It Applies to Healthcare Workers

Type of Respirator	Advantages	Disadvantages/Limitations
Air purifying Particulate Nonpowered Half face "Filtering face piece-type," e.g., N95 "disposable"	Disposable Small size Light weight Simple design — easily understood by wearer Can be reused (short-term) Unrestrictive mobility Inexpensive over short-term Requires no cleaning Requires no maintenance APF of 10 Easy to breathe through (low breathing resistance) Allows good peripheral vision Can be fit tested Allows easy communication Nonthreatening to patient Units without exhalation valves allows use in sterile field Meets CDC criteria for protection against TB and other bioaerosols	Negative pressure device increases inward leakage Facial hair or scars or certain face types will interfere with facial seal Higher costs over long term (vs. nondisposables) Reusable for short-term periods only Units with exhalation valve allow contamination in a sterile field Difficult to fit check No eye/face protection

Air purifying Particulate Nonpowered Half-face Elastomeric face piece type Particulate filter cartridges	Can be reused (long-term) Can be cleaned Unrestrictive mobility Most models are light weight Moderate cost over long term APF of 10 Most models easy to breathe through (low breathing resistance) Most models light weight Low profile models allows good peripheral vision Easy to fit check Can be fit tested Meets CDC criteria for protection against TB and other bioaerosols	Negative pressure device increases inward leakage Facial hair or scars or certain face types will interfere with facial seal Moderate initial cost Presence of exhalation valves and will allow contamination in a sterile field Requires disinfection between use Requires routine maintenance No eye/face protection Communication more difficult than for disposables Some models have a slightly higher breathing resistance Some models are heavy Some models affect peripheral vision More threatening to patients (the more industrial it looks, the more intimidating)
Air-purifying Particulate Nonpowered Full face Elastomeric face piecetype Particulate filter cartridges	Can be reused Can be cleaned Unrestrictive mobility Provides eye/face protection Moderate cost over long term More protective APF of 100 Most models provide adequate peripheral vision	Negative pressure device increases inward leakage Facial hair or scars or certain face types will interfere with facial seal All units have exhalation valves and will allow contamination in a sterile field Higher initial cost than half face piece type Prescriptive glasses cannot be worn due to interference with facial seal

Continued.

Table 5.6 Advantages and Disadvantages of Various Types of Respirators as It Applies to Healthcare Workers (Continued)

Type of Respirator	Advantages	Disadvantages/Limitations
	Easy to fit check Can be fit tested Slightly higher breathing resistance Meets CDC criteria for protection against TB and other bioaerosols	Requires special prescriptive inserts Heavier and bulkier Decreased comfort level Requires routine maintenance Requires disinfection between use Communication more difficult Some models provide limited peripheral vision More threatening to patients
Air-purifying Particulate Powered Loose fitting facial seal e.g., face piece/visor- type Powered Air-Purifying Respirator (PAPR)	Allow persons with beards and those unable to fit standard respirators to be protected Moderate cost over long term Unrestrictive mobility Low breathing resistance Provides cool air Comfortable face seal Allows wearing of prescription glasses Can be reused Can be cleaned More protective than half face piece devices APF of 25 Built-in eye/face protection Meets CDC criteria for protection against TB and other bioaerosols if fitted with HEPA filter(s)	Allows contamination in a sterile field since exhaled air exists around the fabric dam of the visor High initial cost Heavier and bulkier Decreased comfort level High level of maintenance Batteries must be recharged and maintained Bulky and noisy (motor) Communication more difficult Better peripheral vision than helmet/hood PAPR Require disinfection between use Cannot be fit checked or fit tested More threatening to patients

Type	Advantages	Disadvantages
Air-purifying Particulate Powered Tight fitting facial seal, e.g., Helmet/hood type Powered Air Purifying Respirator (PAPR)	Can be reused Can be cleaned Moderate cost over long term Unrestrictive mobility Low breathing resistance Provides cool air More protective than full facepiece nonpowered devices APF of 1000 Allows fit checking Allows fit testing Built-in eye/face protection Meets CDC criteria for protection against TB and other bioaerosols if fitted with HEPA filter(s)	High initial cost Facial hair or scars or certain face types will interfere with facial seal Prescriptive glasses cannot be worn due to interference with facial seal Units with front-mounted exhalation valves will allow contamination in a sterile field Heavier device Peripheral vision inferior to loose fitting PAPR High level of maintenance Batteries must be recharged and maintained Bulky and noisy (motor) Communication more difficult Requires disinfection between use More threatening to patients
Air-supplying Half-face piece or full-face piece, e.g., Supplied-air respirators	Can be reused Can be cleaned Moderate cost over long term if air supply readily available Low breathing resistance Provides cool air	High initial cost Facial hair or scars or certain face types will interfere with facial seal Requires source of quality breathable air Restricts mobility due to presence of airline Units with front-mounted exhalation valves will allow contamination in a sterile field

Continued.

Table 5.6 Advantages and Disadvantages of Various Types of Respirators as It Applies to Healthcare Workers (Continued)

Type of Respirator	Advantages	Disadvantages/Limitations
	More protective than full face piece nonpowered devices APF of 1000 Allows fit checking Allows fit testing Built-in eye/face protection for full face piece devices Meets CDC criteria for protection against TB and other bioaerosols	Prescriptive glasses cannot be worn due to interference with facial seal Communication more difficult Requires disinfection between use Complex maintenance More threatening to patients
Air-supplying Full face piece only e.g., self-contained breathing apparatus (SCBA)	Highest level of protection APF of 10,000	Impractical for health care Very high costs — initial and long-term Highly skilled, technically trained staff required

are not NIOSH approved. They are designed to be used for protection against droplets and splashes and to minimize contamination of a sterile field. In comparison with protection factors obtained with N100s, the protection factors ranged from 3.5 to 4.5 for the Stryker and 2.5 to 3.0 with the Stackhouse.

Respiratory Protection — Selecting the Appropriate Device

Prior the mid-1990s, many healthcare practitioners were inexperienced with respiratory protective devices. They saw them only as devices designed for general industry. In fact, in hospital settings the word *respirator* is more likely to suggest a device for providing respiratory support to a patient than a device for protecting the health care worker. Many in the healthcare industry view surgical masks as providing respiratory protection for the wearer. This belief continued as recently as the SARS outbreak in March 2003.

The selection of a respirator for protecting the healthcare worker from exposure to pathogenic bioaerosols, should follow fundamental occupational hygiene principles based on the risk management paradigm — risk identification, risk evaluation, and implementation of risk control measures. The decision framework used for airborne chemical toxicants as prescribed by NIOSH in its 1987 document entitled *Respirator Decision Logic*[70] has been suggested as an appropriate model.[71] That is, one specifies an acceptable risk of infection (analogous to setting an occupational exposure limit for a chemical) and estimates exposure intensity and duration based on the pathogenesis of, and infectious dose for, the organism based on establishing virulence, infectivity, potential for transmission by inhalation, viability of the organism when present in respiratory droplets of various sizes, and population susceptibilities.

Where it is established the organism presents a risk to human health through respiratory tract exposure, protection should be considered. A respirator would be selected with an average penetration value sufficient to reduce exposure to meet the acceptable risk criterion established through the risk assessment process. For example a N95 may provide an adequate level of protection for pathogenic agent "A" but a N100 is required for agent "B" since it is assessed as presenting a higher risk of infectivity and/or virulence. Table 5.6 summarizes the factors to consider when choosing respiratory protection for healthcare workers.

For example, a N95-rated respirator is considered an appropriate device by CDC and NIOSH for protection against bioaerosols containing *Mycobacterium tuberculosis* since, in part, N95s are worst-case challenged tested to aerosols with aerodynamic diameters averaging 0.3 μm. A single tubercle bacillus measures around 0.8 μm[71] and this study found that 42 CFR Part

84 rated respirators offer much greater efficiency than their 30 CFR Part 11 predecessors, particularly at higher flow rates (moderately high respiratory flows — 85 L/min).

No government or other agency has yet specified an acceptable occupational risk of a *Mycobacterium tuberculosis* infection, the organism most often studied in relation to occupational risk of infection. The same is true for other pathogenic agents, although the scientific literature presents a number of articles that describe in detail risk models.[72, 73, 74, 75, 76] Infectivity data, is available for the Coxsackie A-21 virus where the aerosol infectious dose for tissue culture has been established as 28 times the $TCID_{50}$ (50 % tissue culture infectious dose).[77]

Barnhart et al.[78] has shown that, for tuberculosis in health care settings, based on the estimated aerosol infectious dose from Nicas,[79] and analysis of TB skin-test conversion rates, the use of respiratory protection is estimated to reduce the risk of skin-test conversion by the following proportions:

- surgical mask — 2.5 fold reduction
- disposable dust/mist/fume (30 CFR Part 11) respirators — 17.5 fold
- disposable HEPA respirators —17.5 fold
- elastomeric HEPA respirators — 45.5 fold
- HEPA-fitted PAPR — 238 fold reduction

Note that the no. 42 CFR Part 84 devices were not available at the time of this study, as they only became commercially available in July of 1998.

The authors based their risk assessment on 130 TB patients who produced an average of 0.25 infectious quanta per hour but with marked variation, ranging from 0 to 60 infectious quanta per hour. An infectious quantum is the number of infectious droplets required to cause infection in a prescribed number of susceptible individuals.[80]

Lee et al. also estimated the risk of TB infection using data from their fit test studies and the cumulative risk of infection estimated on a Poisson probability model in a manner that incorporated the rate of successful fit tests of the various brands of respirators for which quantitative fit tests were conducted.[81] Cumulative infection rates were calculated for *M. tuberculosis* infection risks as follows:

- With no respirator use
  - low risk scenario produced 1-yr and 5-yr cumulative risks of 0.0133 and 0.0648, respectively
  - high risk scenario produced 1-yr and 5-yr cumulative risks of 0.0522 and 0.235, respectively.

- With a respirator with a pass rate of 95%
  - low risk scenario produced 1-yr and 5-yr cumulative risks of 0.0007 and 0.0036, respectively
  - high risk scenario produced 1-yr and 5-yr cumulative risks of 0.0029 and 0.0141, respectively.

Application of the Science of Respirator Protection to the Healthcare Setting

As discussed in the previous section, respirators were adapted from industry to health care and initial testing was based on industry standards. Questions have been raised as to whether there is a relevant model for health care regarding respirator use.

Have Respirators Been Evaluated Under "True" Workplace Conditions?

A series of articles (several predating NIOSH and 42 CFR 84 standards) by Brosseau et al. examined the performance of several respirators and surgical masks when challenged with *M. abscessus* aerosols (to mimic TB exposures).[82, 83, 84] Unlike methods used in many other bacterial challenges, Brosseau ensured that the bacterial aerosol went through a drying process such that the majority of particles were individual bacterium, not large water droplets. The authors concluded that nonbiological particles such as polystyrene latex or dioctyl phthalate (DOP) with an aerodynamic particle size similar to the bioaerosol of interest appeared to be an appropriate challenge particle. The investigators also examined the recovery of organisms captured by filters as viable organisms released to the environment after reentrainment from masks. In general, organisms were found to be nonviable when reentrained from masks. Importantly, the authors demonstrated that any facial leakage negated the increase in filtration efficacy gained with N95 masks (the importance of a good facial seal has been discussed in the previous section of this review). These articles confirmed that biological models to assess the efficacy of respirators are possible, and if carefully designed to ensure worst case challenges, may be more representative of actual working conditions than traditional industry models.

A series of articles by Coffey et al.[85,86,87] examined the role of fit testing and respirator performance under simulated conditions. The articles discussed the sequential development of a model to assess quantitative fit testing methods, evaluate the fit-testing methods and examine different test aerosols and their accuracy in assessing fit testing. Importantly, the investigators used a simulated workplace environment to conduct their studies. Subjects donned a respirator and conducted a user seal check prior to an evaluation of total penetration of particles during a series of maneuvers. Simulated testing demonstrated that fit testing gave better protection by screening out poorly fitting respirators.

Finally, Huff et al.[88] clearly illustrated the importance of wearing of a fit-tested particulate face piece respirator in conjunction with the use of simple body substance isolation techniques. The authors tracked the dispersal of radioisotope technetium (Tc^{99m}) during pulmonary function testing. Personnel were evaluated for contamination on clothing, hair, and airways (nose swabs). Laboratory coats and latex glove were the only PPE provided in the first part of the study. In the second part, personnel wore surgical masks, cover gowns, and head covering. For the third part, personnel were fitted with dust/mist/fume respirators designed for protection against radionuclides, gowns, and head coverings and had been trained in infection control procedures. The respirator used was of the face piece type with an elastomeric liner around the periphery of the device to create a good face-seal.

Results for Part 1 and 2 demonstrated levels as high as 11,000 disintegrations/min in the nasal passages of personnel, indicating that surgical masks were ineffective in reducing respiratory tract deposition of technetium. When fit tested respirators were worn, the levels were measured at 50 disintegrations/min or less. One worker, who had not been properly fit tested, had readings exceeding 1000 disintegration/min. This individual was subsequently retrained and retested — a reduction in contamination was subsequently noted, illustrating that the wearing of a fit-tested dust/mist/fume face piece respirator significantly reduced exposure levels to aerosols.

The study concluded that proper infection control techniques (e.g., hand hygiene) and wearing the appropriate PPE (head coverings, surgical cover gowns) resulted in a significant reduction in deposition of the radioisotope onto the body and the lab coats worn under the gown. The study clearly demonstrated that fit testing of N95 respirators significantly reduced exposure levels to the technetium compared to surgical masks. This article is one of few actual workplace evaluations, and it provides a potential model for real-time evaluations while offering a method of sample collection.

Fit Testing versus Fit Checking

Fundamental to the fit-testing process is the educational component; that is, teaching the worker to select the correct mask for best facial fit and to perform a fit check each time a respirator is worn. Hannum et al. examined the effect of three different methods of respirator training on the ability of healthcare workers to pass a qualitative fit test.[89] Employees were divided into three groups: Group A received one-on-one training and were fit tested as part of the training; Group B received classroom instruction and demonstration by infection control nurses in the proper use of respirators but were not fit tested; and Group C received no formal training. Participants then went onto a subsequent qualitative fit test using irritant smoke to check for their ability to correctly adjust the respirator. Location or professional status did not affect fit test pass rate but prior experience wearing respirators did. When the study groups were compared after stratifying for prior experience, there was no difference between Groups A and B but significance difference between the latter two groups and Group C. The authors concluded that fit testing as part of training marginally enhances the ability of HCWs to wear respirators properly and pass a fit test.

Protecting the Eyes of Healthcare Workers

The published literature on the role of eye protection in protecting HCWs from injury and disease is limited. Those studies which have been conducted generally relate to the use of eye protection in the context of dental infection control practice,[90, 91] in reducing the risk of splashes from blood during operative procedures [92, 93] or the protective effect of goggles in protecting against traumatic[94] or chemical injuries.[95, 96, 97] Significantly, no studies were found that measure actual facial/ocular/nasal exposure to bioaerosols and how or what types of eye protective equipment are effective in reducing exposures. The literature reviewed does not address putting on and taking off (donning and doffing) of face shields, goggles, and safety glasses to prevent autoinoculation. Nor does it address the efficacy of manufacturers' protocols for care, sterilization, cleaning, and storage of the equipment. There are no standards specific to the use of face shields and eyewear for protection against bioaerosols.

The need for facial protection in healthcare is suggested by studies such as Kouri and Ernest[98] who examined the perceived and actual face shield contamination during vaginal and cesarean delivery. They found that in 50% of cesarean deliveries and 32% of vaginal deliveries, there was measurable contamination of the face shield surface that was not detected by the physician. This occurred 92% of the time for cesarean delivery and 50% of the time for vaginal delivery. Similarly, Leese et al.[99] measured surface contamination of face shields and goggles resulting from manual dumping of medical waste. Twenty-two percent of face shield and goggle samples were found to be contaminated.

Giachino[100] reported on a study of macroscopic contamination of the conjunctiva of orthopedic surgeons by body fluids. All members of the surgical team at a hospital wore high impact polycarbonate glasses during 60 consecutive orthopedic surgical procedures. In 37 cases both the lenses of the surgeon and his assistant were contaminated by body fluids from the patient, resulting in 59 contaminations, but the significance of these results are unclear due to the uncertainty of the ability of blood borne pathogens to be transmitted through the intraocular route.

The few studies which have looked at the effectiveness of eye protection have found mixed results. Davies et al. collected sera from 50 practicing dental surgeons and 50 control subjects matched for age and sex.[101] Questionnaires from the dentists detailed information relating to protective work-wear and other cross-infection control measures employed within the surgery. The sera were examined by complement fixation tests for antibodies to influenza A and B, respiratory syncytial virus, and adenovirus. The dental group had a significantly elevated prevalence of antibodies to influenza A and B ($p < 0.001$) and respiratory syncytial virus compared with the controls. Wearing of masks or eye protection did not markedly reduce infection with these viruses among the dentists. The authors conclude that dentists are at occupational risk of infection with respiratory tract viruses, and that mask- or spectacle-wearing affords little protection. Using face masks and eye glasses was not correlated with the prevalence of nasal irritation, runny eyes, and itchy skin symptoms in a group of dental hygienists.[102]

Despite this lack of evidence for the efficacy of eye protection, this has been included in formal recommendations to protect HCWs from SARS.[103] Given the documented ability for viruses in the size range of the SARS-CoV to be transmitted via hand to eye contact, this would seem reasonable. However, there is an urgent need to identify the additive benefit of the addition of goggles to other measures designed to reduce exposure to infectious agents among health care workers.

Effectiveness of Interventions in Protecting Healthcare Workers and Preventing Transmission of Respiratory Infections in Healthcare Settings

In order for infection control guidelines to be successful in protecting healthcare workers (HCWs) and patients from SARS, a good understanding is required of what procedures, and specifically personal protective equipment (PPE), are most effective. In addition, organizational and environmental factors, and worker characteristics, influence the ability and willingness to comply with these procedures. A theoretical model which can account for these factors, stems from the PRECEDE (Predisposing, Reinforcing and Enabling Factors in Educational Diagnosis and Evaluation)

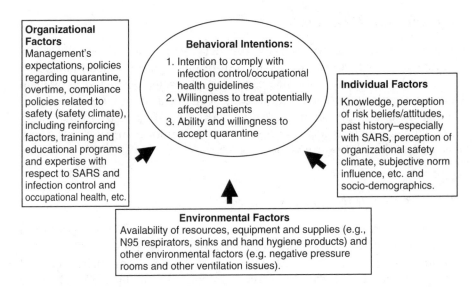

Organizational Factors
Management's expectations, policies regarding quarantine, overtime, compliance policies related to safety (safety climate), including reinforcing factors, training and educational programs and expertise with respect to SARS and infection control and occupational health, etc.

Behavioral Intentions:
1. Intention to comply with infection control/occupational health guidelines
2. Willingness to treat potentially affected patients
3. Ability and willingness to accept quarantine

Individual Factors
Knowledge, perception of risk beliefs/attitudes, past history–especially with SARS, perception of organizational safety climate, subjective norm influence, etc. and socio-demographics.

Environmental Factors
Availability of resources, equipment and supplies (e.g., N95 respirators, sinks and hand hygiene products) and other environmental factors (e.g. negative pressure rooms and other ventilation issues).

Figure 5.2

model of health promotion as developed by Green and colleagues[104] and as modified by DeJoy[105] for application to self-protective behavior at work. Predisposing factors can be seen as the *characteristics of the individual* (beliefs, attitudes, values) that facilitate self-protective behavior. Enabling factors can refer to the *environmental factors* that block or promote self-protective behaviour, including the skills, knowledge, as well as availability and accessibility of PPE and other resources. Reinforcing factors involve the *organizational factors*, such as communication, training, performance feedback, social approval or disapproval from coworkers or management, and other safety climate dimensions. These factors can be seen to interact in the following manner.

There are an increasing number of studies highlighting the importance of a multidimensional or systems approach to worker health and safety, including considering job/task demands; worker characteristics; and, especially, environmental and organizational factors.[106–111] It can be concluded from this body of literature that "compliance" cannot be fully understood by examining each of these factors in isolation, but rather how they relate to each other. This critical appraisal of the literature encompasses all three factors, reviewing information from the SARS outbreak, from other respiratory pathogens which threaten the well-being of HCWs, and from the general health and safety literature in healthcare, and finally in workplaces generally.

Organizational Factors

Evidence pertaining to the key organizational factors from the SARS outbreak are still subject to debate, but research into the determinants of general infection control and health and safety in healthcare as well in other workplaces provide a great deal of relevant information. Organizational factors of importance include both general organizational culture and climate, such as leadership style and institutional mission and goals, as well as specific policies and procedures.

Evidence from the SARS Outbreaks

Five descriptive[112–116] and five analytic studies[117–121] have been published on the hospital-associated outbreaks of SARS in the spring of 2003. Other information sources included letters to the editor, editorials, personal commentaries, and a variety of infection control guidelines.[122,123,124] Some of these reports analyzed organizational factors in terms of their importance in preventing SARS transmission, but the quality of evidence presented varies markedly.

Lau et al.[125] conducted a case-control study of 72 hospital workers who developed SARS in Hong Kong, along with 144 matched controls. They found that having an inadequate amount of infection control training was associated with a higher risk of SARS infection. Specifically, 50% of healthcare workers who developed SARS had not received any SARS infection control training, versus 28% of the controls. Interestingly, the authors found no significant differences between the cases and controls with respect to performing high-risk procedures, incurring minor PPE problems, or having social contact with SARS-infected individuals. In the final multivariate mode, perceptions of an inadequate PPE supply, infection control training less than two hours, and inconsistent use of PPE were significant independent risk factors for SARS infection. The issues related to PPE supply are further discussed in the section on environmental factors below.

Scales et al.[126] described the consequences of a brief, unexpected exposure to a patient with SARS that resulted in sixteen intensive care staff being put at risk of exposure. Of these 16 HCWs, 7 developed the disease. Three of those affected were present in the room for more than 4 hours. A further three of five people who were present during endotracheal intubation developed infections, including one worker who wore gloves, gown and an N95 respirator. The authors discussed the approach to quarantine, emphasizing the desirability of not quarantining more people than necessary but emphasizing that the consequences of missing the diagnosis of SARS for even a relatively brief period can have disastrous consequences, and that therefore a wide net is needed.

The CDC also emphasized the importance of formal respiratory protection programs as well as ensuring that workers understand the correct order to remove PPE.[127] This study noted that many healthcare workers became quite fatigued and recognized that there were momentary lapses where they forgot to put on their goggles, or forgot to change their mask. One editorial suggested that only the most experienced personnel should be involved in high-risk procedures such as intubation.[128]

Organizational interventions which were actually applied in the hospital-associated outbreaks of SARS included temperature checks on hospital staff,[129] quarantine,[130] limiting visitors,[131] hospital closures,[132] and limiting the number of HCWs present during aerosol-generating procedures.[133] None of these interventions, however, have been tested with respect to their ability to prevent SARS transmission.

The study by Park and his coworkers, which retrospectively reviewed HCWs who had been exposed to those American patients with laboratory evidence of SARS-CoV infection, provides some interesting observations on compliance with infection control guidelines.[134] Sixty-six HCWs reported exposure to a patient who was coughing and later found to be SARS positive, yet 40% did not use a respirator. Despite being exposed and developing symptoms, 10 of 17 HCWs were not furloughed. However, none of the HCWs became ill and no local disease transmission occurred.

Evidence from Other Nosocomial Infection Studies and Workplace Health and Safety in Health Care

Much of the evidence that is most relevant to "protecting the faces of HCWs" comes from studies of other infectious diseases transmitted to HCWs or patients.

Specific Policies and Procedures

Studies on the effectiveness of infection control practices for other respiratory viruses have shown that organizational factors can be important determinants of limiting disease transmission. Isolation, or cohorting of patients, restricting visitors and screening admitted patients for respiratory syncytial virus (RSV), have been shown to be more effective in reducing nosocomial spread of RSV, than the use of specific PPE, alone.[135, 136, 137] Outbreaks of parainfluenza virus have been controlled in bone marrow transplant units and neonatal ICUs by application of contact precautions using gowns, gloves, isolation, and cohorting of nurses.[138, 139, 140]

The most important determinants of successful general nosocomial infection control programs in hospitals have been understood since the mid-1980s, when the Study on the Efficacy of Nosocomial Infection Control (SENIC) was published.[141, 142] The following organizational factors were found to be important in determining effective infection control and lower rates of nosocomial-transmitted disease: having one infection control practitioner per 250 acute care beds, having at least one full-time physician interested in infection control, having an intensive surveillance program for nosocomial diseases, and having intensive control policies and procedures. However, in a recent survey of 172 hospitals in Canada, only about 60% of hospitals had evidence of compliance for each of the SENIC factors. The number of institutions that had all four factors was likely much less.[143]

General infection control procedures are focused on protecting patients and the public, while occupational health practitioners are charged with protecting the workforce. While studies have been conducted related to resource requirements for infection control, no similar studies have been conducted regarding resource requirements for occupational health resources. The American Medical Association in 1989 in their publication *Occupational Health Services: A Practical Approach* stated, that "for industries lacking exceptional physical or chemical hazards," the following guidelines are appropriate: for the first 300 employees one full-time occupational health nurse (OHN), and an additional OHN for every 750 employees.[144] In regard to occupational physicians, they state a full-time physician is needed if there are greater than 2,000 employees. It is well recognized that health care does have exceptional hazards, in most, if not all, areas, not the least of which relates to occupational infections. While there have been no studies as to the current levels of occupational health resources in Canadian hospitals, it is clear that it is well below the appropriate levels.

Communication, Training, and Feedback — There is considerable literature with respect to adherence to standard precautions (SP) and measures to prevent the spread of TB. Most of the studies are observational and it has been noted that there is a dearth of controlled intervention studies, but the importance of good communication is a major theme that emerges. A study of 451 nurses employed in a large U.S. hospital center found that organizational factors were the best predictors of adherence to SP.[145] Although the variance in adherence predicted by the model was modest, the factors that predicted adherence to SP included whether compliance was seen as a job hindrance, the availability and accessibility of PPE, and whether feedback on compliance was given. This study, however, did not look specifically at the type of feedback or communication used. Other studies in healthcare and correctional facilities have had similar findings.[146,147]

There is very little information that directly touches on what formative training and continuing education strategies are most effective in implementing and maintaining good infection control practices, nor on what methods of feedback are best. An intervention that was found to improve compliance with barrier precautions (use of cap, gown, mask, gloves, protective eyewear) was prenotification of emergency room staff, [148] which resulted in an increase in compliance with barrier precautions from 63 to 92%. In another study, an educational intervention consisting of lecture and practice sessions for operating room staff was shown to increase compliance with use of protective eyewear from 54 to 66% and double gloving from 28 to 55%.[149] It was unclear, however, how much of this effect was due to awareness by staff that they were being observed.

In a study conducted to analyze the effect of organizational safety climate in health care (discussed further below)[149] in nurses working in a high-risk environment, job hindrance were found to be the strongest predictor of compliance. This suggests that training programs must focus less on knowledge-based training and more on helping workers overcome or reduce the barriers associated with compliance. Task analysis, critical-incident techniques, and focus groups could inform the information base for such training programs.

Most of these studies used self-reports as their measure of compliance. This likely overestimates compliance as studies that have used direct observation have found lower compliance. The act of observing staff also may affect compliance with precautions, such that true compliance is likely considerably lower than either observed or self-reported compliance. Compliance has been generally observed to increase over the course of a study, consistent with a Hawthorne effect. However, what appears to be a methodological weakness may also be an indication of what is required to improve compliance with precautions. The presence of an observer may constitute a very "soft" form of feedback. The optimal form of feedback has not been determined from the literature. It does appear that feedback must be given on an ongoing basis.

A study of Thai healthcare workers[151] demonstrated higher compliance with glove use and hand washing during a peer feedback intervention (83% compliance vs. 49% compliance at baseline). However, compliance fell to 73% in the postintervention phase. The authors noted that other techniques, including in-service educational sessions, computer-assisted learning, as well as provision of education and group feedback by researchers also failed to show long-term effectiveness. The authors noted the importance of cultural sensitivity in how feedback is given, but regardless, emphasize that ongoing observation and feedback is needed, as the effectiveness of programs diminishes over time. They suggested that adjunct measures and more research are needed as to how best to maintain a long-term effect.

Safety Climate

A component of organizational culture is the "safety climate," which refers to the perceptions that workers share about safety in their organization. The importance of the safety climate is increasingly being recognized in health care, as more emphasis is placed on productivity and performance. Hospital-based healthcare workers are having to work faster and harder than ever, in an environment of higher patient acuity, increased patient turnover, and with less time for training and education.[152, 153, 154, 155, 156] To compound the complexity of an analysis of organizational factors in health care is the reality that in most healthcare settings, groups of specialized and interdependent workers interact with each other and with various types of equipment and devices, such that safety performance can decline in a nonlinear fashion as total group workload and situational demands increase. Results of several studies suggest that adherence may often be poorest when the risk of exposure is highest.[157] As discussed below, identification and analysis of special compliance requirements and high-risk task situations should be an important feature of a comprehensive infection-control program. Specifically, there is growing evidence to indicate that it is both incorrect and unfair to assume that healthcare workers have total control over their own compliance behavior.[158]

Although the precise nature of safety climate requires further clarification, there is general agreement that the safety-related attitudes and actions of management play an important role in creating a good or bad safety climate.[159,160,161] Zohar established a 40-item measurement model for assessing perceived safety climate in workplaces.[162] Brown and Holmes in attempting factorial validation of Zohar's eight climate determinants, concluded that an employee's previous experience and, specifically, having incurred work-related injury or disease, may influence employees' perceptions,[163] and therefore urged that longitudinal assessments of climate relative to the onset of physical trauma (in our case, SARS) is needed.

Studies of safety program effectiveness in non-health-care settings have repeatedly shown that a positive or supportive safety climate is an important contributing factor to good safety performance.[164, 165, 166] Specifically, it is known that as safe behaviors are adopted throughout an organization, increasing pressure is put on noncompliers to come in line. As noted by Gershon et al., early research identified management's involvement in safety programs, safety training, and safety communications programs, orderly operations, good housekeeping, and an emphasis on the recognition of good performance rather than on punishment or enforcement as important determinants of workplace safety.[167]

A number of studies have examined the role of safety climate in health care in general[168] and several studies have examined standard precautions with respect to blood and body fluid exposure, in particular.[169] It has been shown that the safety climate has an important influence on the transfer of training knowledge.[170,171] White and Berger[172] insist that it is the interactions amongst workers making decisions that is particularly important; direct feedback on the consequences of use/nonuse of appropriate procedures; information received from the media, professional literature, and other sources; and messages from the organization such as policy and procedure statements, training programs, protective equipment availability and choices, and feedback from supervisors.

Using a 13-item scale to measure safety climate, Gershon et al.[173] found that respondents who perceived a strong commitment to safety at their institution were over 2.5 times more likely to be compliant than respondents who did not perceive a strong safety climate. Consistent with the general hypothesis of the study, job/task and organizational-level factors were the best predictors of adherence. Using the results from the study, a three-pronged intervention strategy was developed that emphasized: (1) the availability and accessibility of personal protective devices; (2) the reduction of job hindrances and barriers; and (3) improvements in safety performance feedback and related communications.

In a separate analysis of 482 nurses in a high-risk environment,[174] job hindrances were found to be the strongest predictor of compliance, and safety climate was the best predictor of job hindrances. Safety performance feedback and availability of personal protective equipment were the strongest predictors of safety climate, together accounting for 30% of the variance.

A later study by the same group of researchers examined the contribution of the predisposing, enabling, and reenforcing factors on compliance with standard precautions in 902 nurses at three large acute care hospitals in different regions of the United States.[175] They found that all three categories of factors influenced general compliance, but predisposing factors were unimportant for compliance with PPE. Their results indicated that a positive safety climate is most likely to increase compliance in HCWs.

DeJoy et al.[176] offered several recommendations: first, safety should be integrated into the management system of the organization. Second, poor safety performance should not be viewed as simply a behavioral or worker-focused problem. Training efforts, which have focused almost exclusively on frontline healthcare workers, should also include supervisors and administrators because they are critical when creating supportive safety climates. Third, safety-related communication and performance feedback systems are needed. These must provide opportunity for two-way communication, which is not the case by simply posting notices or conducting training sessions. Participatory strategies including involvement of safety committees and offering performance feedback was suggested. They also note that certain worker groups, most notably physicians, cannot be allowed to be "outside the loop" in terms of regular safety communications and feedback.

An earlier paper by DeJoy,[177] also recommended providing workers with as wide a variety of personal protective equipment options and choices as possible, training workers in the proper use of the PPE linked to specific job tasks, and attempting to reduce the costs and barriers associated with PPE use. They noted that similar studies that have been conducted with respect to hearing protectors, protective footwear, and other types of protective equipment.

Gershon et al. reported the results of another study on hospital safety climate and its relationship with safe work practices and workplace exposure incidents.[178] A 20-item hospital safety climate scale was extracted through factor analysis from a 46-item safety climate survey. This new scale subfactored into six dimensions: (1) senior management support for safety programs; (2) absence for workplace barriers to safe work practices (3) cleanliness and orderliness of the worksite; (4) minimal conflict and good communications among staff; (5) frequent safety-related feedback and training by supervisors; and (6) availability of PPE and engineering controls. Senior management support was found to be the especially significant with regard to both compliance and exposure incidents. Worker feedback and training were also significantly related to workplace exposure incidents to blood and body fluids.

Rivers et al. recently published the results of a survey of 742 nurses regarding predictors of nurses' acceptance of an intravenous catheter safety device.[179] They too concluded that a positive institutional safety climate was more important than individual factors, and recommended high quality training but also an atmosphere of caring about nurses' safety.

Gershon's group recommended that a safety climate survey be administered in hospitals using the safety climate scale, sponsored jointly by the infection control and occupational health and safety committees. They recommended that the survey be anonymous but be distributed to everyone, and preferably distributed at departmental meetings with a preaddressed in-house envelope. (Nonanonymous but confidential questionnaires would be preferable if there was sufficient trust to allow this.) They recommended the results of the safety climate survey be used in several ways. First, scores on the six dimensions can be ordered from high to low with the dimensions with the lowest score targeted for improvement. Second, safety climate can be measured before and then after any major organizationwide safety initiative. Third, the safety climate can be used to compare departments in the hospital, again to identify areas that require special attention. Fourth, this survey could be used to trend improvements in the overall safety program over time and fifth, the safety climate survey can provide management with valuable employee feedback to address barriers.[180] None of the recommendations from any of these studies, however, has been evaluated in terms of its ability to improve worker safety, once applied.

Environmental Factors

The recently published studies on the hospital-associated outbreaks of SARS in the spring of 2003 have all concluded that direct contact or close exposure to a SARS patient is generally required to transmit the virus, although important exceptions exist.[181]

Evidence Derived from the SARS Outbreak

In some circumstances aerosol-generating procedures have resulted in spread beyond that which is expected by droplet transmission. Further, there is some evidence that fomites on surfaces in hospitals may be able in some instances to transmit disease without direct patient contact. This is also the conclusion of a recent WHO consensus document on the epidemiological features of SARS.[182] Understanding the mode, or modes of transmission is key to designing effective environmental control practices for hospital-acquired infections.

Physical Space Separation

During the SARS outbreaks in Singapore, Taiwan, Hong Kong, Hanoi, and Toronto[183,184,185] a number of different physical space interventions were applied. These included separating triage patients in waiting rooms for emergency wards and other hospital departments; isolating suspected SARS patients in single rooms in emergency departments, general medical wards, and intensive care units, and using anterooms to separate donning and doffing from patient care activities.

In examining the evidence for the transmission route of SARS, Varia et al. found that the risk of developing SARS in Toronto healthcare workers and family members was graded by distance, with exposures less than 1 m from a case being highest risk.[186] Risk decreased sequentially with exposures less than 3 m from a case or greater than 3 m, and whether they took place with or without cough-inducing or aerosol-generating procedures. This implies that physical separation of SARS patients from other patients and staff, should have some effect on preventing transmission of SARS. However, this intervention has not been evaluated formally.

Transmission appears to only occur from those who are symptomatic with the disease.[187] Further, three recently published seroprevalence studies of healthcare workers in the United States, Hong Kong, and Vietnam have shown that asymptomatic infection does not appear to occur [188, 189] Therefore, directing infection control measures against those patients who have symptoms compatible with SARS should be an effective means of controlling the outbreak. This was, in fact, the case in all of the outbreaks in 2003. Once the disease was recognized and appropriate infection control measures put into place, the numbers of new infections declined rapidly.

Engineering Controls

Limiting the generation or dissemination of infectious particles from patients can be seen as a means of controlling the source of a hazardous occupational exposure. Early infection control guidelines for SARS[190, 191] suggested placing surgical masks on suspected patients in triage or while being transported in the hospital in order to reduce infectious exposures. Early presentation to the hospital of symptomatic patients limits exposure of the community to SARS and can be seen as another means of limiting exposures to hospital staff because viral shedding appears to be maximal in the second week of illness.[192] No published studies have evaluated source control as a means of preventing transmission of SARS.

Some procedures, such as intubation, the use of continuous positive pressure ventilation, or nebulizer therapy seemed to result in the generation of finer infectious droplets from SARS patients which could travel farther than those generated spontaneously from patients. Such aerosols seem to be responsible for some episodes for spread at distances greater than those commonly found with large droplets and some instances of failure of infection control practices to prevent transmission.[193,194] Therefore recommendations were made to avoid aerosol-generating procedures, such as nebulizer therapy, and procedures to limit the generation of infectious aerosols during intubation were also developed.[195] Similar recommendations for using closed ventilation systems for intubated patients were also made. Loeb, in a study of ICU nurses in Toronto, did find that assisting with intubations, suctioning before intubations, and manipulating oxygen masks on SARS patients were practices which increased the risk of acquiring SARS.[196] The effect of avoiding these procedures has not been evaluated in terms of preventing disease transmission.

SARS infection control guidelines also recommended that patients be cared for in negative pressure rooms with six to nine air exchanges per hour. These recommendations would not likely be effective in reducing SARS transmission, above that of caring for patients in a single room, if indeed, large respiratory droplets are the primary means of transmission. However, in theory, negative pressure would have the added benefit of reducing exposures to finer droplets produced by aerosol-generating procedures. It is worth noting, however, that in Vietnam, the first affected country to successfully control the spread of SARS, negative pressure rooms were not available in either affected hospital.[197]

The importance of having appropriate ventilation systems in place was shown by the "superspreading" phenomenon seen in Hong Kong, where the index patient in the Prince of Wales Hospital transmitted SARS to 47 healthcare workers. Later studies of the ventilation system revealed that the patient's cubicle was under positive pressure relative to the rest of the ward and the hallway.[198] Furthermore, this problem with the ventilation system appeared to be more important than the use of nebulizer therapy in determining transmission patterns.

An analysis of a large outbreak in the Amoy Gardens apartment complex in Hong Kong concluded that the aerosolization of SARS from fecal material by flushing toilets allowed spread of disease through the building's ventilation system because of improper seals around floor drains.[199, 200] This again resulted in transmission that ranged farther than could be explained by respiratory droplets.

Other engineer controls such as filtration of exhaust ventilation, ultraviolet germicidal irradiation, or increasing ambient air humidity were not included in most SARS infection control guidelines and have not been evaluated.

Environmental Decontamination

Cleaning and disinfecting surfaces was recommended as a means of preventing SARS transmission early in the course of the epidemic. This was supported by the observation that the SARS CoV could survive on plastic surfaces for up to 48 hours.[201] The virus has also been shown to be able to survive up to two days in stool and up to four days, if the patient was experiencing diarrhea.[202] Further the hypothesis that the virus could be transmitted by fomites on surfaces was supported by the observation of Ho et al., that three hospital cleaning staff became infected with SARS, despite having only exposures that involved cleaning a room which was previously occupied by a SARS patient.[203] Similarly, one of the infected HCWs in Seto's cohort did not have an exposure to a SARS infected patient, but was classified as probably being exposed outside the hospital.[204] However, environmental decontamination has again, not been formally evaluated as a control measure for SARS.

Hand washing can also be seen as a similar type of environmental decontamination, which is recommended in all basic infection control guidelines. Seto et al. showed that HCWs who reported hand washing during patient care experienced a lower risk of developing SARS in univariate analyses. However, this effect was not seen in the multivariate analysis.[205] No other evaluation of hand washing has been reported.

Specific Personal Protective Equipment

Controversy arose over whether surgical masks or N95 respirators were required to protect HCWs from SARS. Both Seto in a study on Hong Kong healthcare workers[206] and Loeb in a study conducted in Toronto[207] found that not consistently wearing either a surgical mask or an N95 mask was associated with developing SARS when compared with consistent use. Seto found no difference in risk of infection whether HCWs were using surgical or N95 masks.[208] It should be noted that one hospital where the source of the outbreak was determined to be a patient who was receiving nebulizer therapy, was excluded from this study as "droplet precautions have never been recognised as an effective infection control measure for such aerosol-generating procedures...." In addition, aerosolizing events were not included. The authors concluded that precautions against droplets and contact are adequate for prevention of nosocomial SARS where no aerosol-generating procedures are used. The surgical and N95 masks were both effective in the above scenarios. The situation is less clear where aerosol-generating procedures are in use.

Loeb et al., in a retrospective cohort study of 43 nurses in two critical care units with SARS patients, examined the risks for disease acquisition and did find a trend toward increased protection from N95 masks compared to surgical masks, but this was not statistically significant.[209] Eight of 32 nurses working with patients became infected. Specifically, three of 23 nurses (13% who consistently wore a mask [either surgical or N95]) acquired SARS, compared to five of nine nurses (56%) who did not consistently wear a mask (p = 0.02). The relative risk for infection was 0.22 (p = 0.06) for nurses who always wore an N95 mask when compared with nurses who did not wear any mask consistently. The relative risk for infection was 0.45 (p = 0.56) for nurses who always wore a surgical mask when compared with nurses who did not wear any mask consistently, implying no statistically significant difference between wearing a surgical mask and not wearing a mask at all. However, the difference in relative risk for SARS infection for nurses who consistently wore N95 masks compared to those who consistently wore surgical masks was also not statistically significant (p = 0.5). The study is one of the most informative coming from the SARS outbreak itself, but suffers from many limitations.

Primarily, the results were not analyzed to correct for possible confounding factors. In addition it did not examine whether fit testing was performed on those using the N95 masks, did not address the issues of potential autoinoculation when removing gear, and suffered from small sample size of the cohort. It is worth noting that in Vietnam, N95 masks were not available until the third week of the outbreak, a factor which did not seem to prevent their ability to control it.[210]

The Seto study also found that regularly wearing gowns was protective in univariate analyses, but that only mask usage was significant in the multivariate analysis.[211] The study by Lau found that inconsistent use of goggles, gowns, gloves, and caps was also associated with acquiring SARS in univariate analyses, but were not also significant in multivariate models.[212] One hundred percent of HCWs used an N95 or surgical mask and no difference was noted in the use of N95s between cases and controls. Again, small sample sizes may have limited the power of these studies to show the effects of these interventions. No other published studies have evaluated the effectiveness of face shields or goggles in their ability to protect HCWs against SARS.

Interestingly, the study by Lau found that HCWs who perceived the amount of personal protective equipment available to be inadequate were at higher risk for developing SARS and this effect remained significant in the multivariate model.[213] The study was conducted in five hospitals in Hong Kong, so the researchers were unable to confirm, which specific items (if any) were inadequately supplied. They note, however, that given the large differences they found (odds ratio>5, $p<0.001$), it is likely that PPE shortages were at least partially responsible for many of the SARS infections.

Christian et al. examined a cluster of healthcare workers after exposure to a patient with SARS during cardiopulmonary resuscitation (CPR).[214] Three of the six nurses present during the intubation developed respiratory symptoms and it was suspected that they had been exposed. HCWs were interviewed, the healthcare setting inspected, and policies and procedures reviewed. The CPR event described took place when protocols for management of patients suspected of having SARS were in place but the use of Stryker T4 Personal Protection Systems was being advocated as an additional protective measure. Nine HCWs were present during the intubation. Six nurses did not wear T4 personal protective equipment while three respiratory technicians and physicians did. In addition, the nurses were exposed to an ambubag that did not have an appropriate filter attached during the initial resuscitation. Three of the six nurses developed symptoms in the week after the procedure, however, only one was found later to have positive serology for the SARS-CoV. It was suggested that T4 PPE was potentially more protective; however, not all the subjects involved in the events underwent serologic testing and the level of exposure for each HCW was likely different, with the nurses likely having higher exposures due to the problem of the unprotected ambubag. The patient was not breathing at the time of the intubation that was performed without difficulty, making the generation of small infectious particles less likely. The study did not allow a clear determination of what mode of disease transmission was the most important in this cluster. Importantly, the appropriate removal of equipment was not discussed and it appears that the nurses were not wearing fit-tested respirators.

The authors point out that the delay in some members of the team to respond to the code was due to the time required to put on the T4. This resulted in a second code blue being called and additional HCWs exposed to the index case and suggests that better PPE may conversely result in increased exposures to infections if it is not well suited to the work environment.

Evidence Derived from Other Droplet-Spread Respiratory Infections

Other viruses which can cause significant respiratory infections and have been shown to be transmitted in healthcare settings include other coronaviruses, influenza and parainfluenza viruses, and respiratory syncytial virus. All of these viruses are transmitted through the spread of large droplets or fomites, similar to the primary means of transmission of SARS CoV. However, there have been no reported instances of spread through smaller respiratory droplets over larger distances due to nebulizer therapy or intubation procedures for these viruses. It is uncertain as to whether this is because it does not occur or because it does occur but the transmitted disease goes unrecognized. Therefore, the evidence related to these viruses may be generally analogous to SARS, except with respect to the "superspreading" instances referred to above.

Other coronaviruses are thought to primarily cause mild disease such as the common cold, accounting for up to one-third of cases. However, outbreaks in susceptible populations such as in neonatal ICU's or in elderly people living in long-term care facilities, have been shown to cause significant lower respiratory disease, leading to hypoxia.[215,216] However, no studies evaluating the effectiveness of infection control practices with respect to other coronaviruses have been published.

Outbreaks of nosocomially transmitted influenza are a common occurrence during the winter months in Canada, causing hundreds of thousands of infections and between 500 and 1500 deaths per year, substantially more than SARS. The primary means of controlling the disease is through vaccination of those members of the population who are at highest risk for disease, as well as those who are in direct contact with this population.[217] The latter group includes healthcare workers, who are often the vehicle through which hospital patients or residents of long-term care facilities become infected.[218,219] Droplet precautions are recommended for pediatric hospitals and some adult hospitals caring for patients with influenza-like illness,[220] but have not been evaluated in terms of their ability to prevent transmission.

Respiratory syncytial virus (RSV) is another common cause of outbreaks of moderately severe acute respiratory infections in healthcare institutions, primarily in paediatric hospitals. Infections are transmitted through inoculation of the nose or eye, rather than the mouth.[221] Studies on the effectiveness of infection control practices have shown that organizational controls such as isolation or cohorting of patients were more effective than the use of gloves, gowns, and masks in reducing nosocomial spread of RSV.[222] Screening all patients with viral respiratory infections for RSV on admission and using contact precautions (isolation without masks, but using gown and gloves) was shown in one study to reduce RSV transmission rates by 39% and save $6 for every $1 spent.[223] Two other studies conducted in adult bone-marrow transplant units found similar evidence of effectiveness.[224] Another study, paradoxically, found an association with wearing gowns and an increased risk of nosocomial transmission of RSV.[225] RSV is only able to survive on surfaces for approximately six hours, much less than SARS CoV.[226]

Parainfluenza viruses are also thought to be primarily transmitted through large respiratory droplets. They appear to be less viable in the hospital environment than SARS, as they survive for only ten hours on surfaces.[227] Outbreaks of parainfluenza have been controlled in bone marrow transplant units and neonatal ICUs by application of contact precautions using gowns, gloves, isolation, and cohorting of nurses.[228]

Evidence Derived from Airborne-Spread Respiratory Infections

Measles and varicella zoster are viruses, which can cause respiratory disease and are primarily spread by the airborne route. However, widespread outbreaks are rarely seen in healthcare settings largely because of widespread immunity to both diseases either as a result of successful vaccination programs (for measles) or because of natural infection (varicella). No studies evaluating infection control measures for these viruses, other than vaccination could be found.

An abundance of studies have been published on the prevention of nosocomial transmission of tuberculosis, but the extent to which this information is relevant to SARS is unclear. TB is spread by small droplet nuclei that can travel large distances while remaining aloft after being produced by infected patients. This is unlike the spread through large droplets, by which the SARS coronavirus is generally believed to be transmitted. However, some controls used to prevent nosocomial TB transmission, have the potential to be useful for the control of SARS with respect to the "superspreading" events where smaller infectious droplets are generated.

In the late 1980s and early 1990s it was recognized that infection control practices were not stringent enough to prevent the occurrence of outbreaks of tuberculosis in healthcare facilities.[229] Consequently, more rigorous guidelines to prevent nosocomial transmission of tuberculosis were developed.[230,231] These have generally been credited with reducing the spread of tuberculosis in healthcare facilities, but it remains unclear which components of the guidelines have had the greatest effect.[232]

Physical Space Separation

The airborne nature of TB transmission means that simply physically separating TB patients from other patients and healthcare will not prevent transmission, as long as the ventilation systems are not separated. However, some TB control plans recommend the separation of procedure rooms and general care rooms, so that aerosol-generating procedures do not result in an increased burden of infectious agents in patient-care areas.[233] Similarly hospital designs could help to reduce the environmental contamination of SARS-CoV if patients requiring intubation and nebulization therapy could be transferred to separate procedure rooms.

Engineering Controls

Antituberculosis therapy can rapidly reduce the production of infectious particles, thus limiting exposures to healthcare workers. If effective antiviral therapies could be developed which could reduce the production of infectious viral particles, these could similarly protect hospital workers, even if they do not improve patient outcomes. Other types of source controls such as masking patients or using closed ventilation systems for intubated patients likely have similar effects on reducing the production of infectious particles, but have not been evaluated with respect to preventing transmission of tuberculosis.

Another method of source control is limiting the movement of patients once admitted to hospital with TB. In a hospital with a large HIV unit in Lisbon, Portugal, restricting patient movements was identified as one of a number of infection control measures which were introduced to eliminate risks for nosocomial transmission of multidrug resistant TB.[234]

Ventilation systems, which generate 6 to 10 air exchanges per hour, and exhaust outside the hospital resulting in the creation of negative pressure environments in patient care rooms, have been shown to remove 99.9% of airborne contaminants within 69 minutes.[235] However, one study of the effectiveness of these systems revealed that 11% of such ventilation systems in three U.S. hospitals were not actually generating negative pressure.[236] Further, 19% of TB patients were not isolated on the first day of admission because the etiology of their problem was not recognized. Similarly, Canadian researchers have shown that inadequate ventilation systems of general patient rooms can lead to increased risks of TB infection for healthcare workers because of patients with unrecognized infections.[237]

Ultraviolet irradiation has been shown to enhance the decontamination of infectious airborne bacteria and viruses.[238] While it has a limited effect on surface contamination, because of poor penetrative ability, and does not work well in instances of high humidity, it could also be of some benefit in terms of decontaminating patient-care rooms where the infectious organism is a droplet-spread virus such as SARS. Filtration of exhaust ventilation of isolation rooms with HEPA filters is standard practice to prevent environmental contamination of tuberculosis, but it has not been evaluated in terms of its ability to actually prevent transmission.

Specific Personal Protective Equipment

N95 respirators have been required to be provided for HCWs who work in the United States since 1994, when the CDC TB transmission prevention guidelines were published. However, studies of actual practice have shown that a range of between 44 and 97% of HCWs use these properly.[239] Thus, it is feasible that the improved efficacy of an N95 mask over a surgical mask may be easily lost, if compliance is poor. No published reports on the effectiveness of face shields or goggles, gloves, or gowns were found with respect to preventing nosocomial transmission of TB.

Individual Factors

Several individual factors may affect the compliance of HCWs to using personal protective equipment (PPE) for protection against respiratory infectious diseases in healthcare settings. The majority of research done in this area has been exploring HCW compliance with standard precautions (SP). SP were introduced in the 1980s in response to the risk of transmission of blood borne pathogens to HCWs from patients, in particular HIV. Though the research does not directly examine the compliance of HCWs with facial protection, the reasons for noncompliance with SP can be extrapolated to any PPE.

Knowledge Acquired Through Training and Personal Experience

Knowledge of the appropriate use of PPE is necessary but not sufficient for HCWs to adopt safe work practices.[240] The study by Gershon et al. from 1995,[241] found that HCWs surveyed had high levels of knowledge regarding UP practices but that this did not lead to high levels of compliance. Compliance was noted to be more correlated with perception of risk. Use of PPE only when there is visible blood may demonstrate that HCWs make personal judgments about their own potential risk instead of following a consistent policy.[242] Repeated exposures without consequences may decrease compliance. HCWs may perceive decreased risk if, while caring for patients, they receive repeated exposures to blood and body fluids (BBF) but are never infected. This may lead to a false sense of invulnerability and therefore increase risk taking.[243]

It is noteworthy that, at least the more recent studies on compliance with standard precautions indicate that HCWs do not appear to dismiss or underestimate their personal risk of acquiring an occupational infectious disease[244,245,246]; in fact HCWs are more likely to overestimate their risk.

Gershon et al. in their 1999 study found HCWs less than 40 years of age more likely to comply with SP.[247] This may reflect more recent training. HCWs surveyed were found to have realistic risk perceptions about exposure to BBF: few were fearful of contagion. Level of experience did not necessarily lead to a lack of understanding of risks involved. Nurses who were educated in a more disease driven infection control model, where precautions were used only when the patient was known to be infected by a given pathogen, were less comfortable in UP model as compared to recent graduates.[248]

Students and other HCWs may look to attending physicians as a role model; poor compliance in these senior physicians may lead to lower levels of compliance in their students. Kim et al. had similar findings.[249] Younger physicians, house staff, and medical students were found to be more compliant with SP than senior physicians. The increased compliance probably reflects more recent training. Another study found that compliance with methicillin resistant *Staphylococcus aureus* (MRSA) precautions (which included use of gloves and gowns and hand washing) was related to occupational group with physicians showing the lowest compliance (22%) and physiotherapists and occupational therapists having the highest compliance (89%).[250] Compliance with gown and glove requirements was 65%, and for hand hygiene, 35%. Gershon et al. stated that physicians are "out of the loop" with regard to safety climate within hospitals. Special efforts need to be made to involve them in training, safety programs, and safety committees.[251]

Angtuaco et al.[252] found that fewer gastroenterologists than GI endoscopy nurses used face shields for all procedures (14 vs. 21%; p = 0.02). Overall compliance with use of barrier equipment for both groups was low.

Prieto and Clark interviewed HCWs regarding their attitudes toward use of PPE.[253] Nurses reported confusion at the ward level and uncertainty about the rationale for the uses of PPE recommended in infection control guidelines. They perceived the existing guidelines as lacking specificity to their practice. They also doubted the effectiveness of isolation precautions to prevent disease transmission and voiced frustration with the lack of adherence by allied professionals. Physicians echoed nurses concerns but also felt that their training inadequately prepared them to implement isolation precautions and relied on the nursing staff to direct them. Jeffe et al. also cited the need to teach medical students the importance of the use of PPE before they become set in their ways.[254] Teaching medical students early in their clinical training about the risk of exposure to BBF and specific prevention measures may be associated with more positive attitudes and better compliance with precautions.

Attitudes and Beliefs

Demographics such as gender, education, shift work, or occupation have not been consistently associated with compliance with infection control procedures.[255] Compliance is more often found to be affected by knowledge, attitudes, and perception of risk. DeJoy et al. found that having a positive attitude toward the patients, lower risk-taking tendencies, and greater knowledge of modes of transmission leads to greater compliance.[256] If HCWs do not understand the risk status of patients or that a single momentary lapse in compliance can lead to serious results, they may be willing to take unnecessary risks when providing care.

Perceived barriers may be one of the most important factors affecting compliance. Godin et al. found that HCW perception of their ability to adopt the use of PPE into their practice affected their level of compliance.[257] If they believe that the barriers to their adherence to recommended use of PPE cannot be circumvented they will not comply. Actual working conditions resulting in overwork, lack of time with patients and having to deal with emergencies were reported to have significant negative affects on compliance. Godin et al. also found that HCW are influenced by the subjective norm, i.e., the perception of social expectation to adopt a given behavior.[258] This suggests that if HCWs believe that key persons in their work and social environment expect them to be compliant with the use of PPE, they are more likely to do so.

As noted above, organizational issues impact individual attitudes considerably. For example, workload issues are thought to affect HCW willingness to comply with recommendations for PPE use. Workers who feel stressed and overloaded at work are much less likely to be attendant to safety needs and precautions.[259] Helfgott et al. found that sufficient knowledge of how to prevent occupational exposure did not appear to correlate with compliance with UP.[260] The most common reasons why HCW did not comply were time constraints, hindrance of performing a specific task, and HCWs presumed lack of risk based on identifying infectious patients. It was also noteworthy that this study also found that level of compliance was inversely proportional to level of experience of the HCW. Reasons for this finding were given as increased level of competence, feelings of invulnerability or just plain laziness.

DeJoy et al. in their 2000 study, cited the importance of easy access to the correct PPE when needed, including protective outer garments, eye shields, and face masks as an influence on compliance.[261] The availability of certain PPE can have a significant effect on the attitudes of HCWs toward using them. The greater perceived availability of PPE may lead to stronger beliefs in their effectiveness for prevention amongst HCWs. Face protection is often less readily available in healthcare settings than gloves or sharps containers.

A significant factor that may influence HCW compliance with PPE use is the perception that their use may lead to a decreased quality in the therapeutic relationship between patients and HCWs. Nickell et al. found that during the Toronto SARS outbreak HCWs found wearing of masks particularly bothersome.[262] A mask made communication difficult, recognizing people difficult, and led to a sense of social isolation. DeJoy et al. found that the wearing of PPE places barriers between two people, negatively altering interpersonal dynamics and complicating the performance of tasks and treatment.[263] Respirators cover the face and mouth hampering communication especially for the elderly and those with hearing loss. Use of respirators may lead to increased isolation and fear amongst patients.[264] Prieto and Clark also cited concerns amongst nurses that isolation of patients could lead to depression from lack of social contacts.[265] In trying to avoid these negative consequences, HCW may choose not to comply with PPE recommendation even though they know they should.

Reduction of job related hindrance through analysis and modification of patient care tasks and development of skills based training may increase compliance. HCWs have adequate information and knowledge but need to enhance skills at practicing the use of PPE.[266] Unfortunately, most studies have found that formal education sessions may have effects on compliance levels, but these improvements are found to be short lived. Improvements in compliance may come from informal point-of use prompts or more formal safety performance feedback, rather than official policy statements.

Previous studies have also showed that health care workers view standard precautions, as adversely affecting job performance and the practitioner-patient relationship.[267,268,269] The most common reasons for lack of adherence were insufficient time, interference with job duties, and discomfort. In the Willy et al. study, interference with the practitioner–patient relationship and decreased dexterity were the most frequently cited reasons for noncompliance. Osborne determined that mean compliance rates among Australian operating room nurses were 55.6% with always double-gloving during surgical procedures, and 92% with always wearing adequate eye protection.[270] The variable that had the most influence on compliance was the perception of barriers to compliance, specifically, that adhering to standard precautions interfered with duties. Nickell et al. found in their study of HCWs during the SARS outbreak in Toronto that the most commonly cited difficulty with complying with precautionary measures, especially masks, was that wearing one for any extended period of time was very uncomfortable.[271]

The Challenge of Changing Healthcare Worker Behavior

An important consideration in defining an approach to the management of SARS and other emerging infectious diseases is that whatever the evidence that emerges the key challenge of changing behavior will remain. In recent years much research has been conducted on the components of a successful strategy but much work needs to be done. This is especially true in the context of SARS where the scientific knowledge base will be rapidly evolving simultaneously with the need to implement change. Bero and colleagues have characterized components of interventions that are likely to be successful or unsuccessful, some of which are listed in Table 5.7.[272] In addition Grol and colleagues[273] have characterized the features that were more likely to be associated with a change in primary care practice. An important finding was that recommendations with a strong evidence base were more likely to be effective than consensus statements.

Gaps in the Evidence

Specifically, the following gaps in our knowledge of protecting the "faces" of healthcare workers were identified through our review of the literature.

Epidemiology, Transmission, and Risk Assessment

1. How do respiratory droplets produced by aerosolizing procedures differ from those produced by more "natural" methods such as coughing or sneezing, in terms of their size, their spread, and their infectivity?

TABLE 5.7 Features that Are Likely to Be Associated with Success in Guideline Dissemination

Consistently Effective
Educational outreach visits
Reminders
Interactive educational meetings
Multifaceted interventions

Interventions with Variable Success
Audit and feedback
Local opinion leaders
Local consensus approach
Patient mediated interventions

Interventions with Little or No Effect
Educational materials
Didactic educational meetings

Note: These data indicate that the mere creation of recommendations within a well-grounded program in knowledge translation will be unlikely to achieve a safer workplace.

2. Studies as to the dispersal of droplets and aerosols in the workplace. These studies are important in examining the role of cleaning; the role of autoinoculation; the need for respirators, filters, and ventilation systems.
3. The relative roles of mucosal contamination (autoinoculation) in disease transmission and how much of PPE effectiveness is related to controlling these exposures.
4. How able are respiratory tract pathogens to cause disease through the transocular route?

Risk Management

1. Minimizing the exposure at the source is a fundamental tenet of occupational health and safety, yet development and assessment of engineering controls in the health care sector are sadly overlooked. In particular research is needed in:
   a. standards pertaining to minimizing infectious bioaerosols at source

    b. rigorous and standardized testing for breathing system filters, pulmonary function filters, and heat moisture exchange filters that are commonly used

    c. design research for anaesthesia machines, ventilators, and other respiratory equipment to minimize aerosol generation

    d. studies on the relative effect of changes in engineering controls, such as the role of increasing relative humidity to maximize particle fall out

    e. defining added benefit of nursing high risk patients in a negative pressure atmosphere over physical isolation and adequate ventilation throughout hospitals

2. There is a lack of information concerning the effectiveness of face shields in providing an individual with facial protection. While a few studies have examined the effectiveness for blood and body fluid splashes, no published studies were found that address the effectiveness in providing facial protection against bioaerosols. Design issues for compliance with eyewear protection (e.g., anti-fogging, comfort) have not been adequately addressed in the healthcare sector.

3. The relative importance of fit testing versus fit checking versus other forms of healthcare worker training on infection control procedures needs further assessment.

Compliance with Infection Control Guidelines

1. How can the safety climate of healthcare institutions be improved in light of other changing factors in the sector such as demands for increased productivity and resource constraints?

2. What training methods are most appropriate to teach infection control practices to staff from all occupational backgrounds?

3. What determines individual workers' belief in the effectiveness of infection control procedures and how can this be facilitated to assist worker compliance?

4. What is the best way to provide feedback on adherence to the required practices and use of PPE?

5. What are the most appropriate infection control practices, taking into account sufficient time available to comply with the required precautions while meeting other workload requirements?

6. Can compliance be achieved without being seen as a hindrance to other aspects of the job such as communication with patients and other staff?

7. Are the required PPE and work practices convenient and comfortable for workers to use?
8. How can the impairment of communication and social interaction associated with PPE be overcome?

Focus Group Analysis

In order to develop effective occupational health and safety and infection control policies and procedures for healthcare facilities, it is necessary to have a good understanding how frontline healthcare workers assess the importance the various components of these issues.

Methodology

In order to do this, we organized a series of focus groups in order to ascertain what environmental, organizational, and individual factors were the most important determinants of successful infection control procedures, in the opinion of selected groups of healthcare workers. The focus groups were conducted primarily in two cities, Toronto and Vancouver, and have involved seven different classifications of healthcare workers: (1) occupational health staff; (2) infection control practitioners; (3) physicians; (4) clinical nursing staff; (5) allied health professionals (including respiratory therapists, laboratory technicians, radiology technicians, physiotherapists and others); (6) support staff; and (7) hospital managers. An additional mixed group of occupational health and infection control professionals was held in Ottawa.

Participants were recruited in three ways for the 11 focus groups conducted in Ontario. In the first instance, letters were written to the Chief Executive Officers of 13 hospitals, 11 in Toronto, which had admitted SARS patients, and two in Ottawa, explaining the study's objective and asking them to identify appropriate participants from their facilities. Second, letters were also sent to the Canadian College of Health Services Executives, Greater Toronto Area (GTA) Chapter, the Ontario Society of Medical Technologists, The College of Respiratory Therapists of Ontario, Ontario Medical Association, The Ontario Nurses Association, the Registered Nurses Association of Ontario, and the Occupational Health and Safety workers identified by the Ontario Hospital Association Human Resources database and the Canadian Union for Public Employees (for support staff). Finally, e-mails were also sent to infection control physicians on The Change Foundation's project steering committee requesting their assistance in forwarding the message to other physicians. All invitation letters requested participants to have direct experiences with SARS. In all, 87 individuals came from 21 different healthcare institutions, organizations, and professional associations to participate in the 11 Ontario focus groups. Two focus groups were conducted in Toronto for occupational health staff and hospital administrators, as the response was larger than expected. Two groups of mixed workers from two different facilities were also conducted.

Several different strategies were used to recruit participants in the four focus groups in Vancouver. Nurses, allied health professionals, and support staff from the five acute care hospitals in greater Vancouver, which had confirmed SARS cases during the outbreak, were recruited through their affiliated unions. Infection control practitioners, occupational health staff, and clinical managers were recruited through letters sent to staff from the five hospitals identified by one of the project's steering committee. We were unsuccessful in recruiting physicians to a focus group; therefore, only the physicians group from Ontario is presented here.

Each focus group was approximately 90 minutes in length. Participants discussed three very broad questions relating to the organizational, environmental, and individual factors and their importance in infection control and occupation health and safety in healthcare facilities. The discussion questions and the examples provided for each question appear in Appendix 5.1. Facilitators read out one question at a time and allowed the group to exhaust its discussion of the subject before moving on to the next question. Facilitators tried not to interfere in the discussion except where clarification was required or if some members of the group were having difficulty entering the conversation. There was also an opportunity for participants to discuss other issues at the end of the session, which were not brought up earlier. The discussion questions were developed based on what research has shown to be important and were piloted with a mixed group of healthcare workers and modified prior to their use in the first sessions in Ottawa and Toronto.

All focus groups were recorded and transcribed. Three members of the research committee then began coding the transcripts according to the a priori specification of variables known or suspected to contribute to the effectiveness of workplace health and safety and infection control programs. Codes were divided into organizational, environmental, and individual factors. All three researchers reviewed the same transcript initially and compared their results, so as to standardize coding procedures for subsequent transcripts which were only reviewed by one researcher each. During the subsequent analysis of all the transcripts, researchers tracked the number of times each variable was discussed and compiled quotations which best represented the discussion. New variables were also identified and tracked. Each researcher compiled a one- to two-page summary of each focus group which synthesized the key points of discussion and important suggestions or novel ideas which were raised. These summaries are found in Appendix 5.2. This narrative summary was written based on the one-page summaries, following a discussion with the three researchers on what codes arose most frequently, what codes were lightly discussed or not at all, and which of the new codes identified were raised by more than one group.

Results

Focus Group Participants

Table 5.8 shows the demographic and work-related information of participants in 14 of the 15 focus groups. One group of approximately eight participants from Toronto did not have this information available. Of the 97 participants where information was available, 80% came from Ontario and 19% were from British Columbia. Over 85% came from healthcare facilities where SARS patients were admitted and 44% of participants reported having had contact with a SARS patient at least once. Thirty-seven percent of participants had experienced quarantine during the outbreak, either at work or at home. Participants were mostly female (78%), reflecting the predominantly female composition of the healthcare workforce, especially in the nursing profession, which formed the single largest occupational group (24% of participants). Clinical managers were the next most represented group (12%), followed by occupational health or infection control managers (11%). The other job categories each formed less than 10% of the total number of participants. Only four physicians were able to be recruited, despite several attempts to recruit more. Two of the mixed groups did have physician participants. The average age of participants was 43.1 years.

Content Analysis

Generally the discussions were free-flowing and the facilitators were not directive in presenting the questions, although this varied somewhat from group to group.

General Comments

Focus groups ranged in size between two and 11 people, with most groups having between eight and 10 participants. The discussions covered the topics mostly from the perspective of what occurred during the SARS

TABLE 5.8 **Characteristics of Focus Group Participants (n = 97)**

Variable	Respondents to Question	Results	
Province	97	British Columbia	18 (19%)
		Ontario	79 (81%)
Sex	94	Male	22(23%)
		Female	73 (78%)
Age	92	43.1 yrs (average; range 26–64)	
Job Category	97	Manager (all)	33 (34%)
		Manager (Clinical)	12 (12%)
		Manager (OH&S, ICP)	11 (11%)
		Registered Nurse	23 (24%)
		Support Staff	9 (9%)
		Medical Technologist	8 (8%)
		Respiratory Therapist	6 (6%)
		Infection Control Practitioner	4 (4%%)
		Occupational Health and Safety	5 (5%%)
		Physician	4 (4%)
		Administration	1 (1%)
		Pharmacist	1 (1%)
		Physiotherapist	1 (1%)
Quarantine	97	Any quarantine	36 (37%)
		Work quarantine	14 (14%)
		Home quarantine	14 (14%)
Institutional experience with SARS	97	SARS in facility	82 (85%)
		SARS in ward	52 (53%)
		SARS in room	34 (35%)
Contact with SARS patients	97	Contact with any SARS patient	43 (44%)
Total number of contacts	41	6.9 patients (average; range 1–35)	
Total number of days in contact	35	19.5 days (average; range 1–72)	
History of SARS infection	97	In Self	1 (1%)
		In Co-worker	30 (30%)

Note: Data from 14 of 15 focus groups.

outbreaks, but included discussions of factors pre- and post-SARS, as well. The opinions presented here may not reflect the views of the majority of healthcare workers as we did not try to quantify the responses; however the points discussed here were the elements where most groups spent significant amounts of time. These comments sometimes reflect a range of opinion, which may conflict, but which was expressed in these groups. This should give policy makers and researchers a flavor for the feelings that healthcare workers express about these issues.

Organizational Factors

Lack of Consistency with Safety Instructions and Frequently Changing Directives

This issue was commented on by nearly every group and was a source of much anxiety for healthcare workers both in British Columbia and Ontario. A support worker from Toronto described it this way:

> There was so much information. The information changed on more than a daily basis, and even the managers, sometimes, I am sure they were confused. Which directives to take? Which ones not to take? And I don't think there was enough time for even the managers to relate all the information to the workers. We were just being bombarded with new directives, on how to do certain things and things changed so quickly … when you are so busy trying to actually do work, you don't have enough time to go sit at the computer and read word by word on what's being directed to you. (lines 112–20, transcript 7)

A clinical manager from Toronto felt this about the changing directives:

> There was always that uncertainty of perhaps, there is information which we don't have. And you're telling me this now but will that change tomorrow? … And I certainly think that that affected the compliance of the staff with following protocols and their own comfort levels….(lines 157–66, transcript 5)

It seems likely that the changing recommendations and guidelines undermined the workers' confidence that any of the guidelines would adequately protect them, thus heightening worker anxiety.

Enforcement by Regulatory Agencies

Related to this issue was that of how external organizations such as the Ministry of Labour in Ontario and the Workers Compensation Board in British Columbia exerted their authority in healthcare institutions. There was some diversity of opinion around these issues, in that while many workers saw the measures imposed as being somewhat Draconian, others saw some measures, such as the requirement for fit testing as long overdue. In comparing the role of the Ministry of Labour in healthcare versus other industries, one occupational health and safety professional had this to say:

> The Ministry of Labour traditionally does not go into healthcare settings, ...They go into (other) industries and they say "Okay, where is your card for your fit-testing performance..." If you don't produce it, the employees can be fined, the employer can also be fined right up to senior management and that does happen. But traditionally, in the healthcare setting, they do not come in. So if they do start coming in, there might be a shift. (lines 597–602, transcript 12)

An infection control practitioner, felt that the new levels of enforcement by the Ministry of Labour interfered with rational infection control practice: " We couldn't use those sound principles because we're told that if it's a directive, you have to apply it." (lines 221–22, transcript 9) There were also general feelings that if new health and safety directives were to be successfully applied, they must come with further funding to make them happen. Similarly, an infection control practitioner from Vancouver stated:

> When any sort of organizational body has such power in an entire province to enforce something that suddenly ... it needs to be done with more planning and certainly much more communication and collaborative dialogue, instead of just imposing it on the entire province. (line 277–80, transcript 10)

Workplace Attitudes Toward Safety

Workplace attitudes toward safety were felt to be important for most participants. Generally there was seen to be a lack of commitment to occupational health and safety in health care both by workers themselves and by management. Management's commitment to worker safety is primarily judged by its actions. This was seen during the SARS crisis in terms of whether management was willing to spend money to buy extra PPE and whether they were willing after SARS to hire more infection control and occupational health professionals. It was also seen in their visibility during the crisis. A support staff worker from Toronto characterized it this way:

> I think … more involvement with the president of the hospital. I think that when that person is speaking to you and addressing the issue, you feel like you are in the loop. When you are getting all this second-hand information from everywhere, you wonder what they are hiding. (lines 290, transcript 7)

In the absence of an outbreak, healthcare often sends mixed messages to its employees. A nurse from Vancouver described this:

> I know that at Hospital B there is a policy now that if you have flulike symptoms, if you have the headache and sore throat, you're not to show up for work. But they're monitoring all the sick time that we're using. Some managers … (are) giving direction to use a LOA (leave of absence), instead of a sick day…. It's talking out of both sides of the mouth. (lines 903–11, transcript 2)

The lack of safety consciousness among healthcare workers, themselves, was an area where workers felt there also needed to be some improvements. While having good peer support and more follow-up to training in infection control were mentioned as means to achieve this, participants generally felt that there should be more enforcement of compliance with infection control, through better supervision on the wards. They suggested having consequences in place for noncompliance and also that supervisors should have mechanisms in place to provide feedback on worker performance in terms of infection control.

The recent downsizing of the workforce and the replacement, in most facilities, of the "head nurse" position, with a "charge nurse" who changes from day to day, has made this more complicated. However, one allied health professional offered this solution:

> I think what ended up happening with the SARS outbreak in our facility is that there would be infection control individuals, who would come in … the ICU and speak with whichever bedside nurse was managing that particular patient on that day. That individual, that nurse, then became the infection control officer for the rest of the shift and for every other individual. (line 256, transcript 4)

However, most nurses did not see this as being a sustainable solution (see below, under "Safety Training.")

Another way in which management displays its concern for worker safety is through the provision of adequate range of choices and adequate supply of personal protective equipment. One occupational health worker saw this as being an important determinant of infection control success or failure:

> You're seeing a resurgence of MRSA because of how we had to deal with supplies, we had to break some of our rules and tell people they had to wear a gown from patient to patient. They had to wear a mask for 12 hours. That's not good practice. (lines 193–96, transcript 12)

Other staff, however, saw the MRSA problem as primarily being one of following proper procedures, and not supply.

The occupational health and safety groups, especially those with experience in other industries, felt that their role was generally undervalued in health care, although this was not highlighted by the other groups. This is perhaps part of the problem, as described by a participant from a mixed group in Toronto:

> On the joint health and safety committee, staff could go to any member of that committee and have an issue raised, if they didn't feel that it was being addressed appropriately. But I'm not sure that we probably did that very well, and I'm not even sure if people knew that we had an occupational hygenist, or what their role was in the institution. They are of great value to the organization, but I'm not sure that we always promoted that within the organization. (lines 196–210, transcript 14)

Evidenced-Based and Practical Infection Control Policies

Having specific policies and procedures for infection control and sufficient resources available to carry out these policies was also identified as a key factor. One of the driving factors behind this was that workers often felt that infection control policies developed elsewhere often had little relevance to their workplace, especially if the institution had not experienced SARS. One of the remedies to this disconnect was to involve frontline workers in setting infection control guidelines and procedures. Some of the participants in the focus groups came from institutions where they felt that good infection control policies were in place, but where the resources applied to make these policies happen were not available.

Most groups mentioned that their institutions did not have an adequate number of infection control practitioners, and some (especially the groups composed of infection control practitioners), cited the SENIC study of the literature review as evidence that they did not have enough.

Other participants felt that basic infection control policies and procedures in their institutions were either not well developed or were not enforced. Identified deficiencies included tracking of who receives training in infection control, to ensure that all those who need training actually get it; consistent policies for quarantining individuals; policies regarding the reuse of masks, and policies regarding which patients require negative pressure rooms.

Yet, workers also feel that they want to have the option to use more protective equipment than the clinical situation may warrant. It appears in some situations that physicians may do this:

> I came head to head with a physician over that because after the SARS precautions had been sort of down-graded ... and the physician walks in with his, you know, fully garbed and I was saying ... we have told all of our team members that they no longer needed to wear all of this ... he's like, " I'm not taking any chances," then I say "It's a consistency (issue), everybody has to follow the standards and believe in them." (line 768, transcript 14)

Generally, however, physicians were perceived to be less compliant with the use of PPE than other healthcare staff (see below). Frontline healthcare workers do not often have this option, as one of the allied health professional from Vancouver mentioned resistance to him wearing a mask in the presence of an MRSA positive patient, despite the fact the patient was coughing. This could also be seen as management listening to the concerns of HCWs and trying to accommodate them where possible.

Many participants described the need to establish a respiratory assessment for "high risk" patients on which workers can rely and that doesn't lead to unnecessary precautions being taken. Ideally this is done by having the adequate number of infection control practitioners, who are familiar with the acuity of the patients and with best practices regarding staffing issues. The latter theme was seen to be especially important in ensuring that the extra burden of applying complete PPE against airborne infections is not borne by the staff unnecessarily. If staff are asked to wear this equipment too often when it is not necessary, then it is quite likely that the "new normal" of hypervigilance with respect to infectious precautions will be eroded.

The need for greater availability of infection control practitioners was seen by both infection control professions and non-ICP staff. Interestingly, both groups saw the importance of ICPs being visible on the wards, but often differed in how they viewed their current visibility. ICPs generally saw themselves spending most of their time on the wards, whereas other health staff felt they were not visible enough.

Consensus was not found among participants on whether it was preferable to cohort nursing staff when caring for highly infectious patients. Some groups saw this as being beneficial, whereas others saw it as overburdening a small number of workers. One group recommended that these decisions should be left up to ICPs and not be a nursing decision alone.

As well, it was felt that institutions need to develop clear policies over which workers should be able to work with these patients and whether issues of personal health or health of a household member or pregnancy are grounds for being able to refuse such work assignments.

Many groups mentioned the lack of infection control guidelines for patients and family members as being a source of frustration.

> Sometimes you have the perception that the hospital is afraid to say no to visitors and that they do their best to accommodate visitors, but sometimes it's at the mercy of health care. It happened during SARS. (lines 1127–29, transcript 15)

> I think we should go back to what we did have at one point: two visitors at any one time between the hours of 3 and 8. Period. No children under the age of 13. Period. (lines 1148–50, transcript 15)

> We need to go back to those restrictions…. Yes, I'm sorry you're ill. I'm sorry you can't see your family, but we don't want you taking whatever illness back to your family. (lines 1154–56, transcript 15)

Another area where infection control policies were found to be lacking was in incorporating effective procedures for the cleaning of portable equipment in different care settings. Another was in establishing which procedures can be classified as "high-risk" and require extra protective measures. One group suggested that there should be a specific policy to ensure that one person on the "code team" on the hospital should be responsible to ensure that all team members are using the proper PPE.

Safety Training

Issues related to training healthcare workers in proper infection control procedures also arose very frequently in the focus groups. Many groups felt that existing programs for training in infection control had been inadequate prior to SARS, in that they were often given only to new employees at the time of hiring and no accommodation for ongoing training in infection control existed. However these problems were compounded when SARS struck and healthcare workers were expected to use new procedures and equipment with which they had no experience. Some workers felt they had no extra training during SARS, at least initially.

> Well there were lots of masks available, but we didn't get instructions on how to use them. Nobody instructed us. We just stuck them on our heads as best we could. There was no person that was designated to teach the staff and it was a bad situation.... (lines 221–24, transcript 1).

Others were being trained but by trainers who did not have much confidence in their abilities. One occupational health and safety professional from Toronto stated:

> I think for me personally the biggest thing was that I had to educate and train other people on practices that I didn't even know myself yet. You're learning and you're trying to teach at the same time that you're trying to absorb it and process.... (lines 1042–45, transcript 12)

In other facilities, health and safety training for SARS was delegated to frontline staff who had more experience in infection control, which led to other problems, as outlined by an allied health professional from Vancouver:

> The problem is with primary instructors, who are also the primary caregivers. They have to determine whether their priorities are going to be teaching all the staff as they're doing their bedside care, or are they going to be taking their focus away from their patient and worrying about all the staff. (lines 286–89, transcript 4)

The lack of flexibility or preparedness to rapidly educate staff during SARS was summarized by a manager from Toronto as "You cannot educate in a crisis" (line 558, transcript 6).

With regard to planning for future training, workers suggested that occupational health or infection control keep records of who has received recent training, so they will know who needs to receive more. Some facilities already have similar systems in place, but there was also a recognition that classroom teaching needs to be followed up on the wards in order to ensure that it is being properly applied. Again, an infection control practitioner from Vancouver:

> If you're teaching somebody something that they're not going to apply for a long time or isn't relevant to them at that particular moment, that's not going to be a useful thing to do. You do kind of have to be prepared to grab those teachable moments. And that also again involves being able to be visible, being available. (lines 711–16, transcript 10)

Also the question of where physicians, residents, and medical students fit into infection control training seems unclear, as observed by a nurse from Vancouver:

> There's all these little in-services from infection control and they are all gathering the nurses around the nurses' station to tell them how to do this and I never see the doctors gathering around and their residents, gathering round and getting an in-service. (lines 481–84, transcript 2)

Communication about Safety within Healthcare Organizations

The pivotal role that internal communication played in the SARS outbreak was best described by a manager from Toronto:

> I think communication is paramount to having any success in implementing any infection control procedures and I think that in some organizations that was a challenge, because how do you, you know, staff work three shifts, how do you disseminate all of this incredible amount of information simultaneously in a timely way, when we had new directives coming down the pipeline every hour sometimes. That was a challenge, I think. (lines 53–58, transcript 5)

Much of the communication issues surrounded the dissemination of the constantly changing directives which were discussed above. However, the best means of communicating these messages varied. Most participants agreed that having visible representatives from the hospital in face-to-face meetings was seen as being very credible, and important in terms of boosting staff morale. As another manager from Toronto put it:

> We had a "town hall" (meeting) between the two sites so that everyday there was communication of information. The staff really did want to see somebody, especially in the areas that were high risk areas — the emergency department, the areas where the SARS unit was. They wanted to see somebody from administration and education actually coming onto their unit because they really kind of felt isolated from the rest of the organization. So that was an important role in communicating with the staff. (lines 117–23, transcript 5)

Despite the lack of a widely disseminated outbreak in Vancouver, some staff did not feel that their institutions communicated with staff very well. An allied health professional from Vancouver stated:

> Communication within the institution is one of the major break-downs in terms of infection control...Changes happen, and we saw that every single day during the SARS outbreak and the standards changed sometimes from hour to hour and it was very difficult to communicate that throughout the facility. (lines 80–86, transcript 4)

The amalgamation of hospitals into larger administrative units was seen as a barrier to having good communication, as stated by a manager from Toronto:

> Most of the decisions are being made at Hospital A and then they had to be disseminated down to the campuses, so what happened at my campus was that the information would some-times come from the media before coming to us. That was very difficult for staff and that led to a lot of talking in the corridors and people getting the wrong information. It's a big problem in a big institution. (lines 62–66, transcript 5).

It was generally recognized that relying on the media as an information source was not desirable from the point of view of healthcare workers.

Other communication strategies used during the SARS outbreaks included e-mail distributions to staff. There was some variability in how useful this was seen by staff. Many felt that because they did not have the time or the access to e-mail at work, that this was not effective. A support staff worker from Toronto:

> It would have been nice to have been informed of the changes right off.... Sometimes that didn't always happen.... (Another

speaker) And I can add to that. I personally think the reason that was, is because it was all done by e-mail and a lot of direct people— housekeeping, nursing, anybody that does direct care, do not sit down at a computer before they start their day. I think that it was not the ideal method. (lines 38–45, transcript 8).

Others felt that it was a useful addition to the other forms of communication. Posters and notices were also widely used, especially as reminders, or environmental cues for infection control guidelines, and to inform the public about the situation on arrival in the hospital.

In addition to better communication from the organization to employees, other participants identified communication problems between employees in the hospital. This was described by a member of one of the mixed groups in Toronto.

Many times the patients arrive and we don't know that they've had a cough or a fever or something where we would have to take precautions, so I think there needs to be better communication between departments. (lines 381–83, transcript 15)

Good communication between occupational health and infection control was generally seen as being beneficial both during SARS and after. A support staff worker from Toronto stated:

I don't think you can have a good health and safety program without having infection control included. And if they are not intermingled, then I think the system breaks down (lines 598–99, transcript 7).

Fit Testing

Participants did spend some time discussing fit testing, but the value of it was not universally accepted, as different institutions used different methods and workers often saw these inconsistencies as sources of concern for the whole process. One of the managers from Toronto had this to say:

We have a few issues around mask fitting. One of the things that was a concern … is it necessary? What's the benefit? Beyond that it's even the process and standardization of fit testing, because I think that depending on the company that you hired to do it, the process is not exactly the same….I think there needs to be some work around coordination and standardizing the fit testing process itself." (lines 448–55, transcript 5)

Even if the fit-testing process was successful, there were no guarantees that the masks available would match those on which the worker had been tested. A physician from Toronto noted:

> I think one of the critical issues during this outbreak as well as any outbreak is not only the availability of N95 masks or higher, but are they available for the ones that you've been fit tested with because right now there's a choice probably of about half a dozen that you might get tested for and find the one that fits you. But the problem is that during a crunch, we went through probably half a dozen different companies that provided masks, so trying to provide one that you've been fitted for is difficult. (lines 247–53, transcript 13)

Other Organizational Factors

The increased worker fatigue, especially when having to use large amounts of PPE in stressful situations meant that productivity fell dramatically. Thus staffing levels on a per patient basis likely needed to be increased in order to compensate, and workers felt this was not adequately addressed. As well, because of the casualization and outsourcing of the labor force, management needs to recognize that many of their workers work in more than one site, and are often not working full-time at any one institution. This has implications for many of the organizational factors discussed above.

Environmental Factors

Participants spend the least amount of time talking about environmental factors, which included the availability of personal protective equipment (PPE).

Physical Space Separation

While participants recognized the importance of physical space separation in assisting infection control in hospitals, there appeared to be a great variation in space available.

Isolation Rooms for Patients with Suspected Communicable Diseases — A member of the occupational health/infection control group in Ottawa stated:

I mean directives came out and said patients presenting to triage with infectious or respiratory symptoms had to be immediately isolated. Well, I mean, they would all be isolated together in the big waiting room, right? Like it couldn't happen. There weren't (enough isolation rooms). I mean we have ten rooms with closed doors on. It's impossible. (lines 733–39, transcript 9)

An allied health professional from Toronto commented:

A lot of our ICUs are open concept with only a select few isolation rooms and there was always an issue of a patient was going sour and we didn't have an isolation room. What are we going to do? You know, and so we were like hunting every-where for an isolation room, and then it had to be negative pressure on top of that, so that put us in another bind.... (lines 872–77, transcript 3)

However, it seems that most of the facilities have adapted to the "new normal," of hypervigilance regarding respiratory precautions. A nurse from Toronto describes the current situation:

Whenever a patient has a temperature, right away the nurses put that patient under fever/pneumonia precautions, so we call infection control and place that room under isolation. If there is a patient in there, we take that patient out so we have to shift the whole floor around and put that patient in a private room.... That will continue and the only person who can take that person off the isolation is the infection control. (lines 515–22, transcript 1)

Anterooms for HCWs to Change into PPE — The same was true for anterooms. A participant from a mixed group in Toronto commented: "As far as an anteroom, we don't have those. They never existed" (line 481, transcript 15). However, many facilities did have anterooms for workers to use, or were developing them.

Negative Pressure Rooms — As Ontario hospitals were directed to provide negative pressure rooms for their patients during the SARS out-break, most facilities had experience with creating them and using them. One manager from Toronto was clearly convinced of their utility:

Initially, when this all started, patients who were being admitted were being admitted to negative pressure, ventilated rooms. There were a number of things that were done though to help create negative flow.... I think also too, when you look at the period of SARS III, what will make the difference, it definitely is, if we create negative pressure rooms in this area. (lines 198–208, transcript 6)

Another manager viewed the negative pressure directive as more of a precaution:

I guess, back to negative pressure, its interesting because in regards to SARS, if its not airborne then that wouldn't have been a necessity, but because as you mentioned earlier, it was the learning process and certainly we all wanted the very best for both our patients and healthcare providers. (lines 43539, transcript 6)

However, this participant also recognized that establishing the negative pressure room was not enough.

I feel that unless you do testing of the rooms once you've put in the unit, you don't have a clue what you have and that's the issue I've been fighting.... You should even have continuous monitoring to see that negative pressure is maintained. (lines 446–61, transcript 6)

Environmental Decontamination

Generally participants felt that most of their facilities had adequate hand-cleansing gel stations, which could compensate for the areas where there might be a lack of hand-washing sinks. A nurse from Toronto observed: I found that (during) SARS in our institution, it was the first time I worked there that they went around and they actually disinfected and cleaned the doorknobs, the handrails, the pillars. I had never seen it before and they did it twice a day. (lines 1095–97, transcript 1)

Availability of Specific PPE

Nearly all groups mentioned the supply problems with N95 masks during the SARS outbreak, as described above. There were also supply problems with face-shields and goggles, leading one member of the Ottawa focus group to comment:

> The problem with the goggles is that ... you have the choice between something that may work and offer you some protection or something that might work better that nobody is going to use. (lines 681–684, transcript 9)

Individual Factors

Knowledge

Certainly, the knowledge of infection control procedures and the rationale behind them was found by most groups to be important. A manager from Toronto had this to say:

> If we're going to expect that staff will want to work in a unit with patients infected with SARS or something similar, then we're going to have to do a lot better by providing cited evidence to support decisions that are being made otherwise.... the word of mouth is just not going to work. There needs to be something to back that up. (lines 608–12, transcript 6)

Another support staff worker from Toronto said:

> There were lots of employees, I found just from chatting back and forth, that if there was another outbreak of SARS in the hospital, they would be gone. They would leave because ... of all that uncertainty and fear. So I think an education for the employees would make a huge difference. If they knew what they were dealing with and if they knew what precautions to take." (lines 371–74, transcript 7)

However, it was also generally felt that knowledge alone was not sufficient in allowing workers to protect themselves from infectious diseases at work.

Attitudes

Attitudes such as decreasing compliance when feeling stressed or overworked, and professionalism, which can lead to the HCWs placing their safety concerns below those of patients who need help, were generally felt to be more important than knowledge. A support staff worker from Vancouver expressed her professionalism this way:

> We work in this field and we know we are going to be exposed to this and we chose this field to work in, so you just have to safeguard and take all the precautions you can.... It's different when you have inexperienced workers coming in. (lines 547–52, transcript 8)

A nurse from Toronto explained it as a mix of both professional ethics and personal empathy for her patients:

> I think in general, the nurses think, oh yeah, I probably can (become infected), but "I decided to be a nurse and I'm going to do it because what happens if we all stop?... What happens to me when that's me the patient?" (line 897–900, transcript 1)

Beliefs

Beliefs were also felt to be important by most participants. One nurse from Toronto described how her experience with SARS undermined her belief in the directives which were designed to protect her:

> I volunteered to work on the SARS unit. I only did it because I knew all the nurses and I thought, "Okay, I'll do it." But about June 5th and you go on the unit and the three doctors who are giving us the education ... then one of these doctors became ill. I thought, "Okay, it's just Russian roulette here" .... Nobody felt safe at all. (lines 760–67, transcript 1)

Yet, generally, the heightened fears of infection with SARS during the outbreak led healthcare workers to be very vigilant for themselves and for their coworkers. An allied health professional from Toronto noted that:

> During the outbreak, really compliance or noncompliance was a nonissue. Everybody just did and there was no question about it. I think the fear of contracting the disease was palpable, very real. Nobody was trying to cut corners. (lines 590–93, transcript 3)

A support staff worker from Toronto stated:

> If SARS were to hit tomorrow and let's say you have a SARS patient that comes ... I would feel a lot more comfortable if I

put on a mask, if I put on a respirator, just because I knew that there was a SARS patient in our facility. (lines 340–43, transcript 7)

In some circumstances, this fear led some workers to refuse to work. A physician from Toronto stated that:

And then you had some people who refused outright. We had one cardiologist at Hospital C who would not come into Hospital D to cardiovert a baby. Absolutely refused to come. And then we had some physicians that just disappeared. They never saw a patient. (lines 575–78, transcript 13)

Past exposures to disease can lead to decreased compliance when experience shows that barriers are not needed 100% of the time. A support staff from Toronto:

I remember when I first started working (at) the hospital, I was ever so careful what I touched and I had my limits. I would never press the elevator button if I didn't have a paper towel in my hand. Now, it's like all those issues they are everyday routine. You don't think about them as much as you used to. I think every once in a while we need to kind of "wake up." (lines 801–805, transcript 7)

An allied health professional from Toronto also recognized the problem:

That's the problem … because you do get, sort of, these people that are put in protective isolation that turn out to be nothing and then after a while people start to ignore the precautions because they think it's going to be another nothing again. So I think it has to be a sort of balance." (lines 349–51, transcript 3)

Impact of PPE on the Job

Time constraints, increased workload, discomfort, and peer involvement were some of the issues HCWs mentioned.

Time Constraints

Participant from the Ottawa focus group:

I think the staff need to have direction on what is required, but it needs to be realistic, because what we've been told is... that (in) triage, you change your goggles, gloves, mask and gown between every patient and its 100% not feasible. It can't be done. Patients would be dying waiting at the triage desk. (lines 792–96, transcript 9)

Increased Workload

An infection control practitioner from Vancouver stated:

Of course, it is a lot of extra work for the staff wearing protective eyewear, wearing an N95 mask, which increases your oxygen consumption, wearing gowns, wearing gloves. It can be very hot, very uncomfortable and that continues to be a barrier. (lines 411–14, transcript 10)

A support staff worker from Toronto found that the discomfort dramatically increased her workload:

I remember going to clean a room and I'm a custodian so I do everything from the ceiling, walls, floor ... I had to wear double of everything except the mask, but I had the shield. All I know is by the time I got out of the room, I could squeeze my clothes. I was so dehydrated. You can't just go back and get a drink. It's too time consuming.... Because just coming out you have to strip and then you have to regown, double of everything and you have to go back in. And the time that it takes to put all these layers on is just so much that you can't be bothered. (lines 398–405, transcript 7)

Discomfort

Many participants felt that wearing the full protective equipment during SARS was quite uncomfortable, as described by a physician from Toronto:

The masks weren't very comfortable.... Obviously, everybody found the respirators, in particular, cramped or irritating too. You sweat with them, so that's going to affect the compliance. (lines 390–94, transcript 13)

A nurse from Toronto said:

We had five or six different masks but it was your choice, whatever felt comfortable to you. There were some very strange in their function and they looked funny and they felt funny and they smelt funny. So sometimes in an evening you might wear three different masks because you're trying desperately to get something that is comfortable and doesn't smell like dill pickles and whatever else. They were awful. (lines 204–209, transcript 1)

Another Toronto nurse said that with regard to the masks: "Our girls complained of rashes and they had to... (use) a lot of different skin care products." (lines 1107–09, transcript 1).

Peer Environment

Many workers found that poor compliance with the use of PPE in role models and coworkers, especially physicians to be quite frustrating. A support staff worker from Vancouver explained:

I think I washed my hands five times every time I came out of a room because you had to wash your hands before you took something else off. So that was one of my big concerns, and the other one — doctors.... Doctors not washing their hands. It doesn't matter if it's a SARS patient or who, doctors don't wash their hands.... Especially when the SARS epidemic was here, people should have been a little bit more diligent in washing their hands and they weren't and that bothered me. (lines 157–63, transcript 8)

An occupational health and safety manager from Toronto described another source of frustration:

People wandering around with gloves and touching elevator buttons. That's what most of our (OH&S) staff get upset about. They feel they are being diligent and donning everything properly and using it when it's appropriate and they see somebody else totally disregarding it. (lines 946–51, transcript 11)

Peer feedback on compliance with PPE was seen to be effective, if it was applied. But it was often left to the nurses to police others coming in and out of the rooms, a role which they did not feel they wanted to take on. A nurse from Vancouver observed:

> I never see the doctors and their residents gathering around and getting an in-service (on infection control).... And then, when you're the police at the bedside "Hey, wash your hands!" "All right. settle down." And you know what, it's the fifth time today that I'm telling somebody to wash their hands. (lines 482–85, transcript 2)

However, sometimes physicians will use nurses as a source of information about proper infection control, as describe by another nurse from Vancouver:

> Some of the doctors ...were better. They came and asked me before they went in (to a SARS patient's room) and they even said ... come with me. And I went. So that was actually the first time, because they usually just go in and out. Some of them were actually a bit concerned. (lines 494–96, transcript 2)

Allied health professional from Toronto:

> If someone didn't comply, everybody else helped them comply. 'Cause we had one person that didn't want to comply and it was just like everybody was on the case of that person and they eventually did. (lines 656–59, transcript 3)

Exhaustion/Fatigue

Many participants mentioned fatigue as a major cause of failing to follow proper infection control guidelines. A nurse from Toronto described her experiences:

> I work 12-hour shifts in emergency, rarely got a break, so we were not permitted to have fluids at the desk. None. None in the care area. So we were going for five or six hours with nothing to drink. We were so exhausted. So at the end of your 12-hour shift by six and seven hours you're so exhausted that you're crazy. That is now leading to sloppy practice. (lines 866–77, transcript 1)

Attitudes of Others

The attitudes of family members can be an important determinant of increased compliance with infection control guidelines, as described by a support staff worker from Vancouver:

My son-in-law was angry (that I was working) but you just reassure them that you're taking a shower and you're taking all the precautions. And my boyfriend was the same way. You make sure that you wear that stuff and take all the safety precautions because he didn't want me getting sick. I think we were more at ease, but our family members were definitely upset. (lines 555–59, transcript 8)

Table 5.9 shows a summary of the key points from the focus group analysis.

Conclusion

The content analysis of the 15 focus groups has shown that frontline healthcare workers see more organizational factors being important in determining the success of occupational health and safety or infection control programs than factors in the physical environment or individual factors. This supports what has been found from the literature review. The fact that healthcare workers feel these factors are important does not mean that they necessarily are the *most* important factors, but it shows that policy makers and researchers must address them if they want to have healthcare worker support in developing their policies and procedures. How these results complement or contrast those discussed earlier and how they can be used to develop priorities for research in this area will be the subject of the final section of this report.

Priorities For Further Research

The priorities of healthcare workers as outlined above included both areas where the literature review found substantial information, and areas where knowledge gaps were identified through the literature review.

Comparing Results from the Literature Review with Those From the Focus Groups

The lack of consistency/changing directives problems, and the enforcement issues from the Ministry of Labour in Ontario and the Worker's Compensation Board in British Columbia were themes that were somewhat unique to the SARS outbreaks. Many of the suggestions that emerged in the focus groups conflicted with one another. While many healthcare workers were frustrated by the frequently changing directives, others felt

Table 5.9   Summary of Key Factors Identified by Healthcare Workers

1. Organizational Factors:
   Lack of consistency with safety instructions and frequently changing
      directives
   Enforcement by regulatory agencies
   Workplace attitudes towards safety
   Attitudes and actions of management
   Safety climate
   Perceived importance of occupational health and safety
   Evidence-based and practical infection control policies
   Participation of front-line HCWs in development of infection control
      guidelines
   Adequate resources for infection control
   Adequate number of infection control practitioners
   Better enforcement of infection control guidelines
   More accommodation of worker concerns
   Infection control guidelines for patients and visitors
   Safety training
   Repeated safety training
   Assess the appropriateness of the "train-the-trainer" model
   Track who has been trained and who needs training
   Develop policies to deal with part-time staff, physicians, residents and
      students
   Communication about safety within healthcare organizations
   Face-to-face "town-hall" meetings are necessary to build confidence during
      a crisis
   A variety of communication media are likely most effective
   Communication strategies need to be adapted for large, multi-centred
      organizations, especially with fewer lower managers
   Communication between employees, units and especially OH&S and
      infection control is important in creating safe workplaces
   Fit-testing
   Other organizational factors
   Worker fatigue
   Casualization and outsourcing of the workforce

2. Environmental Factors:
   Isolation rooms for patients with suspected communicable diseases
   Anterooms for HCWs to change into PPE
   Negative-pressure rooms
   Environmental decontamination

Table 5.9   Summary of Key Factors Identified by Healthcare Workers *(Continued)*

Availability of specific PPE:
Masks
Face shields or goggles

3. Individual Factors:
Knowledge of infection control procedures and the rationale behind them
Attitudes such as professionalism
Beliefs in effectiveness of infection control guidelines, as modified by past
  experiences.
Impact of PPE on the job
Time constraints
Increased workload
Discomfort
Peer environment
Peer compliance
Peer feedback
Attitudes of family members

that officials were not forthcoming enough with new information. The differing views on the implementation of rules requiring fit testing for healthcare workers were particularly noted.

Clearly, if the safety climate within healthcare was better and workers had more confidence in their employers' commitment to worker health and safety, employees would have more confidence in the messages and directives they received during a crisis situation such as SARS. The relatively low profile of occupational health and safety within healthcare is perhaps best reflected in the observation that very few focus groups, aside from those containing health and safety professionals, seemed to be aware of occupational health and safety professionals at all. Tasks such as fit testing of respirators often fell to infection control practitioners, not to occupational health and safety professionals as it would have in other industries (although this appears to vary from facility to facility). Certainly more research on what levels or standards are needed to promote effectiveness in occupational health, similar to the SENIC studies for infection control, is needed.

Another suggestion that emerged from the focus groups was to involve experienced and credible frontline healthcare workers in formulating the infection control guidelines and occupational health directives. In most cases these guidelines are developed only by experts who are very well versed in the science behind the guidelines, but may be less informed on how to best translate the science into practice. Allowing some form of adaptation of guidelines by local policy makers may help in this regards, but it is also likely that allowing significant variation in guidelines from facility to facility would increase the uncertainty in their reliability. Similarly, the suggestion to allow workers to use more personal protective equipment than a clinical situation may warrant, could lead to a lack of confidence in the guidelines in general. It is difficult to balance many of these issues and operational research in these areas could greatly inform the discussion.

The suggestion to develop stronger infection control guidelines for patients and visitors is also an area where policies could be developed immediately. The most recent Health Canada guidelines on SARS do include references to visitors and patients, but their description is scanty and mechanisms for ensuring compliance are not developed in most institutions.

The importance of training in infection control is obvious, but the literature review and the focus groups agreed that one-off didactic sessions are unlikely to be the key to ensuring that workers practice appropriate infection control. Again there was inconsistency in how workers viewed the "train the trainers" model for infection control training, and the decision to use this versus other models may be based on trying to balance the concerns of frontline workers who already feel overburdened and other methods. Certainly, finding innovative ways to ensure that physicians, residents, part-time staff, and students receive annual and ongoing infection control training and feedback should be seen as a priority. Nurses do not want to be, nor should they be, seen as the "infection control police" for other health professions.

Equally, the importance of good communication within healthcare organizations is self-evident. The strategies that provide the most effective communication within organizations are not clear and are an area in which research can inform greatly. No single strategy seems likely to meet the communication needs of any organization; therefore it is more a question of what mix of strategies works best for which messages.

Participants did spend significant time discussing fit-testing, but the value of it was not universally accepted, as different institutions used different methods and workers often saw these inconsistencies as sources of concern for the whole process. The fact that prior to SARS, fit testing had not been a requirement in healthcare facilities likely contributed to this perception. The questions raised by workers during the focus groups regarding fit testing appeared to be addressed by the literature (see literature review section). As noted, fit testing is helpful in reducing exposures, but whether this is attributable to the training that accompanies fit testing or because fit testing by an expert leads to improved seal between face and respirator, is unclear. This has been identified as an area in need of further research.

Other organizational factors that should be addressed with more research would be the role of how workforce changes such as the increased use of casual workers and outsourcing of some basic services such as cleaning and laundry affect the health and safety of all workers.

The low visibility of occupational health and safety in healthcare is perhaps also reflected in the relatively low priority that focus group participants gave to environmental controls. In general occupational hygiene and engineering controls are seen as being the preferred starting point in reducing risk of injury or illness in workers, but this has received relatively little attention in health care in relation to the use of personal protective equipment. Negative pressure rooms were discussed at length by the focus groups but the added benefit of negative pressure, above that for isolation with adequate ventilation systems throughout the facility, was already identified as an area requiring further research. Certainly policy changes to include infection control considerations, such as physical space separation, ventilation systems, and environmental decontamination issues when designing new facilities or renovating old ones, could already be considered by policy makers at this point.

How specific knowledge, attitudes and beliefs around infection control and occupational health can be improved in individuals, is largely mediated through the organizational factors identified above. The full PPE required for use with SARS patients was found too cumbersome, uncomfortable and imposed additional time-constraints and workload on healthcare workers. However, part of these findings could have been influenced by the fact that many people were being introduced to the use of this equipment during a crisis when normal coping strategies may not have been functioning. In hospitals where N95 masks had been introduced for general use in the past, it seemed that there were fewer complaints from the staff with their use. Trying to define precisely who needs to use this equipment and when, and what amount of protection is afforded by it, were identified as priorities from the literature review. Another research priority from the point of view of healthcare workers would be in designing protective equipment that provided the most protection and least discomfort.

The importance of peer feedback and peer compliance had been previously identified in the literature review as being key determinants of safety training success. Certainly the attitudes of family members and society in general to the SARS outbreaks greatly influenced healthcare workers in their attitudes to practicing infection control during SARS and this probably should be addressed in terms of infection control policies on deciding which healthcare workers should be working in high risk areas.

Further Research Priorities

The following criteria were used in identifying research priorities:

1. Degree to which the knowledge gained from exploring the "gap" would reduce risk to health care workers (i.e., how big is the gap in our knowledge and does additional knowledge provide any significant benefit to protecting HCWs?).
2. Ease with which a research study could be designed and answered.
3. Whether research is currently underway in this area.
4. Cost and feasibility of the proposed research and/or of the intervention.
5. Stakeholder interest.

We have divided the priorities into three groups of research, with the following order of priority:

1. Improving the workplace health and safety through organizational factors: (i.e., how best to bring about meaningful knowledge translation).
   a. How can the safety climate of infection prevention and occupational health of HC institutions be improved? What approaches best facilitate an organizational culture that promotes safety?
   b. What are the best mechanisms to provide communication to front line workers in order to ensure appropriate infection control practices?
   c. What are the best mechanisms to provide feedback to frontline HCWs in order to ensure infection control measures are practical and feasible while still enhancing safety?
   d. What are the best ways to train HCWs on appropriate use of personal protection equipment?
   e. What are the health and safety effects of the recent changes to the healthcare workforce, in terms of increased casualization and increased outsourcing of services?
   f. What key components of an infection prevention and occupational health program are needed to improve or maintain worker health and safety in healthcare facilities?
2. Epidemiology and transmission of respiratory pathogens:
   a. How do respiratory droplets produced by aerosolizing procedures differ from those produced by more "natural" methods such as coughing or sneezing, in terms of their size, their spread and their infectivity? This question is key because it addresses

the issue of the hierarchy of precautionary measures (i.e., are the same level of precautions required for situations that do not generate aerosols by mechanical means?)

b. Do infectious organisms survive on barrier equipment and clothing and for how long? The implications for this are for environmental decontamination, reuse of barriers versus the use of disposals and to assess the potential importance of auto-inoculation through contaminated PPE.

c. How are respiratory tract pathogens able to cause disease through the transocular route?

3. Risk reduction through engineering controls and personal protective equipment:

a. What is the relative effect of engineering controls to maximize particle fall out or decrease viability of organisms, e.g., temperature, air exchange, relative humidity?

b. There may be simple yet effective measures to decrease these aerosols that could have significant impact on reducing the risk of exposure.

c. What design criteria are required to minimize generation and dispersal of infectious aerosols in medical equipment such as anaesthesia machines, and ventilators? This question addresses the relative importance of decreasing aerosols at source — is it effective in practice?

d. What is the added benefit of nursing high risk patients in a negative pressure atmosphere over physical isolation and adequate ventilation throughout hospitals? There has been a great emphasis placed on hospitals improving access to this technology, yet evidence to support their use is lacking.

e. What is the effectiveness of facial protection against bioaerosols?

f. In conjunction with question 2c above, answers to this question will clarify the relative importance of full facial protection, versus eye protection, versus nose and mouth protection.

g. What is the relative importance of fit testing versus fit checking of respirators? The reasons for selecting this as a priority is less an issue of burden of disease but because of stakeholder interest, the implications for where resources are expended and the potential extrapolation of this data to other airborne illnesses.

Appendix 5.1

Focus Group Discussion Questions

This focus group will discuss four broad questions related to different factors which influence the success or failure of infection control and workplace health and safety in health care facilities. Each question will be given a fixed amount of time for discussion. The entire exercise should take less than 90 minutes.

1. Organizational factors
   Some examples of workplace organization and hospital culture include:

   - How your place of work is organized to function on a day-to-day basis
   - Committees, protocols, and programs in place that address infection control and occupational health and safety
   - Communication within the institution
   - Perception that employers adequately respond to concerns of their employees
   - Perceived commitment of administration to infection control and workplace health and safety and availability of training programs

   "How do workplace organization and the hospital culture influence (a) the implementation of sound infection control practices, in general (b) the use of facial protective equipment; in particular and (c) occupational health and safety initiatives?"

2. Environmental factors:
   Some examples of the physical environment include:

   - Availability of negative pressure rooms, hand-washing sinks, and appropriate space to allow separation of patients who may have contagious diseases
   - Availability of surgical masks, N95 masks, gowns, facial shields, goggles etc.

   "How have these factors affected your ability to practice safe infection control and, in turn, did you feel comfortable that the environment you worked in was safe?"

3. Individual factors:
   Ultimately it is the individual who makes the decision whether to use or not to use a particular piece of protective equipment or to follow (or not) an established protocol. Examples of factors which vary from person to person in the same workplace include:

   ■ Perceived likelihood of catching the disease and the severity of the disease
   ■ Personal knowledge about infection control guidelines
   ■ Confidence in the effectiveness of infection control guidelines
   ■ Family life circumstances
   ■ Ease or difficulty of incorporating infection control into daily work.
   ■ Preference for particular types of protective equipment

   "What individual factors have influenced you in practicing safe infection control and occupational health?"

4. Other factors:
   "Are there other factors, not already discussed that you feel are important in determining the success or failure of infection control procedures or occupational health and safety initiatives?"

Focus Group Summaries

Nurses — Toronto — November 25th, 2003 (Nine participants)

Key Points:

■ Impact of frequently changing directives: During the SARS outbreak there were directives coming from the Ministry that were frequently changing, translated into institutional directives that were frequently changing — raising fears among HCWs, particularly when directed to discontinue use of PPE.
■ Communication from organization to HCW: Related to frequently changing directives, communication was recognized to be a problem, with there being difficulty for HCW in finding out what the current guidelines were.
■ Fit testing: With the institutional requirement for fit-testing, participants voiced concerns that the supply of the particular mask with which they'd been tested was not always available — with accompanying fear of incomplete protection when using a different mask.

Suggestions

- During an outbreak, there should be a coordinator or responsible person who could coordinate the dissemination of information to HCW
- Infection control practitioners need to be more visible at education sessions for HCW
- The number of infection control practitioners needs to be increased
- The management/organization needs to listen to the concerns of HCW and accommodate them where possible

Nurses — Vancouver — December 12th, 2003 (Seven participants)

Key Points

- Necessity of advanced planning for emergencies: Facilities do not have the resources to deal with emergency situations.
- Delivering safe care in emergency situations: Staff take on themselves the complexities of trying to deliver safe care in emergency situations. Many examples were cited where nurses wanted to rush in to assist a patient in crisis and the difficulty of doing this with proper protection on.
- Consistency of policy and practice: Nurses strongly perceived an inconsistency in policy and in the application of policy throughout a individual facility and in the community. Examples include when to wear and how to use PPE and when quarantine is required.
- Development of infection control guidelines for patient behaviors and compliance by patients: There are perceived to be no policies directing infected patients behaviors. If an infected patient was not bedridden, their access to the facility was not constrained leading to concerns that they could be spreading infections to staff, visitors, and other patients.
- Contracted out staff: The nurses have been told not to give direction to contracted out staff. This presents an ethical and practical dilemma for nurses when they see a staff person not complying with safe infection control practices. Further, the style of cleaning where one individual does one task and another individual does a separate task is counter to trying to minimize exposure to infectious agents.

- Minimization of staff concerns: Staff felt that their concerns were minimized and suggestions for practice were overruled. For example, when one nurse wrote that staff should use a mask when caring for a MRSA patient because she was producing sputum and was coughing a lot, the Infection Control Nurse (ICN) just overruled her without discussion. According to the ICN, a mask was to be used only during suctioning.
- Compliance with infection control practices: Nurses don't want to be the infection control police but often find themselves in this position.
- Development of protocols for pressurization of rooms: It was reported that there was a lot of problems with practices surrounding the negative pressure rooms — who turned on the pressure, alarms, who monitors the pressure in different situations (when more than one room was being used).
- Leave management programs and use of sick leave: Management is concerned about how much sick time is used by staff but this should be balanced by the need to keep workers with infectious diseases away from the workplace.

Suggestions

- Assignment of staff to SARS patients should be informed by infection control professionals who are familiar with the acuity of the patients and with best practices regarding staffing issues. Nurses question the practice of assigning the same staff to SARS patients versus sharing responsibilities.
- There should be criteria for identifying conditions that make staff vulnerable when caring for highly infectious patients — pregnancy, undergoing chemotherapy, or having a partner undergoing chemotherapy, etc.
- An emergency plan should be developed to minimize dealing with emergent issues on the fly.
  - Basic infection control policies and procedures were either not developed or were not enforced. There needs to be consistent policies and procedures that are enforced and monitored. These include:
    - Methods for monitoring and enforcing compliance
    - Consistent policies for quarantining individuals
- Policies regarding the reuse of masks.

- Tracking of training to ensure that all those who need training actually get it.
  - Policies regarding negative pressure rooms.

Allied Health Professionals — Toronto — November 25th, 2003 (Seven participants)
Key Points

- Leadership, communication, coordination, and the involvement of frontline workers in decision making are key factors in gaining the trust of health care workers and making them feel they are working in a safe environment. The feeling of safety is much more than the provision of personal protective equipment. The more management was visible on the floor, and the more workers were engaged in discussion and understood the development of policies, the higher degree of confidence in safety measures was felt.
- The identification of patients as high risk unnecessarily leads to compliance fatigue. Resorting to a perceived type of "universal respiratory precautions" is not productive.
- There is a stark difference between ICU and ER. ICU workers generally knew what they were dealing with (although not at first). ER workers don't know what to expect when they approach a patient. There are fewer resources such as isolation/negative pressure room in the ERs. Some feel these conditions lead to laxness on the part of ER workers.
- There was no agreement on the identification of high risk procedures and how to deal with them or how to clean portable equipment in different work settings.

Suggestions

- Involve frontline workers in policy setting.
- Establish a respiratory assessment for high risk patients on which workers can rely and that doesn't lead to unnecessary precautions being taken.
- Establish effective procedures for the cleaning of portable equipment in different care settings.
- Establish protocols for high risk procedures — when, where, how and by whom procedures should be done.

- Establish with certainty the mode of transmission and the efficacy of PPE.
- Evaluate experience after an outbreak to assess the effectiveness of policies and practice and make improvements.

Allied Health Professionals — Vancouver — December 12th, 2003 (Five participants)

Key Points

- There is a great desire for standardized infection control policies and procedures that are enforced by individuals specifically assigned this task as part of, not in addition to, their regular duties.
- Within the need for standardization, professionals want to be given the ability to make choices about the appropriate use of personal protective equipment when their assessment shows a need for its use. Infection control professionals should support the judgment of professionals who have direct contact with infectious patients on a daily basis. For example, choosing to use a mask with an MRSA patient who is productive should not be discouraged.

Suggestions

- When dealing with an infectious disease that little is known about but that clearly can result in serious illness or death, maximum precautions should be taken at first, followed up by a tapering off of precautions as more is known about appropriate guidelines.
- Standardize infection control policies and procedures and allow staff to use their judgment in certain situations.
- Especially where there are shortages of PPE supplies, those in high risk situations should be given the equipment in priority.
- Medical surveillance programs should be reinstated.
- Training programs should be delivered at times convenient for all staff who need to attend. They should be delivered by staff as part of their job not in addition to their job.
- Training needs to happen more than once in order to ensure that staff remember how and when to use PPE and proper infection control procedures and techniques. The site should be prepared for disease outbreaks and not scrambling when they occur. Procedures, repeated enough, will become routine.

- Assessment protocols on admittance should be developed to ensure that no one who needs to be isolated is missed and there is minimal isolation of those who do not need it.
- Wearing PPE for long periods can lead to exhaustion. Where this occurs, there should be time for a break before proceeding to the next task.
- Especially in emergency situations with an infectious individual, there should be a member of the response team that is charged to consider infection control issues and who could be the individual who helps people dress and undress appropriately. As procedures become more routine, this may become unnecessary.
- Emergency departments should all have isolation rooms.
- Multitasking should be considered in situations where it is appropriate to limit the number of individuals needing to don PPE to do certain routine tasks.
- Storage and availability of PPE must be considered to avoid having to spend time looking for equipment.
- The availability of the appropriate PPE on carts presents a problem that need to be addressed.

Managers — Toronto Group 1 — November 26th, 2003 (Ten participants)

Key Points

- Communication that is timely, ongoing, consistent, and reaches all staff is paramount to having any success in implementing a good infection control program. This is a great challenge in health care due to shift work and the size of facilities. Mixed messages led to staff feeling very insecure about directives and policies.
- The lack of infection control practice leaders was a factor in the difficulties faced in some institutions to get buy-in from staff on infection control directives. Since the SARS outbreak organizations have hired more infection control professionals but are concerned that lack of funding will result in these individuals being cut.
- Cutback of support staff has raised concerns about the adequacy of cleaning being carried out.
- Casualization of the workforce may be leading to increased potential for exposure as staff move from work site to work site in order to get an adequate number of work hours.

■ Emergency situations pose particular problems for staff members who have to deal with the balance between providing care in a safe manner and providing care in a timely manner.

Suggestions

■ Consistent and effective screening protocols are necessary to ensure proper identification of potential infectious diseases.
■ There is a need for ongoing education outside periods where there is no outbreak of an infectious disease. There needs to be an identification of the content of any program and the most effective way to provide this education.
■ There must be an identification of the proper PPE required at different stages in the care continuum, including the use of disposable versus nondisposable equipment.
■ There needs to be identification of what is meant by an isolation room, a negative pressure room and the appropriate use of each at different stages in the care continuum.

Managers —Toronto — Group 2 November 26th, 2003 (Seven participants)

Key Points

■ Establishment of standard protocols, that are widely and well communicated, with consistent follow-up/enforcement were significant themes for this group. There was support for a comprehensive infection control program developed and implemented by staff who are given clear role definitions and areas of responsibility. Confusion and uncertainty needs to be addressed by the development of evidence based information that forms the basis of standard protocols.
■ Guidelines and standards aid in inspiring confidence in staff but also aid in assisting in getting senior management to fund infection control initiatives.
■ Staff members are more likely to be compliant with protocols when they understand the principles and the evidence supporting the protocols. This group of managers felt that evidence based standards are easier to communicate to staff since managers themselves have more confidence in the information they give out.

■ Despite the desire for standards, staff members need to be given permission to use their own judgment in cases where they determine a higher level of protection may be warranted.

Suggestions

■ Development of standards for number of negative pressure rooms per patient population and having at least one room per facility.
■ Development of a computer program that aids implementation of infection control programs.
■ Establish best practice for storage of alcohol based hand wash gels.
■ Have an infection control professional on the design team for new or renovated buildings or areas.
■ The infection control team should have an individual with engineering expertise.
■ Develop a cost effective way to retrofit rooms to provide negative pressure environment.
■ Since infection control practitioners are in short supply, organizations should pool resources to create tools available to everyone in order to minimize wasted time and efforts.
■ Establish protocols for wearing or not wearing uniforms to and from work. There is a need to reestablish the practice of changing into uniforms and work shoes before work and changing back after the end of the shift.
■ Working with infectious patients while wearing PPE can be exhausting. Break times must be established to keep staff from burning out.

Support Staff — Toronto — December 10th, 2003 (Eight participants)

Key Points

■ Communications to staff about infection control procedures: There was a tension evident in the discussion between concern over frequently changing directives leading to confusion and uncertainty as to what to do, and the need for rapid dissemination of information so that support workers could be kept up-to-date. Support workers appeared to feel out-of-the-loop with respect to safety information.

- Distrust of management: This theme was prevalent, with a fear that they were not being told the entire truth by management, and that they would more highly trust the same information coming from a peer (interestingly, it was mentioned that there was trust of the president of the hospital).
- Fear of infection: Due to lack of knowledge regarding the mode of transmission, there was fear that their required activities were placing them at risk. This is also related to the two other points above: that they feared they were not being told when they were at risk. Examples included fear of catching SARS from fellow HCWs in the cafeteria, or a mail-room staff worried about contact with mail coming down from the floors.
- Lack of compensation/danger pay: There was a lot of dissatisfaction with the lack of compensation for support staff who worked in the same physical areas as frontline nursing staff who were compensated. A perceived lack of recognition for their service.

Suggestions

- Have one person in charge of communication instructions/directing staff every shift regarding new policies (i.e., during outbreak).
- Need for increased communication regarding infection control procedures (and changes).
- Every department should have a training program in order to keep up to date on new policies and procedures.
- Cleaning staff should not be delivering food after cleaning washrooms.
- There should always be a supply of masks and respirators for which employees have been tested on the units at all times.
- There should be extra breaks when PPE is required due to discomfort associated with prolonged use of PPE.
- There should be clean gowns to put over one's uniform prior to entering the cafeteria (because of fears of catching SARS from other staff).
- An essential core of staff should be trained to take over right away in case of an outbreak (like code teams which respond to cardiac arrests).
- There should be annual in-services and education for IC policies and procedures.
- Always have one person in every department who can act as an "IC Steward" who is trusted, who can convey concerns about IC to the organization.

Support Staff — Vancouver — November 12th, 2003
(Three participants)

Key Points

- Professional Commitment: The workers displayed pride in their work, and that they were thorough in cleaning SARS rooms in order to protect others. There was also a sense that they were fully compliant with all infection control procedures when other HCW were not. The theme of "we chose to do this job knowing all the risks involved" was raised on several occasions.
- Infection control practices of other HCW: The workers portrayed themselves as the "conscience" of the units by pointing out when others were not complying fully. There was discussion that physicians in particular were not compliant and there was concern that they were spreading infection.
- Organizational valuing of support staff: There was the impression that housekeeping staff did not have policies in place to protect them (i.e., from dirty linen) and that priority was placed on nursing and frontline staff. Along the same lines, communication of their concerns to management was a problem.
- Changing directives: Again perceived as a problem.

Suggestions

- A protocol for the transport of garbage/linen from patient care floors is needed (some concern over practice of popping holes in sealed garbage bags in order to compress).
- There is a need to be frugal with PPE supplies for isolation rooms because otherwise they are wasted unnecessarily which is expensive.
- Stricter policies on limiting the access of visitors (especially children) to infected patients are needed.
- There should be stricter policies on the circulation of infected patients in the hospital.
- Patients actively coughing should be masked even when in own room.
- There should be a plastic barrier/seal around the bed of infected patients, especially if coughing.
- Lab techs should have a small set of phlebotomy supplies they take to the bedside of infected patients and then discard (rather than carrying tray of supplies from room to room).
- Need for more education on hand washing.

Infection Control Practitioners/ Occupation Health and Safety Professionals — Ottawa —November 23th, 2003 (Six participants)

Key Points

- Perception of strong safety climate: There is not a strong safety climate in health care in general, both from workers who are expected to apply infection control guidelines and management, who must provide adequate leadership and funding of occupational health and safety programs.
- Safety-related attitudes and actions of management: A key measure of the importance of safety in their place of work was whether management take actions and direct resources to occupational health programs and infection control. While this was not seen to be a priority before SARS, a the time of the SARS outbreaks in Ontario, resources were mobilized quickly to assist with the new Ontario Ministry of Labour directives on fit testing and hire more safety officers, for example.
- Purchasing policies with respect to safety: Related to the above factor. Another way in which hospitals displayed their concern over safety in the workplace was how rapidly and willing they were to purchase a variety of personal protective equipment for healthcare workers during the time of the SARS outbreaks in Ontario.
- Lack of consistency with safety instructions and recommendations from outside agencies: Related to the individual factor discussed below about the lack of confidence in infection control guidelines, participants felt that the rapidly changing guidelines and directives which they received from authorities hindered their efforts to protect workers in that this undermined their credibility.
- Individual beliefs that guidelines are not relevant: Guidelines may not be relevant to HCWs in their place of work, predominantly because no cases of SARS presented to their institutions, but also because of the rapidly changing guidelines and directives which they were given.

Suggestions

- Current infection control measures rely too heavily on the use of personal protective equipment and did not make enough use of other means of protecting workers and patients, such as source controls, engineering controls, and the design of physical space in hospitals.

- Many organizational structures in hospitals are important determinants of workplace health and safety with respect to infectious diseases such as where OH&S professionals fit in the administrative structure of the institution.

Infection Control Practitioners — Vancouver — December 9, 2003 (Three participants)

Key Points

- Safety training: Training in this context relates to training HCWs in infection control practices. Participants spoke at length about training activities in their facilities both for new staff and in-service training for currently employed staff, and how important they felt this was to protecting workers and patients. Training also included instruction on fit checking of masks and follow-up to training with healthcare workers in their place of work.
- Communication about safety from the organization to employees: Part of the follow-up to safety training included ways of communicating with hospital staff. The primary means identified was by the use of posters and signs to remind staff about the need for PPE use. As well, participants felt that the use of e-mail contributed to the dissemination of infection control information, especially when combined with printing out hard copies and posting for those without e-mail access.
- Availability of infection control practitioners: This was seen in terms of needing more staff to follow-up with healthcare workers on the wards, to conduct in-service trainings and to review patients placed on specified precautions (airborne, droplet, contact, etc.) in a timely fashion. Being infection control practitioners, they were aware of current recommendations regarding the number of ICPs based on the number of acute care beds and recognized that they were understaffed.
- Policies and protocols for infection control: Clear infection control policies and guidelines greatly facilitated the practice of good infection control and helped to protect healthcare workers. This included not only when to place patients under specific infection control precautions, but also when to follow-up on patients to ensure that precautions are not applied for an unnecessarily long time.
- Lack of consistency with safety instructions: Changing information contained in repeatedly updated infection control guidelines

undermined the confidence that healthcare workers had in their effectiveness.

■ Availability of negative pressure rooms: Participants mentioned the use of negative pressure rooms as a part of controlling respiratory infections, while recognizing that there is great variation in their availability.

Suggestions

■ Participants from one hospital noted that the creation of a separate cost-center for SARS greatly facilitated internal dissemination of PPE to protect HCWs and allowed them to more directly measure the cost of the outbreak to their institution.

Occupational Health and Safety Professionals — Toronto —Group 1— November 26th, 2003 (Eleven participants)

Key Points

■ Importance of OH&S: The general opinion is that OH&S is undervalued compared to infection control, and that there generally is a lack of integration between OH&S and IC (where integrated, it works well). Also, it is difficult to find personnel with experience in both infection control and occupational health and safety.

■ Safety-related actions and attitudes of leaders: Having a CEO who is involved in safety issues proves that the organization is committed to safety, and fosters trust among employees for management. When the CEO is not supportive, managers felt resentful and unsupported.

■ Merit of keeping nonessential staff out of the workplace during outbreaks: This was done in different institutions with differing results. Pros: reduces possible exposures, eliminates personnel who may get in the way when there are increased demands on patient care because of PPE and ICP (e.g., researchers). Cons: creates a double standard, staff shortages result in change in duties.

■ Adequacy of PPE Supplies: The importance of having a centralized distribution system for supplies was recognized, with lack of a good supply system leading to stockpiling and lack of supplies for high-risk institutions. The importance of having at least a two-week supply on site was recognized, with some discussion as to the benefit of having storage of supplies on every patient care unit.

- Methodology and resources used for fit testing: Discussed in detail.
- Masks: The discomfort associated with masks was seen as the greatest individual factor influencing compliance with PPE.

Suggestions

- Each hospital should have a manager of OH&S services in order to advocate for OH&S and give it the importance it deserves in the workplace.
- Each hospital should have all nonessential personnel (to clinical care) work off-site and discharge patients from hospital whenever possible.
- Each hospital needs to have policies regarding personnel who fail fit-testing: duties to accommodate, find alternate work, compensation if they cannot work.
- There is a need for adequate room for storage of PPE supplies on-site both within institution and on each clinical care unit.
- There is a need to study the question of whether successful fit testing on one occasion persists (i.e., is the success of the fit maintained with prolonged use?).

Occupational Health and Safety Professionals — Toronto — Group 2 — November 26th, 2003
Key Points

- Composition of decision-making team: A key theme with this group was the lack of OH&S professional involvement in decision making at top levels in the province. Directives came down from decision makers who were not conversant with the issues of frontline staff, including OH&S and infection control professionals who were responsible for implementing the directives.
- Directives must come with resources: The directives did not come with the resources necessary to carry them out. There was a huge shortage of trained and experienced OH&S and IC practitioners. There was a huge shortage of PPE, especially masks.
- Good infection programs and protocols must be in place to ensure adequate level of readiness for next crisis: The weaknesses in the system and the shortages that the crisis identified can be linked to a lack of attention to good infection control and health and safety programs and practices for the past decade at least. Though there are references to infection control in regulations in Ontario, there

is no attention paid to health care by the regulator and practices have been very lax. When the crisis came, the system had to move too far too fast and couldn't cope.

■ Infection control practices already going back to pre-SARS levels: The above observation is linked with a concern expressed that the state of infection control and occupational health is already going back to pre-SARS practices. For example, during the crisis, facilities were looking for professional staff to assist them through it. Now that it is over, these staff are being let go without consideration of what is necessary to maintain an effective prevention program in order to ensure an effective program and to be ready for the next crisis.

■ Ministry of Health needs to resource their standards whether they are called directives or guidelines: There is a fear that the Ministry of Health has downgraded "directives" to "guidelines." This was interpreted to be the Ministry's attempt to get off the hook for providing resources that should come with directives.

■ Protocols must be standardized and resourced: The group emphasized the need for standard protocols and for support for organizations trying to implement these protocols in a crisis situation. The stress on OH&S and IC was enormous and little support was provided to them. One issue that caused considerable stress was the fact that IC personnel were asked to educate and train staff when they were unsure of themselves of the directives or of proper techniques such as fit testing.

Suggestions

■ OH&S professionals must be part of the team making decisions and setting policy related to infectious diseases.

■ Capital projects such as building new facilities or redeveloping old ones must be reviewed taking into consideration infection control requirements. The funding must be in place to incorporate needs identified by this assessment.

■ All negative pressure and isolation rooms should have glass in them so patients can be observed without have to go into the room.

■ Ventilation must be monitored to ensure that it is functioning properly.

■ Respiratory technologists have to be part of the decision-making team in facilities.

■ Clear roles and responsibilities have to be assigned to individuals within an organization so there is no confusion. Special consideration has to be given to how compliance is enforced; is enforcement strictly a management issue or not?

■ In the era of nursing shortages, are nurses more likely to prefer a facility with tough standards and enforcement protocols or one that is lax? There is a balance between allowing nursing to make judgment calls and requiring them to follow appropriate infection control protocols.

Physicians — Toronto — November 25th, 2003 (Two participants)

Points of Interest

■ Commitment to early training: Training in the use of PPE and IC protocols needs to start in medical school, accompanied by a system for fit testing for such "transient" HCWs.

■ Environmental factors: This group discussed many environmental factors in detail, such as negative pressure rooms, sink/rinse availability, and availability of masks (particularly correct masks based on fit testing). More emphasis was placed on environmental compared to individual or organizational factors.

■ Fear of transmitting infection: Described as a factor affecting willingness to work and possibly compliance.

Suggestions

■ IC and OH&S should be unified or same division.

■ Equip entire wards such that they can be rapidly converted to negative pressure when needed.

■ Reinforce the importance of doffing equipment when leaving patients' rooms (because of concern regarding contamination of common surfaces and equipment).

■ Monitoring of compliance/auditing of HCW with infection control is important.

■ Fit testing should be more systematic, and should also be done for "transient" HCW such as medical students and residents.

■ Staff should be advised not to wash hands in the patients' bathrooms/washrooms (often this is the only sink available).

■ Need to start infection control training in medical school.

■ Should approach all respiratory secretions as being potentially infectious (analogous to experience with blood and body fluids).

- Rewarding HCW for "100% attendance" is a bad idea as it encourages HCW to come to work when sick.
- Quality control for negative pressure rooms needs to be improved.

Mixed Group 1 — Toronto — November 26th, 2003 (Nine participants)

Key Points

- Organizational decision making: Having a centralized decision-making process for infection control issues allowed for rapid consensus and facilitated communication of directives to employees.
- Education: Education and training of employees was seen as key to ensuring compliance and appropriate use of infection control procedures.
- Cohorting of infected patients: During the SARS outbreaks, patients with SARS were placed in negative pressure rooms located all over the hospital. The disadvantages of this meant that there wasn't a team of employees looking after SARS patients, and that employees could be looking after both SARS patients and non-SARS patients with the possibility of nosocomial spread.
- Compliance with IC procedures in medical and nursing leaders: Physicians in particular were seen as idiosyncratic in their use of PPE and were not consistent in following guidelines, with a negative impact on other employees who were expected to behave differently.

Suggestions

- Education: New approaches to education and training of HCW in infection control should be adopted, that are collaborative, interactive, and based on high quality material — that can be used across the province.

Mixed Group 2 — Toronto — December 10th, 2003 (Ten participants)

Key Points

- Supervision and screening of non-HCW: There was a lot of concern regarding the lack of screening of visitors to the hospital during the SARS outbreaks, and also the lack of enforcement of IC precautions among visitors. This was felt to pose a danger to HCW. There seemed agreement among members of the group that visiting hours, and numbers of visitors, be restricted as they have been in the past.

■ Contamination of surfaces in the hospital: In conjunction with the above point, there was a fear that common areas were contaminated (e.g., common bathrooms without automatic taps and with a lack of paper towels for turning off faucets).

■ Differential treatment from other HCWs: The makeup of this focus group appeared to be largely support staff, or non-patient-care staff. The group felt that they were treated differently from patient care staff in terms of communication of information regarding infection control procedures. They also seemed to feel that their concerns were not listened to — that procedures and policies that would protect them from infection were not in place (i.e., transporting soiled laundry). This led to a lack of trust in the management of the organization.

■ Communication: The participants in this group felt that communication of infection control policies and procedures needed to be improved.

Suggestions

■ Temperature logs for inpatients should be scrutinized by HCWs for the previous twenty-four hours (concern that elevated temperatures on other shifts were being missed).

■ Have standardized screening tools for infection in the hospital (during outbreaks).

■ There should be screening at all entrances (not just the emergency room) for all persons entering the hospital.

■ There should be tighter restrictions on visitors' access to hospital, perhaps even banning all visitors altogether (no children < 13, 2 visitors at a time, set hours).

■ Need to publish an analysis of the SARS outbreaks and circulate to staff, create up-to-date policies and procedures for IC.

■ In-services should be short, pertinent, in unit of work, and with sufficient notice.

■ An annual course in IC is needed with requirements for certification.

■ Fit testing should be done at the start of employment for all new employees, and orientation for new employees should include a section on IC.

- Need for reeducation of PSAs in IC procedures.
- Pedestal sinks are preferred, or sinks with automatic sensors (where do not need to touch handles).
- Build a new hospital only for infectious diseases, that is well-equipped, and designated as a "respiratory hospital" or an "infectious diseases" hospital.
- "Somebody unplug the public purse": health care is expensive and needs to be funded appropriately in order to prevent future outbreaks.
- Measures for danger pay/compensation should be consistent across all hospitals.
- Danger pay should not be "blanket" but tailored to risk and exposure (should be similar to overtime instead of double or triple time; or use other options such as days off with pay). There is a need for ongoing education for the community, and orientation for patients admitted to the hospital IC.

Participating Institutions
Ontario

Children's Hospital of Eastern Ontario
Credit Valley Hospital
Lakeridge Health Corporation
MDS Laboratory Services
Markham-Stouffville Hospital
Ministry of Labour
Mount Sinai Hospital
North York General Hospital
Ontario Nurses Association
Orthopaedic and Arthritic Institute
Scarborough General Hospital
Scarborough Grace Hospital
St. John's Rehabilitation Hospital
St. Joseph's Health Centre
St. Michael's Hospital
Sunnybrook and Women's College Health Science Center
The Ottawa Hospital
Trillium Health Centre
University Health Network
West Park Healthcare Centre
William Osler Health Centre

British Columbia

Providence Health Care
St. Paul's Hospital
Surrey Memorial Hospital
Vancouver General Hospital
UBC Hospital

Chapter 6

Ventilation Systems for Handling Infectious Diseases in a Healthcare Environment

Stephen J. Derman

CONTENTS

Introduction

"The thing that kills women with (childbirth fever) … is you doctors that carry deadly microbes from sick women to healthy ones," said Louis Pasteur during an 1879 seminar at the Academy of Medicine in Paris.[1] Though Pasteur was referring to the lack of hygienic practices as a cause of disease, a practice that continues to this day, an inordinate number of diseases are attributable to nosocomial infections — those that originate or occur in a hospital or hospital-like setting — including airborne infectious diseases.

Appropriate ventilation in hospitals and healthcare facilities is a key component for protecting patients and healthcare workers from the transmission of airborne infectious diseases. Tuberculosis, SARS, and avian influenza have received attention due not only to their ability to be transmitted in the air but also for their virulence. In the event of an epidemic or pandemic, where a hospital or other healthcare facility will receive a large influx of patients, those facilities will likely be unable to safely treat such patients or to protect other patients, healthcare workers, and visitors.

In the healthcare environment, as well as in non-health-care environments, a disease can be associated with a specific mode of transmission. Infectious diseases in the healthcare environment are generally transmitted via airborne mechanisms that make physical contact with surfaces that harbor infectious organisms. Control of exposure by physical contact is usually accomplished by an attempt to isolate the source, utilizing antimicrobial agents on the host and the receiver, and by attempting to isolate the host from the receiver.

According to Eric Toner, droplet spread is the primary route of transmission for influenza, a typical airborne infectious disease that is transmitted by droplet nuclei. The virus can be transmitted by large droplets, which generally travel three to six feet, and by small droplet nuclei and aerosols thjat can remain suspended in the air for longer periods of time and travel greater distances from their source.[2]

For several reasons, hospitals are incapable of providing sufficient quantities of negative pressure rooms for controlling airborne transmissions through their central heating, ventilation, and air conditioning (HVAC) systems should a surge of patients occur. In addition, they have a very limited number of isolation rooms that are used to safely house those patients. Several hospitals recently have started developing plans to accommodate a greater number of patients, yet most plans are inadequate for five to five hundred patients. In addition, unless required by certain local or regional requirements, hospitals are required to have only one isolation room (considered adequate to house one patient). During an epidemic or pandemic one isolation room per facility would be insufficient. The building codes continue to allow older facilities to follow more relaxed requirements than newer facilities.

The Department of Homeland Security has been prompting hospitals to reevaluate their emergency preparedness, primarily for a terrorist threat. Biosafety issues, including bioterrorism, are included in this appraisal. The Joint Commission on the Accreditation of Healthcare Facilities (JCAHO), also responding to emergency preparedness concerns has been evaluating hospitals' emergency preparedness with greater scrutiny.[3]

Compliance with many existing standards at many hospitals and healthcare facilities could be difficult to achieve. Infection control procedures must be diligently followed. Staff must properly don and utilize personal protective equipment. The proper use of respiratory protection as part of a comprehensive Respiratory Protection Plan still continues to haunt healthcare facilities.[4] Effective in the second half of 2004, employers using respiratory protection have been required to annually fit test their employees. The standard also calls for medical evaluations, training, and workplace surveillance. According to a healthcare safety and health expert, many hospitals would not be in compliance with several components of the standard.[5] In 2003, lack of preparedness in several Asian countries and Canada caused the SARS outbreak to escalate in healthcare facilities.

The Centers for Disease Control (CDC) states that healthcare facilities should allow adequate time for the air handling system to clean 99% of airborne particles from the air (describing more infectious types of procedures or situations). The time period to achieve that safety factor ranges from 46 minutes at 6 air changes per hour (ACH), to 23 minutes at 12 ACH. Fifty percent more time is required for 99.9% removal efficiency. It should be noted that these assumptions assume perfect air mixing and no aerosol-generating source: impossible scenarios in a healthcare environment with patients. The CDC continues that with an infectious patient in a room, coughing, breathing, or sneezing, perfect mixing does not occur. Caution should be exercised and a healthcare worker should allow additional time prior to reentry before the air can be cleared of the infectious agent(s).[6] One must contemplate how much time a healthcare worker will allow before either entering such a room, bearing in mind that the HCW will already be stressed with an increased workload and sometimes have had at best questionable training,. During a surge or of patients, it is conceivable that some procedures may not be followed.

The Problems Including the Surge Scenario

The list of diseases that are transmitted in the air is extensive: they include the common cold, SARS, aspergillosis, valley fever, pneumonic plague, measles, Hansen's disease (leprosy), polio, anthrax, influenza, Legionnaire's disease, and tuberculosis. Avian influenza and the hemorrhagic fevers are believed to also be spread through the air, though the circumstances could be more extenuating. (There are several modes of transmission of avian influenza: the primary source is believed to result from direct contact; however, airborne exposures are also possible. Marburg, Lassa, and Crimean–Congo hemorrhagic fever viruses can spread from one person to another, once an initial person has become infected, at times from another host).[7] Though the etiologic agents, modes of transmission, symptomatology, and outcomes vary, the fact remains that these and other airborne diseases continue to pose a risk to the general population.

The healthcare environment is inherently risky for patients, visitors, and staff. There are over 100,000 deaths per year in the United States attributed to nosocomial hospital infections.[8] Most patients generally do not come into a facility while they are healthy. Thus, when they do enter the hospital their immune systems may well be compromised and the individual will be be more prone to becoming infected. They may have an infectious disease and infect others, including other patients, visitors, and healthcare workers. These individuals could in turn pass the disease to not only individuals in the healthcare environment, but to friends, family members, and other people they associate with, either formally or casually. And the infectious disease will continue to spread.

During an epidemic or pandemic, a surge of patients is expected to arrive at a healthcare facility. They will first be triaged in an area that normally is not controlled for pressure differentials; therefore, migrating viral or bacterial particles will begin to spread through the facility. In several large cities, emergency rooms have at best only one to two negative pressure isolation/trauma rooms; that is insufficient to control the volume of patients needing treatment.

During major epidemics influenza hospitalizations for high-risk persons may increase between two- and fivefold,[9] placing healthcare workers at increased risk of infection. Small infective doses are thought to be responsible due to the rapidity with which the disease spreads through a population. Couch et al. studied natural airborne transmission of respiratory infection with Coxsackie A virus type 21. Using two groups of adult volunteers — one infected with the virus and the other noninfected and antibody free — separated by a double walled, wire screen four feet wide, transmission of infection was demonstrated on day six. [10]

Coughing patients can produce high amounts of viral or bacterial particles. The 2001 Guidelines for Design and Construction of Hospital and Health Care Facilities states that emergency waiting areas, bronchoscopy suites, and triage areas be under negative pressure and have a minimum of twelve air changes per hour.[11] Prior to the publication of this standard, the requirement was six air changes per hour. Because this standard became effective in 2001 and most hospitals are more than four years of age, it is likely that they do not provide higher rates of dilution ventilation to greatly decrease transmission rates. Therefore, ERs and other treatment and diagnostic areas could be excellent vectors of transmigration. Once patients are seen and admitted to medical wards, there will likely be an acute shortage of holding, diagnostic, and treatment rooms.

The Association of Heating, Refrigeration and Air Conditioning (ASHRAE) Standard calls for bronchoscopy suites to be negative with 12 air changes an hour. How many in the nation are actually up to code is questionable at this time.[12] The 2001 American Institute of Architects (AIA) Guidelines and 2005 ASHE Proposed Ventilation Standard call for radiology (x-ray) rooms to have between 6 and 15 ACH. Prior to 2001 the Standard was 6 ACH for treatment rooms; the newer standards clarify 15 ACH in surgery, critical care, and catheterization rooms. Given these disparities in pressure controls, hospitals older than 2001 could act as transmission vectors for diseases and could increase the amounts of transmissions to patients and staff.

Pressure relationships in healthcare traditionally were focused on fire and smoke control. Plenums and elevator shafts, stairwells, smoke dampers and barriers have been tightly controlled and inspected by regulatory agencies over the years. However, pressure control as a method of reducing and controlling for nosocomial infection or emerging infectious diseases has been underutilized both as theory and practice. Very few healthcare facilities in the United States are equipped with adequate numbers of negative pressure rooms with antechambers. The numbers of hospitals capable of using their central HVAC system to provide additional negative pressure at 0.01 inches of water gauge is unknown.[13]

Maintenance of the central HVAC systems is also crucial to proper functioning and providing safety factors. Streifel has commented that the maintenance of central HVAC systems is often questionable and leads to malfunctioning. Room leakage, fan belt slippage, filters unchanged at regular intervals, water pooling in fan rooms creating mold intake, systems not cleaned at regular intervals, all lead to potential down stream problems. He adds, "The management of the mechanical systems in health care becomes critical for the safe environment of care."[14]

During the TB outbreak in the 1990s, California hospitals did create negative pressure rooms as per the CDC guidelines. However, when they were tested for functionality, a good percentage were found to be not really negative to corresponding corridors. The California Department of Health Services study found a lack of full compliance with CDC guidelines at all hospitals studied. The study showed that many of the rooms tested were actually under positive pressure, that the pressure differentials were undocumented, and that work practices varied among hospital staff.[15]

Transmission

Because of the potential for a moderate to a large influx of patients with any number of infectious diseases, facilities must develop contingency plans for effectively dealing with this situation. Several types of controls must be considered and implemented when dealing with this and other types of emergencies. Plans need to be developed with state and local public health agencies and other healthcare organizations. Each facility must develop a coordinated multidisciplinary response plan to deal with an emergency. All affected departments should be involved.

Human sources of the infecting microorganisms in hospitals may be patients, family members, visitors, or other healthcare personnel. These individuals could be infected with an acute disease, they could be in the incubation period, they can be infected but have no apparent disease, or they can be chronic carriers of an infectious agent.

Once a patient enters the health care environment, they are evaluated and a course of treatment is prescribed. The CDC has categorized the risk that patients could pose to others into two categories: standard precautions and transmission based precautions. They are described below.[16]

Standard Precautions

Standard precautions combine the major features of universal (blood and body fluid) precautions (designed to reduce the risk of transmission of bloodborne pathogens) with body substance isolation (designed to reduce the risk of transmission of pathogens from moist body substances). Standard precautions apply to (1) blood; (2) all body fluids, secretions, and excretions except sweat, regardless of whether or not they contain visible blood; (3) nonintact skin; and (4) mucous membranes. Standard precautions are designed to reduce the risk of transmission of microorganisms from recognized and unrecognized sources of infection.

Transmission-Based Precautions

Transmission-based precautions are designed for patients documented or suspected to be infected or colonized with highly transmissible or epidemiologically important pathogens for which additional precautions beyond standard precautions are needed to interrupt transmission in hospitals. There are three types of transmission-based precautions: airborne precautions, droplet precautions, and contact precautions. They may be combined for diseases that have multiple routes of transmission. They must be used in addition to standard precautions.

In addition, there are specific syndromes in both adult and pediatric patients that are highly suspicious for infection and identify appropriate transmission-based precautions to use on an empiric, temporary basis until a diagnosis can be made. These precautions also must be used in addition to standard precautions.

There are five main routes that microorganisms are transmitted; they are contact: droplet, airborne, common vehicle, and vectorborne.

Droplets are generated from the source primarily during coughing, sneezing, and talking, and during the performance of certain procedures such as suctioning and bronchoscopy. Transmission occurs when droplets containing microorganisms generated from the infected person are propelled a short distance through the air and are deposited on the host's mucous membranes. Because droplets do not remain suspended in the air, special air handling and ventilation are not required to prevent droplet transmission. Please note that droplet transmission must not be confused with airborne transmission.[17]

Airborne transmission occurs by dissemination of either airborne droplet nuclei (small-particle residue {5 μm or smaller in size} of evaporated droplets containing microorganisms that remain suspended in the air for long periods of time) or dust particles containing the infectious agent. These microorganisms can be dispersed widely by air currents and may become inhaled by a susceptible host within the same room or over a longer distance from the source patient, depending on environmental factors; therefore, special air handling and ventilation are required to prevent airborne transmission. Microorganisms transmitted by airborne transmission include *Mycobacterium tuberculosis* and the rubeola and varicella viruses.

Isolation precautions are designed to prevent transmission of microorganisms by these routes in hospitals. Because agent and host factors are more difficult to control, interruption of transfer of microorganisms is directed primarily at transmission. The recommendations presented in this guideline are based on this concept.

Placing a patient on isolation precautions usually presents certain difficulties to the hospital, patients, personnel, and visitors. Isolation precautions may require specialized equipment and environmental modifications that add to the cost of hospitalization. Isolation precautions may make frequent visits by nurses, physicians, and other personnel inconvenient, and they may make it more difficult for personnel to give the prompt and frequent care that is sometimes required. The use of a multipatient room for one patient uses valuable space that otherwise might accommodate several patients. Moreover, forced solitude deprives the patient of normal social relationships and may be psychologically harmful, especially to children. These drawbacks must be weighed against the spread of serious and epidemiologically important microorganisms.

Solutions

Contact and droplet precautions have been shown to be effective in preventing transmission of several respiratory pathogens, including influenza. Hospitals should consider institution of control measures including masking all patients with respiratory symptoms in emergency departments, admitting areas, and waiting rooms, and initially managing all inpatients with febrile respiratory illnesses with droplet/contact precautions until they know otherwise.

Lessons learned from the SARS model are currently considered a model for patient care and preventing an outbreak of highly communicable diseases in the healthcare environment. The reader is encouraged to review appropriate infection control procedures in addition to the ventilation controls described here.

1. Determine where and how patients will be triaged, evaluated, diagnosed, and isolated.
2. Only admit patients when medically indicated or if appropriate community isolation is not possible.
3. Determine where suspected SAS patients will have respiratory specimens collected. These areas should have negative air pressure and the capacity for decontamination and disposal of waste.
4. Identify appropriate paths, segregated from main traffic routes as much as possible, for movement of patients. Determine how they will be controlled (e.g. dedicated patient corridors, elevators).
5. Optimize necessary patient transport by:
   a. Ensuring that transport staff use infection control precautions.
   b. Ensuring that receiving locations are prepared for the arrival of those patients.
6. Ensure that patient flow is unidirectional to prevent higher risk patients from going back through waiting areas.
7. Ensure that staff flow will prevent healthcare workers from crossing between high and low risk areas.
8. Determine how to cohort potentially exposed or symptomatic patients in the event that significant disease transmission has occurred in the facility.
9. Ideally, patients with suspect or probable highly infectious diseases should be admitted into airborne infection isolation rooms (AIIRs).

However, if these rooms are spread out through the facility or if there are a large number of patients, facilities may choose to cohort patients onto specific nursing units that have been modified to accommodate those patients. Cohorting patients, rather than placing them in AIIRs throughout the hospital, has the advantages of physically isolating these patients from nonspecific communicable disease patients; this may make dedication of staff easier. Additionally, experience in Taiwan and Toronto has shown that cohorting SARS patients is a highly effective method of interrupting transmission. However, even when a dedicated unit is created, there are some cases in which AIIRs are preferred:

a. Patients who are known to have infected other people should be placed in AIIRs given the potentially increased risk of transmission.
b. Patients in whom the risk of SARS is being assessed should be housed in AIIRs rather than on the SARS ward to minimize the risk that they will acquire SARS. They can be moved to the SARS ward if they are deemed likely to have SARS.[18]

Ventilation Controls

Cody and Fenstersheib further described the following engineering controls in attempting to control SARS:[19]

1. Patient room engineering controls
    a. Determine current capacity for isolating SARS patients for medical, pediatric, and ICU settings (e.g., AIIRs).
    b. Determine how rooms designated for SARS care will be modified to achieve appropriate airflow direction or air exchanges.[20]
    c. Determine how airflow/negative pressure will be verified and monitored.
    d. Unit and facility engineering controls
        i.   Establish and maintain a pressure gradient with airflow from the "cleanest" (nurses' station) to the "least clean" (patient room) area.
        ii. If possible, nurses' stations should be configured so that HCWs do not need to wear complete PPE.
        iii. Determine the best location in the hospital and how to modify existing rooms/wards/floors to develop a "SARS Ward" where patients and the staff caring for them could be cohorted within a healthcare facility. Ideally, this location would have the following characteristics:

2. An air supply that is separate from other areas of the facility and one where a negative pressure ward could be created.
   a. Rooms that could also be converted to negative pressure to the hallway.
   b. Identify a separate designated space for a SARS evaluation center ("Fever Clinic") associated with the healthcare facility. This center may be a temporary structure or utilize existing structures. The purpose of this center is to separate potential SARS patients from other patients seeking care at a healthcare facility. Determine needed ventilation, restroom facilities, and water supply. for the center. Also determine appropriate patient traffic routes for patients who must be taken from the evaluation center to the healthcare facility.
3. Develop plans for alternative measures for containment and engineering controls if the number of SARS patients exceeds the isolation capacity (e.g., portable HEPA filtration units).
   a. Depending on the number of SARS patients, disease acuity, importance of cohorting, and availability of airborne isolation rooms (AIRs), consider constructing new AIIRs or temporary anterooms.
   b. Consider cohorting admitted patients onto designated SARS units depending on personnel and availability of AIIRs.
   c. Construct barriers to prevent movement and access.
4. Prior to entering the isolation ward, healthcare workers should exit through a buffer change room and don the appropriate PPE ensemble (Gown/Tyvek with hood, gloves, shoe covers, face shield/goggles) prior to entering the isolation ward and the isolation room.
5. Prior to opening the isolation ward, a comprehensive commissioning of the ventilation system should be conducted; this should include a complete test and balance of the system to verify the design parameters are met, and to measure and record exhaust and supply volumes, damper/louver positions, fan operation, and indicator panel status. Additionally, the final acceptance should include visual verification of negative pressure in the following sequence of increasingly negative pressure areas: nurses station; corridor; isolation room. Visible smoke should be used to ensure that airflow is in the proper direction in all areas (from clean to "less" clean areas) prior to accepting the system.
6. Healthcare workers, custodial staff, and maintenance personnel should be trained on the proper function of the negative pressure system, including the effect of opening doors, blocking vents, etc. on ventilation performance. Infection control procedures and the

ventilation system are complex, and require strict adherence to good work practices in order to maintain the integrity of the controls.[21]

According to the CDC,[22] the following protocols describe additional ventilation controls:

Airborne Infection Isolation Rooms (AII)

Ventilation for airborne infection isolation rooms shall meet the following requirements:

1. Provide a continuous differential air pressure monitor that will alert clinical staff when differential pressure is not maintained.
2. All air from the airborne infection isolation room shall be exhausted directly to the outside.
3. All exhaust air from the airborne infection isolation rooms, associated anterooms, and associated toilet rooms shall not be combined with any other nonairborne infection isolation exhaust system.
4. Exhaust grilles or registers in the patient room shall be located directly above the patient bed on the ceiling or on the wall near the head of the bed.
5. Outside air intakes for air handling units shall be located a minimum of 25 feet from contaminant sources.[23]

Protective Environment (PE) Rooms

Ventilation for PE rooms shall meet the following requirements:

1. Provide a room envelope that is well sealed.
2. Provide a continuous differential air pressure monitor that will alert clinical staff when differential pressure is not maintained.
3. Air distribution patterns within the protective environment room that conforms to the following:
   a. Supply air diffusers shall be generally above the patient bed.
   b. Return/exhaust grilles or registers shall be located near the patient room door.[24]
4. Protective environment rooms shall remain under positive pressure with respect to all adjoining rooms whenever an immunocompromised patient is present. Protective environment rooms shall be tested for positive pressure daily when an immunocompromised patient is present. When HEPA filters are present within the diffuser

of protective environment rooms, the filter should be replace based on pressure drop.[25]

Airborne Infection Isolation (AII) Rooms

Airborne infection isolation rooms shall remain under negative pressure relative to all adjoining rooms whenever an infectious patient is present. They shall be tested for negative pressure daily whenever an infectious patient is present.[26]

Local Exhaust Ventilation

Portable air handling units have come on the market quite recently. Their primary advantage is that they can supply additional air changes, filtration, or help create a negative pressure room in a room that was not designed for that purpose. Charney tested portable air HEPA units at San Francisco General Hospital during the TB outbreak in San Francisco and found the portable units to be efficient at providing both negative pressure and additional air scrubbing.[27] The HEPA technology, now upgraded and supported by ULPA filtration, has been available for many decades, used by NASA, and approved for many different uses within health care. These units were challenged, tested, and showed very efficient scrubbing using a surrogate aerosol as a test agent. The portable machines can be fitted into the existing exhaust plenum of the hospital or exhausted out of a patient room window, as the exhaust side of the filter showed no breakthrough of test agent aerosol during testing. Given a median size patient room of 100 cubic feet and six air changes an hour with neutral pressure (before installation), a 500 cfm capacity portable scrubber can achieve a more desired 12 to 15 air changes of scrubbing, thereby decreasing the airborne microorganisms.

The latest challenge test results published by Mead et al.[28] demonstrated that the best performing designs showed no measurable source migration out of the inner isolation zone and had mean respirable particle counts up to 87% lower in healthcare worker positions. Investigators concluded that with careful implementation under emergency conditions in which engineered isolation rooms are unavailable, expedient methods can provide affordable and effective patient isolation while reducing exposure risks and potential disease transmission to healthcare workers, patients, and visitors.

Conclusion

At the present time hospitals and their current ventilation systems are not prepared for a virulent emerging infectious disease influx. All areas are problematic, as healthcare facilities are not equipped to safely contain a surge paradigm. Healthcare facilities have the potential to be vectors for transmission rather than control areas, Therefore, transmissions will not only occur from patient to patient but from patient to healthcare worker as viral or bacterial particles will transmigrate throughout the facility. Most HVAC systems are not designed to have scrubbing potential or highly efficient air changes per hour that could dilute at 99% of aerosolized organisms within a rapid rate to prevent transmission. Hospitals at this point are underdesigned for ventilation for an emerging pathogen, and too little attention is being paid to this deficiency.

Air handling systems can be designed to not only provide comfort (for both temperature and humidity) but also provide controls to reduce and eliminate contaminants in the air (chemical and microbial, as appropriate). This chapter was not designed to regurgitate the existing standards, but to describe best current practices used in the control of infectious diseases; readers are encouraged to review the most current documents on this subject including those described in the endnotes.

DETERIORATION OF THE PUBLIC HEALTH SYSTEM AND THE SQUANDERING OF LIMITED RESOURCES

IV

Chapter 7

Infectious Diseases, A Resurgent Problem: Developing Effective Public Health Responses

George Avery, Ph.D., MPA

CONTENTS

Since the 1960s, the United States, like much of the rest of the world, has seen a decline in the ability of its public health system to address the threat of infectious disease.[1] Historically, policies to address infectious disease have been motivated by fear.[2, 3] In the late twentieth century, fear of infectious disease was lost to advances in medicine and public health. These advances resulted in the elimination of smallpox, the discovery of effective antibiotics and vaccines, and sanitation improvements that together reduced the death rate due to infectious disease by 90%. However, as Senate Majority Leader Bill Frist noted,

> To have believed with the Surgeon General forty years ago that the great advances of biological science were capable of permanently suppressing infectious disease was to have been unaware that these triumphs were appropriate to only one phase in the life of a continually evolving enemy whose natural rate of evolution and adaptation is far greater than our own. [4]

The decline of emphasis on infectious disease coincided with severe deterioration of the U.S. public health infrastructure. By the late 1980s, the public health service was subjected to significant social and cultural discounting, with subsequent neglect and deterioration in its capabilities.[5,6] By the early 1980s, states were ceasing tuberculosis surveillance at a time when drug resistant strains were arising. Over the next two decades infectious disease mortality rates, after discounting for the AIDS epidemic, rose 22%.[7] In 1988, the Institute of Medicine (IOM) warned of the risks of deterioration and suggested a need to rebuild the system.[8] By the 1990s, the system had deteriorated to a point where responses to infectious disease threats could best be characterized as uncoordinated and "jury-rigged."[9] State public health laboratories declined to the point of having difficulties with small outbreaks,[10] and disease surveillance systems were facing similar weaknesses.[11] In 1992, the IOM further warned that the public health system had continued to deteriorate, and stressed that emphasis needed to be placed on the prevention of infectious disease.[12]

Infectious Disease: A Resurgent Problem

Since the mid-1970s, the emergence of a number of new pathogens and reemergence of older diseases has highlighted the fact that, contrary to expectations, epidemics of infectious disease remain a problem of public health concern.[13,14,15] Infectious diseases remain the largest global cause of death.[16] The United States suffers 325,000 annual hospitalizations and 5,000 deaths from food-borne illness alone.[17] Social and climatic changes in the twentieth century have resulted in changes in disease ecology allowing penetration of infections such as AIDS and West Nile fever to new niches.[18, 19, 20, 21] Antibiotic prescribing patterns have resulted in the emergence of new strains of common bacteria resistant to available treatments.[22] Syphilis rates have begun to rise, particularly among homosexual men, raising concerns of a new increase in the incidence rate for HIV.[23] The emergence of a highly pathogenic avian influenza strain in Southeast Asia, which has been compared to the 1918 "Spanish" influenza strain, warns of the potentially devastating impact of a new influenza pandemic.[24,25,26]

Responding to an epidemic disease crisis requires not just a public health infrastructure, but a public health *system* capable of providing a coordinated response involving federal and state agencies, health and nonhealth agencies, as well as public health departments and private health care providers.[27] Lessons learned from the 1999 West Nile fever emergence provided evidence that an effective response required laboratory diagnostic capabilities, a surveillance system involving cooperative private practitioners, infrastructure for field epidemiological investigations, and the capacity to mobilize resources for treatment and preventive interventions.[28]

The current public health system has significant deficiencies that prevent it from adequately responding to either bioterrorism or other epidemics. According to the General Accounting Office, the basic capacity for infectious disease surveillance is lacking.[29] Time lags, communication difficulties, and personnel shortages render the system unable to respond rapidly.[30] Systematic problems are seen in terms of funding levels, workforce training, and adequacy of public health laws.[31] Thirteen states lack an epidemiologist and 18% of public health laboratory positions are vacant. Over 40% of those epidemiologists lack training in the field.[32] One-third of states lack the staff to investigate outbreaks of food-borne diseases.[33] Resources entering the public health system for modernization are largely from the federal level, and heavily targeted at the bioterrorism issue. Fiscal difficulties are resulting in reduced support for core public health tasks.[34] In Colorado, the state cut all funds for local health departments as a result of the influx of federal bioterrorism money, leaving some departments with funds for bioterrorism experts, but no budget for routine public health activities.[35] Over a quarter of local health departments in the United States have outsourced their communicable disease control functions.[36] Local jurisdictions are having difficulty in gaining involvement from hospitals and medical personnel in training and planning activities.[37] These deficiencies were observed during the 1999 West Nile fever outbreak in New York where a relatively small outbreak taxed the resources of the largest state, local, and federal public health departments.[38]

The Problem of Coordination

Developing effective policy responses to national problems often depends on coordination across not just agencies but also, in a federal system, across political jurisdictions. The problem of coordination arises because, as Pressman and Wildavsky note, "Everyone wants coordination — on his own terms."[39] Rational participants in the intergovernmental policy arena seek to maximize the utility of public programs according to their own local and individual needs.

March described a model for a "garbage-can" process wherein organizational actions occur in an environment characterized by a changing array of actors with a multiplicity of individual objectives.[40] Such processes are commonly found in areas where organizations have ambiguous processes, participation, and presences.[41] Within the framework of these processes, the attention given to a particular decision is dependent upon both the nature of the decision and alternative claims on the decision makers. Given the differing, and changing, perspectives on actors, decision making is often decoupled from the properties of the decision circumstance.[42] Solutions and problems exist simultaneously, and are only coupled when an opportunity arises that forces or allows the organization to do so. The organization will choose a satisfying solution simply because an optimal option is temporally dislocated. Likewise, problems are often identified only because a solution exists. Garbage-can processes within different structure have been found to produce differing performances and results. Structural features and organizational norms produce systematic bias in decision-making processes.[43] The rules and operating procedures that form the structural aspects of the organization are typically developed to reduce the ambiguity and attention demands of organizational processes. These routines bridge the gap between organizational structure and action by constraining the possibilities faced by the organization.[44]

In terms of complex system theory, a garbage can is viewed as a stable and adaptive self-organizing system. Such systems are generally found in an area of a second-order phase transition between ordered and chaotic systems. A physical analogy would be the phase transition state between a crystalline solid such as ice and liquid water, or a spin glass. Rigid systems fail to adapt because they are effectively unable to utilize information about their environment. Chaotic systems are so reactive to the environment that the ability to organize is compromised by information overload. In the transition state, the system has the flexibility to adapt, but a framework in which the ability to utilize information is maximized. [45]

A complication arises in complex organizations when conflict occurs between component groups. When conditions of a perceived need for joint decision making and a difference in goals and problem perception exists, the situation is ripe for conflict. When the pressure for joint decision making is unilateral, resistance typically occurs, which leads to conflict.[46] This conflict takes place because of differences in organizational paradigms. What may be a pressing need for one agency may not fit within the worldview of another. In a practical sense, the implementation of an existing solution is dependent on the acceptance of all agency partners. Solutions outside of an agency paradigm will not permeate the organizational information filters. Thus, "jointness" depends on a shared perception of the decision environment.

Game theory would assume that competing agencies would, through rational actions, adopt a course of action designed to minimize the costs to both parties.[47] Arthur argues that under conditions of complication, rationality breaks down. Human logical capability ceases to cope. Human agents cannot rely on perfect rationality and thus are forced to guess the behavior of colleagues, and shared assumptions cease to apply. This results in a loss of problem definition and a switch to inductive behavior. Problems are translated into temporary models, which are adapted according to environmental feedback. Agents linger with their most believable local hypothesis until it is no longer functional.[48] Cosmides and Tooby have argued that such specialized tools often function in a manner better than deductive rationality. Cognitive specializations trigger which inferential model will be used and which information will be communicated and accepted.[49]

In a crisis atmosphere, operational and strategic levels of decision making exist. The levels differ in their outlook, interests, policy orientations, and distance from the problem they share. The attendant administrative tension helps shape the degree of decentralization. Time pressure at the operational level and overload at the central level invite a formal decentralization of decision-making responsibility. Bureaucratic politics ensures that a sole effective center of decision making rarely exists.[50]

Graham Allison, in his study of decision making during the Cuban Missile crisis, examined the problem of an organization in which the various actors form a constellation of allied agencies that act according to these individual biases. Coordination of the behavior of these actors requires a set of ground rules, or standard operating procedures. Each agency functions within a bounded rationality defined by the circumstances of the individual agency and its routines. Each has separate criteria of acceptable performance, and responds within their individual perspectives. Agencies try to avoid uncertainty by negotiating the environment, and react to unavoidable uncertainty within the framework of the agency culture. Dramatic change only occurs when forced by failure or resource constraints, or through resource excess that allows a widening of the attention sphere.[51]

Public Health and the Coordination Problem

Leadership and coordination problems, as much as funding, are responsible for the decline in the public health system. In the United States, weak leadership and resource scarcity aggravate the coordination problem. Scutchfield et al. have found that leadership and the ability to form partnerships outside the agency are key determinants of public health agency performance.[52] From the mid-1970s, the ability of public health advocates to actually influence policy has steadily eroded.[53] At the state level, agencies are criticized for failure to take leadership on health issues, excessive caution in making decisions, and frequent leadership turnover.[54] Similar criticism has been made regarding federal officials.[55] Traditional public health has sunk into a distant second place role in the provision of health care, a position that hinders efforts to raise public health issues to importance as policy issues.[56] A further complication in developing a coordinated response to public health emergencies is that the system is legally and structurally fragmented. Public health powers are derived from state police powers, with only weak ability for the national government to coordinate policy development and implementation.[57]

The weakness of public health leadership has hindered the ability to develop a system capable of meeting infectious disease threats. Kingdon describes a policy process where discrete streams of problems, policies, and political actors are coupled in "policy windows," or critical moments when policy entrepreneurs can successfully match problems to policy solutions.[58] By the mid-1990s, public health advocates had developed solid technical proposals to upgrade the public health system, but lacked sufficient political strength to create windows for implementing these solutions.[59] The window had yet to open.

By the late 1990s, the policy stream for advancing these goals contained two proposed policy solutions, revolving around different issues with the potential for exploiting public unease to advance rebuilding the public health infrastructure. One solution, based on previously unknown diseases such as AIDS and Ebola, which emerged while rates of known epidemic diseases such as tuberculosis were growing, emphasized the need for non-disease-specific improvements in the infrastructure for surveillance, research, training, and prevention of infectious diseases.[60,61,62]

The second strategy focused on using a program to respond to threats from biological weapons as a vehicle to obtain structural improvements to the public health system.[63, 64] In 1999, prompted by the 1995 use of the nerve agent Sarin by Aum Shinrikyo in Tokyo and subsequent revelations that the group had attempted to use anthrax spores, the U.S. government launched an Antibioterrorism Initiative. This program addressed issues related to infectious disease control and surveillance, but in the context of responding to an attack with infectious agents that had historically been developed as military weapons.

The Association of State and Territorial Health Officers has argued that an effective bioterrorism response requires general improvements in the public health system.[65] Effective postattack management of bioterrorism involves traditional public health responses to the event of recognition, intervention, prevention of further casualties, and public reassurance. Joseph McDade of the Centers for Disease Control argues that improvements targeted by the U.S. Department of Health and Human Services bioterrorism program benefit the ability of the public health system to manage other infectious diseases by rebuilding capacity that eroded in the 1960s and 1970s.[66] Other advocates make the same argument.[67]

Although adoption of this strategy has resulted in the allocation of significant resources, questions remain about whether it is the appropriate way to address the problem of preparedness for epidemic emergencies. Critics argue that the trickle-down effect from the focused bioterrorism program is less effective than general infrastructure improvements, and may divert resources from general public health efforts that have greater impact.[68,69,70] Much of the appropriations are diverted to programs of dubious value in protecting public health; program focus is targeted on traditional military agents of biological warfare, and the program definition as a "security" issue rather than a public health issue has resulted in response leadership and a secrecy culture that have hindered, rather than aided, the response to real incidents.[71,72,73]

A large part of the blame can be placed on the patchwork manner in which preparedness policy has been developed and implemented, which is largely due to the direction of the policy stream. Drexler[74] notes that the definition of the problem as a security rather than a public health issue defines the nature of the political debate, leading to "budget brinkmanship" games rather than a reasoned analysis of the problem and potential solutions. The framing of a policy issue results in the privileging of some strategies over others, as was also seen in the 2001 foot-and-mouth disease epidemic in the United Kingdom.[75] While entrepreneurs have been successful in spurring action, the mismatch between the problem and the solution has resulted in most new resources going to unintended uses. Thus, less than 1% of the $9.7 billion fiscal year 2001 terrorism budget went toward the public health infrastructure. As one prominent advocate of preparedness programs notes, most of the funded projects for bioterrorism preparedness would actually play no role in a real incident.[76]

The focus on bioterrorism results from political pressures that create a climate where political actors must act to show that they are addressing a threat.[77] This creates an environment where security and law enforcement agencies are given an outsized role in addressing public health problems, but may have organizational priorities incompatible with health needs. Thus, interorganizational conflict is almost a certainty in a crisis, and response pathways are shaped by conflicting demands from these agencies. Thus, it can be anticipated that the ability of the program to respond to public health problems will be handicapped and constrained by structures imposed by security concerns. As Garrett has noted, as a result of the bioterrorism focus, public health has "boarded the train, but only as a passenger."[78,79]

An example of the distortion of public health planning imposed by the bioterrorism focus can be found in planning to deal with an influenza pandemic. Prior to the 2001 anthrax attacks, planning for an influenza epidemic was considered as a good model for bioterrorism preparedness. After the attacks, the situation reversed, despite the fact that an influenza pandemic would likely arise on a far greater scale than a biological warfare incident.[80]

The definition of a problem shapes the perception of it, the agencies involved, and organizational perceptions of what procedures are appropriate to manage it. A definition as security issue typically would involve more agencies than a public health problem definition, and those organizations have different assumptions, as was seen in the 2001 anthrax event. In that case, failure to prioritize public health aggravated the problem.

Among other issues, security and law enforcement agencies tend to compartmentalize information, which may keep data from those managing consequences. The Department of Homeland Security, with a key role in bioterrorism planning and civil defense, is already showing tendencies toward compartmentalization, excluding from planning many experts who lack existing security clearances, and demonstrating little understanding of the integration of civilian entities into operations.[81] As Moynihan noted a culture of openness in government, even in security affairs, improves efficiency by emphasizing analysis and full exploitation of information.[82] Similarly, t'Hart and colleagues argue that in crisis situations and in preparing for a crisis, centralized control is often nonoptimal, reducing the ability to apply expertise to the control center.[83,84] In bioterrorism, and even more so during a natural epidemic, the proper management of health aspects of consequence demands information sharing in order to include those with health expertise in the decision process.

The first test of the current preparedness program occurred with the small-scale anthrax attacks in 2001. Significant problems appeared, despite involving an agent that the system is specifically designed to respond to. Cooperation between the FBI and CDC was limited, with the CDC never having access to the contaminated letters, resulting in improper control measures at the Brentwood, NJ post office that led to additional casualties. The FBI had the agent analyzed by the U.S. Army Medical Research Institute (USAMRIID) rather than the Centers for Disease Control, the agency primarily responsible for the public health aspects of a bioterrorism, and not until further infections had occurred.[85,86]A contributor to the problem was that the FBI, lacking public health expertise, did not respond properly to a situation that did not fit its script of how a terrorist incident looked.[87] This was not a new issue. During the 1994 Mirage Gold nuclear terrorism exercise, the FBI agent in charge of the response was unwilling to share information with other agencies, and the FBI failed in 1995 to notify local authorities of a nerve agent threat against Disneyland.[88] The Topoff exercise identified coordination as a problem in bioterrorism response capability. Lack of a clear chain of responsibility seriously hindered public health officials in dealing with the epidemic side of the scenario, resulting in further casualties and social disruption.[89] The response to a real incident, a year later, revealed that the lessons had not been absorbed.

Healthcare Cost Containment and Provider Cooperation

A key to an effective public health response to epidemic disease is the participation of local healthcare providers.[90] The public health system depends on the overall healthcare system for data to identify outbreaks, as well as for the treatment of patients and preventative measures such as vaccinations. Access of patients to health care and cooperation by providers is critical to preparedness.[91,92]

Coordination with local healthcare providers is complicated by problems arising from the cost-containment environment in which they operate. The incentive structures found in managed care organizations create little incentive for concern with public health issues.[93] For example, control measures for sexually transmitted diseases receive low organizational priority in managed care organizations in the Medicaid program, despite high incidence rates, as treatment is often viewed as cheaper than prevention.[94] Healthcare price controls are known to result in lower quality services and care.[95] The cost of preparedness can be high, and largely exists as an unfunded mandate. Hospital costs for preparedness in Kentucky, for example, are over ten times the amount appropriated by public authorities.[96]

The result is a significant lack of capacity and cooperation to deal with public health emergencies. Hospitals lack surge capacity to deal with an emergency. By the late 1990s, the patient load from a routine influenza season was overtaxing primary care and emergency departments.[97,98,99] The General Accounting Office, reviewing public health preparedness, found that only 11% of hospitals surveyed had the ability to increase isolation capacity, and five out of seven states reviewed lacked enough healthcare workers to deal with an emergency.[100] Economic factors, coupled with the regulatory requirements of the Clinical Laboratory Improvement Amendments of 1988, have resulted in a significant decline in the use of microbial studies to provide a definitive diagnosis of infections, with an adverse effect on the surveillance and control of infectious diseases.[101]

Building surge capacity is complicated by the fact that the vast majority of healthcare facilities in the United States are privately owned and therefore not subject to direct control by public health officials. As a result, cooperation is voluntary. In many cases, cooperation is not forthcoming. Wetter and colleagues, studying hospital emergency departments, found that less than 20% had done any planning for bioterrorism or other public health crisis.[102,103] At the same time, hospitals have seen cuts in infection control programs.[104] Anecdotal evidence indicates that some hospitals actively avoid becoming involved in a public health crisis, including discouraging personnel from smallpox vaccination.[105] Such lack of cooperation between providers and public health personnel can hinder an appropriate response and extend the effects of an epidemic, as occurred during a pertussis outbreak in Arkansas in 2001 and 2002.[106]

At the physician level, three factors work against effective integration into the public health system. Unless forewarned, primary care physicians are not prepared to identify the unusual disease that presents as similar to the usual disease, and workloads in general make education in that area a low priority.[107] There are weaknesses in the public health infrastructure that raise concerns about the ability to change this situation. Currently, only about half of states offer such continuing training to healthcare providers, and only 10% require it.[108] Despite funding through the bioterrorism program, less than 40% have effective communications systems to link healthcare providers to the public health system.[109] At the same time, compliance problems exist with physician participation in existing mandatory disease reporting systems due to both ignorance of and antagonism to requirements.[110, 111] Specialists, who may be better trained to recognize unusual diseases than generalists, are less likely to participate in reporting systems.[112]

Public Health Law, Response Coordination, and Public Leadership

Public health law in the United States is shaped by the American federalist system and the high valuation placed by the country on self-determination. The result is devolution of policy control to the local level.[113] While this can create the opportunity for development of tailored local responses, it increases the problems associated with coordination of a response by increasing the number of involved actors.[114]

State and local laws differ between jurisdictions, and may be inconsistent with emergency needs. Minnesota, for example, requires separate court orders for each person quarantined, a major hurdle in the event of an infectious disease epidemic.[115] Uniform public health powers, suggested from the perspective of bioterrorism preparedness, include censorship of the media, liberalized search and seizure rules, and the power to compel civil servants and medical personnel to work, which are considered by many to threaten the balance between security and liberty.[116] The Model State Health Powers Act proposed by the CDC[117] with a focus on bioterrorism drew significant criticism for exceeding reasonable bounds in compromising civil liberties.[118] In the aftermath of AIDS, questions remain as to whether the coercive approach outlined in the Supreme Court's *Jacobsen v. Massachusetts* decision and in the Model Act remains a workable solution.[119,120,] Privacy rights, rights to assembly, and other impingements on individualism are a much greater barrier to public health intervention than at the time of Jacobsen.[121] Potential public distrust as a result of a compromise of civil rights is cited as a threat not just to bioterrorism programs, but public health in general. Etzioni[122] criticizes the Model Act for emphasizing uniform distribution of burdens, noting that, regardless of how distressing differential impact is, effective response needs to be adapted to the situation, and the burden resulting from an incident necessarily will not fall equally on the population.[123]

Distrust, therefore, means that preparedness needs to consider not only legal and medical issues, but also public relations and communications. In the event of an epidemic, public agencies can face a significant challenge to their perceived legitimacy, and hence to their authority.[124] Additionally, psychological trauma and distress may be a consequence of the outbreak as much as the disease itself. Strategies need to be planned to ensure that the population is informed of the nature of a problem[125] and the reason behind interventions that may impose on cherished civil liberties. In order to exercise such leadership, partnerships are needed with media and communications outlets.[126, 127] The value of favorable public opinion was demonstrated in the Arkansas pertussis outbreak, where responses were more effective in counties with greater trust between the public health and general communities.[128]

Surveillance — A Partial Success Story

Currently, the biggest strides in coordinating the public health and healthcare systems have occurred in the area of disease surveillance.[129] It is argued that the best tool for the control of public health threats is a broadbased program improving the disease surveillance system.[130] This is fortunate, as addressing a crisis requires first that the problem be recognized. Disease surveillance systems in the United States have traditionally relied upon passive reports of predetermined diseases from private practitioners to state public health agencies.[131] By the early 1990s, these systems were severely strained.[132] Internationally, surveillance is in even worse shape.[133, 134] Since the late 1990s, progress has been made in improving the United States surveillance systems, including new techniques such as syndromic surveillance from insurance and healthcare databases.[135,136,137,138,139] Cooperation between managed care organizations, with comprehensive databases on patient encounters, and public health organizations offer opportunities for improving surveillance efforts.[140,141] Similarly, additional agencies, such as the Central Intelligence Agency, have been utilized to provide sentinel information.

Although improvements in the system are being made, the General Accounting Office has determined that significant gaps remain.[142] Problems exist, such as balancing the Health Insurance Portability and Access Act (HIPAA) privacy rights of patients with public health needs[143, 144] and evaluating the effectiveness of such systems.[145] Despite the rapid increase in bioterrorism funding, overall funding for upgrading surveillance has remained flat.[146] It is a credit to the public health system that advances have been made despite these problems.

Surveillance, however, is an aspect of preparedness that remains malleable to management at higher levels. Because infectious disease threats arise at a local level and, due to resource distributions, must be responded to at a local level, the emphasis in preparedness needs to be to improve the infrastructure at the state and local level. This means that more jurisdictions need to be involved in preparedness programs, increasing the coordination problem. Lipsky and Hjern[146,147] suggest that successful policy implementation requires attention to the street-level operators and their active participation in planning and development. Centralized control is an example of what is described as a system where:

> By consistently taking the power to make decisions about the ways to innovate, adapt, and coordinate efforts away from those who are directly affected, policymakers have created institutions that are less able to respond to the problems they were created to address.[148,149]

Thus, a successful program must ensure that trained personnel are available at the local and state levels, communication between local, state, and federal public health agencies is adequate, and planning and coordination with local and state emergency responders and healthcare workers occurs. Local health leaders are critical due to their ability to work with and gain the cooperation of local health and government agencies.[150] Integration of and cooperation with local healthcare providers will be far easier if the training and planning that occurs has more relevance to day-to-day tasks. Unfortunately, local health officials are more often than not excluded from the planning process.[151] For example, blame for failure to obtain cooperation by healthcare workers in the voluntary smallpox vaccination program is placed on the failure to involve local healthcare leaders in the planning process.[152] Even within state health departments, cooperation between operating units may be problematic. For example, one study has found that state public health laboratory directors are, in general, unclear of their role in the agency response in the event of an emergency.[153]

A proper communication strategy can relieve a problem that arose from the emphasis on bioterrorism. The use of the rhetoric of bioterrorism to gain public health resources has a potential to create new problems, in the form of hysteria and hoaxes. Over 200 hoaxes were logged between 1997 and 1998, of which 13 involved more than 200 potential victims. These were blamed on the effects of media coverage and the rhetoric of government officials. Each involved significant opportunity costs to the individuals and agencies involved.

Research, Drugs, and Vaccinations

Preventive measures such as vaccination have proven to be among the strongest tools in the public health arsenal against infectious diseases, including the elimination of smallpox outside of the laboratory.[154,155] Vaccines have eliminated from 87 to 100% of the annual morbidity of vaccine-preventable childhood diseases in the United States.[156] Vaccines, however, have not been a high priority in the health policy arena.[157] In recent years, however, regulatory and liability issues have raised barriers to the development of new vaccines and the production of existing ones.[158,159] Since 1998, shortages have caused delayed or missed vaccinations for nine out of twelve routine childhood vaccines as well as the 2003 to 2005 influenza vaccine shortage.[160] Currently, worldwide production capacity for influenza vaccines is less than anticipated needs in the event of a new pandemic.[161]

A number of factors impacting the vaccine market have caused a decline in the number of manufacturers willing to enter the market. Economically, there are few incentives for the creation and production of vaccines. The market is small compared to other drugs, and dominated by a single purchaser, with almost 60% of the supply purchased by the federal government through the Vaccines for Children program, which creates a cap on the price of the vaccine. Both the federal government and private insurers limit reimbursement to physicians at rates insufficient to pay for the infrastructure for a proper vaccination program.[162] Panic over the availability of the antibiotic ciprofloxacin during the 2001 to 2002 anthrax incident, led the federal Department of Health and Human Services to threaten to abrogate the manufacturer's patent rights in order to obtain a lower price, creating new barriers to the prospects for the development of new vaccines and treatments.[163] Similar threats are seen with antiviral drugs for treating influenza.[164]

At the same time, liability and regulatory issues have increased the cost of producing a vaccine. An errant British study on the side effects of pertussis vaccination led to widespread lawsuits against manufacturers, driving three of four manufacturers from the market. Although the federal government offered some relief through the National Vaccine Injury Compensation Program and the Homeland Security Act, opt-out provisions and lack of coverage for non-childhood vaccines have limited its utility.[165, 166, 167] The Food and Drug Administration (FDA) functions under incentives to prevent harm from drug side effects, even if the drug benefits exceed the potential harm.[168] Hypersensitivity to side effects on the part of FDA has resulted in costly delays in approval, while timidity by public health agencies has further delayed getting vaccines into circulation, in at least one case prompting the manufacturer to withdraw a vaccine.[169] Such timidity can heighten public fears of vaccination, reducing the vaccination rate and effectiveness in developing herd immunity.[170] Fear of adverse events, for example, had a negative impact on efforts to use vaccination to control the 2001 to 2002 Arkansas pertussis outbreak.[171] Constricting the pipeline for development of new antimicrobial drugs is costly in terms of public health. Increasingly, common infectious bacteria such as *Streptococcus sp.* are demonstrating resistance to antibiotics. The Institute of Medicine estimates that by 1998, the annual cost of resistant infections was $4 to 5 billion, with 19,000 deaths due to hospital acquired resistant strains.[172] At the same time, no new class of antibiotics was introduced between the 1970s and the approval of the glycycline tygecycline in June 2005, and only a handful of vaccines for biological warfare agents are under development.[173]

Replacing private sector development with government drug and vaccine development is unlikely. The total National Institutes of Health budget for infectious disease and bioterrorism research is just over a billion dollars, and on average a new drug requires $500 million in development costs.[174] Some promising vaccines for SARS, tuberculosis, and West Nile virus are in the federal pipeline, along with new derivatives of antituberculosis drugs.[175] Even if federal resources for development were adequate, the issue of production capacity would remain. Worldwide, the vast majority of pharmaceutical production capacity remains in the private sector. Development and production of vaccines and drugs thus will remain dependent on the private sector. Ridley notes that significant progress has been made in leveraging government research efforts through public–private partnerships, but greater effort is needed to ensure coordination of these efforts with health system and public health needs.[176] Not just partnerships are needed, but careful attention to the consequences on development and production, in the form of incentives and disincentives, arising from policy decisions.

If adequate supplies of drugs are secured, there remains the problem of assuring that they are distributed properly. Current planning relies on the use of centralized prepackaged stockpiles, based on anticipated bio-terror agents, distributed to the scene of an outbreak. Although federal officials claim the ability of the bioterrorism initiative to assemble Disaster Medical Assistance Teams and drugs from the National Pharmaceutical Stockpile on-site in 12 to 24 hours, the best estimates are that in reality this would not occur for 48 to 72 hours.[177] There would then be additional delays in utilizing the aid for treatment. The multiagency response mechanism diverts resources from frontline responders like hospitals to centralized teams, at the cost of preparedness for those responders. Preparedness based on the security/law enforcement/natural disaster model, disproportionately emphasizes the "siren" responders relative to the hospital and surveillance components that more likely will detect and deal with an infectious disease threat. National Guard Rapid Assessment and Initial Detection (RAID) response teams, for example, cost enough to equip nearly 50,000 hospitals with decontamination facilities. Funds might be better spent to ensure local medical and drug supplies shelf-life extension and local "bubble" programs.[178, 179] Even a relatively small increase in local reserves and hospital resources can buy an additional 12 to 24 hours for federal emergency assistance to arrive in a crisis. While national stockpiles may be very useful in a large-scale event, they may not be needed in a higher probability small incident, and are not immediately available when a crisis is identified.

Conclusions

AIDs, SARS, drug resistant tuberculosis and other bacteria, hantavirus, West Nile fever, all offer clear and recent evidence that infectious disease remains a serious threat to population health, and that the public health system needs more attention than it receives. The United States is in urgent need of improvements to ensure that its public health infrastructure is able to meet threats such as emergent diseases, diseases evolving to resist treatment, and epidemics created out of human spite. Preparing for these threats, however, requires appropriate policies in place for dealing with them. Although progress has certainly been made in the recent past, significant hurdles remain to having an effective system capable of controlling infectious diseases.

The biggest problem facing the system is coordination and direction. The U.S. federalist system assures that public health remains a divided responsibility. Separation of public health responsibilities from the healthcare system further splits these responsibilities. As a result, a significant coordination burden exists. Preparedness for epidemic disease requires cooperation between the public and private sectors and across jurisdictions. Because different organizations have different values and view the world through different priorities, a considerable effort is necessary to ensure that their strategies are capable of interacting to identify and manage a public health crisis.

The biggest challenge the public health system faces in obtaining this coordination is leadership. The public health system is notable for having weak leadership, high rates of turnover in senior agency positions, and consequently a poor capability to frame issues, compete for resources, and secure cooperation from other organizational actors. Core public health issues such as infectious disease control place second to issues raised by outside activist groups, such as healthcare financing or anti-smoking activities. Budgets are strained due to ineffective advocacy, and new sources of revenue are often offset by cuts in state and local support. Much of the emphasis in the schools of public health is on ancillary issues to the traditional public health core, which further complicates the leadership issue. Lack of leadership credibility impairs the ability to convince outside organizations to set aside their individual needs and cooperate in developing programs to assure that a system is in place to identify and control epidemics.

Organizations decide to adopt strategies based on what is needed to meet what they perceive as important goals. Coordinating these strategies requires credible leadership capable of convincing organizations that the larger goal is important enough to modify strategies, and leadership capable of identifying and obtaining sufficient resources to ensure that the strategy can be implemented. The need for such leadership is evident not just in preparing for new epidemics, but for other catastrophic emergencies as well. The well-known consequences of the failures of federal, state, and local leadership in Louisiana during the 2005 Hurricane Katrina disaster, highlight the weaknesses of the system, particularly when compared to the stronger response by authorities in neighboring Mississippi. Failed leadership and uncoordinated planning and response can have lethal consequences for those who have the misfortune to be victims of a disaster.

Overcoming thirty or more years of neglect is not a task that can be accomplished overnight, but with effective leadership the system can be repaired. Strong building blocks exist — the United States possesses what may be the most technically proficient health system in the world, a medical and biological research capacity second to none, a first class government and university research infrastructure, dominance in communications and information management technology, and a strong and wealthy economy. The challenge for public health leadership is to use these raw materials to create a system capable of effectively utilizing them to protect the health of the public against the threat of epidemic disease.

Portions of this article were previously published in Avery G. "Bioterrorism: Fear, and Public Health Reform: Matching a Policy Solution to the Wrong Window." Public Administration Review 2004; 64(3): 275.288 (Blackwell Publishing).

Chapter 8

The Fallacy of Bioterrorism Programs: A Catastrophe for U.S. Public Health

Hillel W. Cohen, M.P.H. Dr. P.H., Robert Gould, and Victor Sidel

CONTENTS

Introduction

Recent bioterrorism preparedness programs that illustrate irrational and dysfunctional responses to inadequately characterized risks should be of urgent concern to all members of the public health community. Since anthrax spores were released in the U.S. mail system in 2001 and caused five fatalities and widespread panic, the spores have been linked to a U.S. military research program, suggesting that the release might not have occurred had the anthrax program never existed. The smallpox vaccination program has also been linked to fatalities and other serious adverse events, although evidence of risk of exposure to smallpox has been minimal. Indeed, the smallpox vaccination campaign may have been motivated by a political rather than health agenda. Continuing bioterrorism preparedness programs are similarly characterized by failure to apply reasonable priorities in the context of public health and failure to fully weigh the risks against the purported benefits of these programs. Such programs may cause substantial harm to the public health if allowed to proceed.

Efforts by the United States to prepare for the use of biological agents in war based on flawed evaluations of risks have had serious health consequences for military personnel. In addition, they have led to significant weakening of international agreements against the use of biological agents. Massive campaigns focusing on "bioterrorism preparedness" have had adverse health consequences and have resulted in the diversion of essential public health personnel, facilities, and other resources from urgent, real public health needs.[1] Preparedness proponents argued that allocating major resources to what were admittedly low-probability events would not represent wastefulness and would instead heighten public awareness and promote "dual use" funding that would serve other public health needs.[2] Public health resources are woefully inadequate, and the notion that bioterrorism funding would bolster public health capability seemed plausible to many, even though we and others have argued that the "dual use" rationale is illusory.[3,4] An evaluation of recent experience concerning anthrax and smallpox can help illuminate these issues.

Anthrax

Despite extensive work on the possible weaponization of anthrax, there has been no example of effective use of anthrax as a weapon of indiscriminant mass destruction. In 2001, shortly after the events of September 11, weapons-grade anthrax spores were mailed to several addressees, but none of the intended targets were injured. Of 11 people who developed inhalation anthrax, 5 died. Of the 12 who had cutaneous infections, all recovered after administration of antibiotics.[5] Thousands of people in potentially exposed areas such as postal sorting centers were advised to use antibiotics prophylactically. Millions of people were terrified, and many thousands in areas where there was no possible risk of exposure also took antibiotics. Congress was closed for days, mail service was disrupted for months, and state and county public health laboratories were inundated with white powder samples that ranged from explicit anthrax hoaxes to spilled powdered sweeteners.

Despite early speculation linking the anthrax release to "foreign terrorists," evidence led investigators to suspect an individual who had been working in a U.S. military facility that may have been in violation of the Biologic and Toxin Weapons Convention.[6,7] Whether or not that specific individual was involved, it appears likely that the perpetrator or perpetrators were associated in some way with a U.S. military program, that the motive for the extremely limited release was political, and that, without the existence of a U.S. military laboratory, the material for the release would not have been available.

This experience supports the view that, as a consequence of the inherent difficulties in obtaining and handling such material, mass purposeful infection is highly improbable and the likely impact on morbidity and mortality is limited.[8,9] However, the nature of U.S. "biodefense" programs may modify this prognosis; such programs may result in dangerous materials being more readily available, thus undermining the Biologic and Toxin Weapons Convention.[10,11,12] Despite an absence of evidence of anthrax weapon stocks posing a threat to U.S. military personnel, and despite problematic experiences of the military anthrax vaccination program, the U.S. government announced plans to spend as much as $1.4 billion for millions of doses of an experimental anthrax vaccine that has not been proven safe or effective and the need for which has not been opened to public debate.[13]

Smallpox

The 2002 to 2003 campaign to promote smallpox as an imminent danger coincided with the Bush administration's preparations for war on Iraq and the now discredited claims that Iraq had amassed weapons of mass destruction and could launch a biological or chemical attack in "as little as 45 minutes."[14, 15] A media campaign describing the dangers of smallpox coincided with the buildup for the war. An unprecedented campaign advocating "preventive" mass smallpox vaccinations, to be carried out in two phases was announced in December 2002.[16] The program involved half a million members of the armed forces and half a million health workers in phase 1 and as many as 10 million emergency responders in phase 2. Before then, the debate on smallpox had been whether the stocks of stored standby vaccine were adequate or whether they should be increased. The World Health Organization (WHO), the Centers for

Disease Control and Prevention (CDC), and virtually every public health official took the position that the vaccine involved too many adverse events to warrant mass vaccination when no case of smallpox existed or had existed for more than twenty years.[17] When the Bush administration announced support for mass vaccinations, WHO did not change its position, but the CDC and other U.S. public health officials and organizations, including the American Public Health Association (APHA), decided to acquiesce.[18]

The coincidence of the Bush war calendar and the smallpox vaccination calendar, while not conclusive, is nonetheless consistent with an inference that the war agenda was the driving force behind the smallpox vaccination campaign. Since the invasion, evidence has emerged that allegations regarding Iraqi weapons of mass destruction were deliberate exaggerations or lies.[19] The evidence is highly suggestive that the smallpox vaccination program was launched primarily for public relations rather than public health reasons.

The vaccination campaign did not proceed as planned. Opposition arose on both safety and political grounds,[20, 21] and most frontline health professionals simply did not volunteer to participate. Of the 500,000 health professionals who were targeted for inoculations in phase 1, fewer than 8% participated.[22] Despite efforts to avoid vaccination of those who might be at elevated risk, the CDC reported that there were 145 serious adverse events (resulting in hospitalization, permanent disability, life-threatening illness, or death) associated with smallpox vaccinations among civilians.[23] Of these cases, at least three were deaths.

Three deaths resulting from thousands of inoculations would have been justifiable in preparation for a real threat of smallpox or in the midst of a smallpox outbreak, when vaccination could have saved many more lives. However, in the absence of any smallpox cases worldwide or any scientific basis for expecting an outbreak, these deaths and other serious adverse events are inexcusable. In August 2003, an Institute of Medicine committee that had been charged with reviewing the vaccination program came back to the position that had been generally accepted before 2002: that mass, preventive inoculations were unwarranted. According to the committee report:

> In the absence of any current benefit to individual vaccinees and the remote prospect of benefit in the future (as such benefit would be realized only in the event of a smallpox outbreak, and the outbreak occurred in the vaccinee's region), the balance of benefit to the individual and risk to others (through contact with the vaccinee or through disruption of other public health initiatives) becomes unfavorable.... In the absence of other

forms of benefit, therefore, offering vaccination to members of the general public is contrary to the basic precepts of public health ethics.[24]

The report further cited "lingering confusion about the vaccination program's aims."[25] We find it difficult to comprehend how a program with confused aims and known serious risks can be viewed as having a positive risk-benefit ratio or how public health organizations could accept such a program without subjecting it to extensive critical examination and debate.

The smallpox vaccinations harmed others beyond those who suffered side effects. Considerable public health resources were used in the campaign. In a climate of state and local budget crises coinciding with the war and occupation, a downturn in employment, and a tax cut for the wealthy, public health services have been cut or are at serious risk. Funding for bioterrorism programs is not correcting the deficit, because such funds have been for the most part specifically earmarked for preparedness efforts and cannot be transferred to other public health programs. In general, federal increases in public health funding are much less extensive than state or local cuts.[26] During the height of the smallpox vaccination effort, a number of state health officials complained that important work, including tuberculosis screening and standard children's inoculations, had to be scaled back.[27] The siren song of dual use — that bioterrorism funding would strengthen the public health infrastructure — has shown itself to be an empty promise, as preparedness priorities have weakened rather than strengthened public health.

Broader Problems

Even worse, bioterrorism "preparedness" programs now under way include the development of a number of new secret research facilities that will store and handle dangerous materials,[28,29] thus increasing the risk of accidental release or purposeful diversion.[30] Reports of accidental leaks and improper disposal of hazardous wastes at the U.S. Army facility at Fort Detrick serve as further warnings,[31, 32] as do revelations of mishandling of biological agents at the Plum Island, New York facility that studies potential bioweapons that affect animals.[33] (Note. Facilities at biosafety levels 3 and 4 are authorized to handle dangerous biological materials. Level 4 facilities may handle the most deadly and contagious pathogens like smallpox and Ebola viruses. Source. Reprinted with permission from the Sunshine Project: available at: http://www.sunshine-project.org).[34]

Most important, the proposed development of "biodefense" programs at sites, such as national nuclear weapons laboratories, that are traditionally secretive in their operations, also provides an impetus for a potential global "biodefense race" that would likely spur proliferation of offensive biowarfare capabilities.[35,36] Accidents or purposeful diversions from these facilities seem at least as likely as terrorist events, and perhaps more so, since the deadly materials are already present. The Patriot Act has greatly expanded the cloak of secrecy that shields these facilities from public awareness and oversight.[37]

In short, bioterrorism preparedness programs have been a disaster for public health. Instead of leading to more resources for dealing with natural disease as had been promised, there are now fewer such resources. Worse, in response to bioterrorism preparedness, public health institutions and procedures are being reorganized along a military or police model that subverts the relationships between public health providers and the communities they serve.

What can we do? Advocacy groups and local coalitions have emerged to oppose the widespread siting of potentially dangerous bioterrorism laboratories and have demanded that such facilities be open to the public. Labor unions that helped resist the smallpox vaccinations can be vigilant against further efforts to enlist health workers in poorly conceived and misguided campaigns that pose unnecessary risks to patients, workers, and communities.

Above all, it is imperative that public health organizations such as the American Public Health Association take a fresh and critical look at the government's biopreparedness agenda and advocate for a comprehensive program that promotes global health security. Such a program would initiate appropriate and focused preparedness efforts only in the context of concerted and cooperative international steps designed to reduce the likelihood of infection from all sources. The modalities employed would range from strengthened treaties to provision of adequate clean water, food, shelter, education, and health care for all.[38] Those of us working in public health can insist on a reevaluation of the entire bioterrorism preparedness agenda and demand a close examination of its goals and consequences before additional resources are invested in programs that so far seem to have done more harm than good.

In light of the daily toll of thousands of deaths from illnesses and accidents that could be prevented with even modest increases in public health resources here and around the world, we believe that the huge spending on bioterrorism preparedness programs constitutes a reversal of any reasonable sense of priorities. While some still feel that bioterrorism preparedness programs will protect us from catastrophe, we agree with David M. Ozonoff, chairman emeritus of the Department of Environmental Health at the Boston University School of Public Health, that these programs represent "a catastrophe for American public health,"[39] and we hope it is not too late to change this dangerous direction.

War, poverty, environmental degradation, and misallocation of resources are the greatest root causes of worldwide mortality and morbidity, as well as ultimately being the underlying causes of terrorism itself. Bringing an awareness of this reality to the public is no easy task. However, one important step will be for the public health community to acknowledge the substantial harm that bioterrorism preparedness has already done and develop mechanisms both to increase our public health resources and to allocate them in a manner that will do the most good for all inhabitants of our increasingly fragile planet.

Acknowledgments

We wish to thank Susan M. Hailpern for assistance in assembling the data and graph for Figure 8.1. Thanks to the Sunshine Project (http://www.sunshine-project.org) for the map of U.S. biodefense facilities shown in Figure 8.2.

Contributors

All authors contributed to the ideas for this article. H. W. Cohen led the writing.

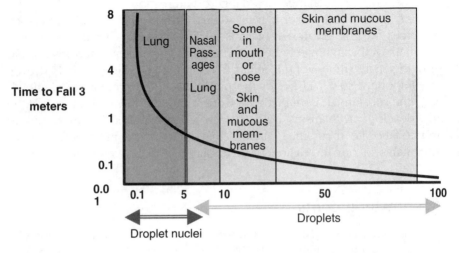

Route of Exposure - Deposition

Figure 8.1

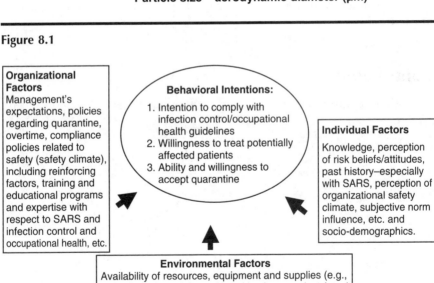

Figure 8.2

INFLUENZA:
THE OLD AND
PRESENT THREAT

Chapter 9

A Killer Flu? Planning for the Devastating Effects of the Inevitable Pandemic

Margaret A. Hamburg, M.D., Shelley A. Hearne, Dr.P.H., Jeffrey Levi, Ph.D., Kim Elliott, M.A., Laura M. Segal, M.A., and Michael J. Earls

CONTENTS

The Real Dangers of a Pandemic

Scientific experts estimate that an "inevitable" major epidemic of a new influenza virus strain could result in millions of deaths if preventive actions are not taken.

The seasonal flu kills approximately 36,000 to 40,000 people and hospitalizes more than 200,000 in the United States each year.[1] Annually, influenza costs the national economy over $10 billion in lost productivity and direct medical expenses.[2] Many view the flu as a relatively predictable and manageable health threat.

Recently, however, health experts worldwide have been sounding the alarm about a different type of flu. They warn of the "inevitable" emergence of a new, severe strain of the flu virus against which people have no immunity to protect them.[3] This could result in a rapidly spreading, worldwide pandemic of this new, potentially lethal strain of the disease.[4] New strains of the flu traditionally emerge in animals, often in poultry and pigs, and then as the disease develops over time, it can become transmitted to humans. The severity of an emerging pandemic would be determined by the particular strain or the new form of the virus and how easily contagious it proves to be in humans.

The World Health Organization (WHO), Centers for Disease Control and Prevention (CDC), and other health authorities believe that the emergence of a pandemic flu could be devastating to world health and economic stability.

- In the United States, projection models predict that a pandemic may cause over half a million deaths and 2 million hospitalizations.[5]
- The estimated economic impact of a pandemic flu outbreak in the United States today, based on projections from the relatively mild 1968 flu epidemic, would be $71.3 to $166.5 billion due to death and lost productivity, excluding other "disruptions to commerce and society."[6]
- The United States would be impacted by the global implications as soon as a pandemic outbreak occurred in any part of the world due to the interdependence of economies. Sectors, such as hospitals and the health care system, which rely on supplies manufactured in other parts of the world, including Asia, would feel

immediate repercussions and supply shortages. Travel restrictions, possible limitations on public gatherings and events, and other measures taken to limit the spread of disease would also have rapid and far reaching repercussions. Since a pandemic could likely result in political and economic destabilization, particularly in developing countries, it poses serious national security concerns for the United States. In a May 2005, *New England Journal of Medicine* article, Dr. Michael Osterholm, Director of the Center for Infectious Disease Research and Policy at the University of Minnesota, wrote that "even a relatively 'mild' pandemic could kill millions of people worldwide."[7]

■ Based on historical trends and predictions, virologists and epidemiologists predict that a new flu pandemic will emerge three to four times each century.[8] Health officials around the world are troubled by the severity of the "avian flu" circulating in Asia (the H5N1 flu strain). They fear this avian flu could become the next pandemic for humans. The regional director of the WHO for the Western Pacific region stated in February 2005 that the "world is now in the gravest possible danger of a pandemic."[9]

■ As of June 17, 2005, this "bird flu" virus had killed 54 individuals and had spread rapidly among bird populations.

■ As of April 2005, the strain seemed to be exhibiting a mortality rate of over 50 percent in humans. Experts are concerned that when the mortality rate decreases, the virus's transmission rate will increase.[10]

■ Health officials are concerned that the avian virus could become more contagious among humans, and that it could remain in a strain against which humans have no natural resistance.[11]

CLSA Asia-Pacific Markets, the Asian investment banking arm of Credit Agricole of France, estimates that avian influenza has already cost the impacted region in Asia $8 to $12 billion, mostly from lost revenue from poultry and related industries.[12] Health and Human Services Secretary Mike Leavitt, said on May 16, 2005, "I am acutely aware of the disaster that a pandemic could cause. Many of us are particularly worried about H5N1 avian influenza virus, and we're right to worry. It has infected at least 89 human beings and killed more than half. There is a chance that this virus could cause the next pandemic."[13]

Major Flu Outbreaks of the Twentieth Century[14]

1918: The "Spanish" flu pandemic killed 500,000 in the United States, 50 million worldwide.

1957–58: An outbreak spread from China across the globe, killing approximately 70,000 in the United States. In April 2005, a company testing laboratory proficiency mistakenly distributed samples of this pandemic strain to laboratories worldwide, triggering global concern until all samples were accounted for and destroyed.[15]
1968–69: The "Hong Kong" flu, the most recent pandemic, affected millions worldwide and disrupted world economies.
1997: The first identification of the avian "bird" flu, which remains active in Asia.[16]

The World Health Organization secretariat reported in April 2005 that:

Since January 2004, events affecting both human and animal health have brought the world closer to an influenza pandemic than at any time since 1968. Whereas past pandemics have consistently announced themselves with an explosion of cases, events during 2004, supported by epidemiological and virological surveillance, have given the world an unprecedented warning that a pandemic may be imminent. They also have opened an unprecedented opportunity to enhance preparedness.[17]

While experts predict a pandemic flu is "inevitable," subsequent deaths in the United States, predicted to be over a half million people, are not. Increased federal leadership, convening national and state pandemic influenza plans into operational blueprints, procuring adequate antiviral medication for treatment, and putting a process in place now for rapid influenza vaccine approval, are all steps that should be taken immediately.

Protecting the United States and the world against the threat of a pandemic would, at the same time, better prepare countries for threats posed by infectious illnesses, including the intentional spread of disease by terrorists. The threat of a pandemic influenza outbreak was highlighted by the U.S. Department of Homeland Security (DHS) as one of 15 disaster planning scenarios for which the United States should be prepared.

In order to help understand the current statues of U.S. preparations and highlight ways to improve them, in the following report, Trust for America's Health (TFAH) provides:

▪ A state-by-state examination of potential deaths and hospitalizations due to a flu pandemic based on model estimates;
▪ A state-by-state examination of capacity to treat citizens with recommended antivirals based on model estimates;

- A review of U.S. and state pandemic readiness, including a comparison to other nations' progress; and
- Recommendations for improved pandemic readiness.

Overall, the report finds:

- Despite the health and economic implications of such an event, pandemic planning efforts are lagging in the United States, especially when compared to the United Kingdom and Canada.
- The United States has not assessed or planned for the disruption a flu pandemic could cause both to the economy and society as a whole. This includes daily life considerations, such as potential school and workplace closures, potential travel and mass transit restrictions, and the potential need to close stores resulting in complications in the delivery of food and basic supplies to people. Daily life and economic problems would likely emerge in the United States even before the pandemic flu hit the country due to the global interdependence of the world economy.
- Aspects of the planning process, such as ensuring vaccine and antiviral capabilities and surge capacity readiness, are incomplete or fragmented.
- The failure to establish a cohesive, rapid, and transparent U.S. pandemic strategy could prove a major weakness against a virulent and efficient virus — putting Americans needlessly at risk.

Dr. Anthony S. Fauci, National Institute of Allergy and Infectious Diseases, National Institutes of Health, has said: "I would like to emphasize that although we cannot be certain exactly when the next influenza pandemic will occur, we can be virtually certain that one will occur and that the resulting morbidity, mortality, and economic disruption would present extraordinary challenges to public health authorities around the world."[18] Dr. Julie Gerberding, Director, CDC notes: "Today, many influenza experts, including those at CDC, consider the threat of a serious influenza pandemic to the United States to be high. Although the timing and impact of an influenza pandemic is unpredictable, the occurrence is inevitable and potentially devastating."[19]

Model Estimates of the Impact of a Severe-Strain Flu Virus Epidemic

In order to illustrate the potential severity of a pandemic outbreak in the United States, the chart below uses one model based on assumptions from the current avian flu outbreak. Scientists have used a number of different models to estimate the scope and impact of the emergence of a new strain of the flu. The basic U.S. planning model is based on assumptions from the 1968 "Hong Kong" pandemic flu, which was considered to be relatively mild. Experts also have predicted higher and lower estimates based on different sets of assumptions. The numbers below adapt the model to reflect moderate assumptions for the current avian flu threat.

The WHO has estimated that there would be a "contraction" rate of 25% for this flu strain. This means they warn that countries should be prepared for approximately 25% of their populations to get sick from the pandemic virus. Other scientists have estimated that up to 50% of countries' populations could become infected.

The current strain of the avian flu is viewed as significantly more lethal than the 1968 pandemic flu strain. A high-level pandemic, such as the 1918 pandemic, is considered to be six times more lethal than the 1968 flu.[20] The projections below reflect a midlevel estimate of a three times higher rate. These numbers are reflected in the "Projected Dead" column in the table below. The range of estimates from low level to high level severity death rates can be found in Appendix A.

Due to the severity of the avian flu strain, experts also believe that it would result in a much higher hospitalization rate than estimates using the 1968 strain. The estimates below, in the "Projected Hospitalizations" column, reflect a midlevel estimate of a three times higher rate. A more virulent strain of flu, changes in medical care and treatment procedures, and an aged population are all factors behind this projection. The range of estimates, from low level to high level severity hospitalization rates, can be found in Appendix A.

Potential Impact: State by State Analysis

To assist state and local health agencies with pandemic readiness, CDC developed a computer model (FluAid 2.0) that generates mortality, hospitalization, and outpatient rates for different ages in the population on a state-by-state basis.[21] FluAid derives its default numbers from the 1968 Hong Kong pandemic, which had a relatively minor impact on the United States. According to Dr. Keiji Fukuda, the Chief of Epidemiology and Surveillance Section, Influenza Branch at CDC's National Center for Infectious Diseases, a high severity pandemic, similar to the 1918 pandemic outbreak, may have a mortality rate of six times the 1968 pandemic.[22] To estimate the potential impact from a H5N1 pandemic on the United States, the following projections multiplied the default FluAid mortality rate for each state and each age group by three (the midpoint between the default numbers and the possible six times mortality rate); hospitalization rates are also three times the default FluAid number.

Projections of deaths and hospitalizations from an H5N1 pandemic are only estimates. Variables, including the virulence of the virus, its attack rate, and the success of preventative and controlling measures (including the use of antiviral medication and the development of a vaccine) would influence the actual total. The mortality estimate provided below — a U.S. death toll of over half a million persons — varies from some other experts' forecasts, all projections agree on a critical point: the risk of a pandemic is serious enough to justify urgent steps to improve U.S. ability to fight this virus if it starts to spread.

It is also important to note that planning and accommodating for the surge of sick patients presents a separate, massive challenge to the health care system — a consideration that the projected death toll should not overshadow. The impact of over 2 million hospitalized patients would test and possibly overwhelm the surge capacity of hospitals nationwide. For instance, according to the American Hospital Association, in 2003 there are only 965,256 staffed hospital beds in registered hospitals.[23]

As of May 2005, the U.S. had stockpiled 2.3 million courses of the antiviral medication Tamiflu®, which could be used as a treatment in the event of an outbreak, and intended to order approximately 3 million more with funds appropriated by Congress to total 5.3 million courses. The WHO is currently estimating that an avian flu epidemic could impact 25% of countries' populations. In the United States this means it could affect nearly 67 million individuals, based on FluAid projections and population numbers. With the current level of the U.S. Tamiflu® order, *over 61.5 million Americans who could be infected would not receive antiviral medication*. If the United States orders additional courses of Tamiflu®, they would not be available until 2007, unless production capacity significantly changes.

In an actual pandemic, there would likely be geographic concentrations of the disease, especially in the initial stages of an outbreak. U.S. government officials may decide to "front-end" target the limited supply geographically in hopes of containing the initial spread. However, it is likely that the pandemic would still spread to the remainder of the country. As a result of the pandemic's national scope and lacking a prioritized distribution plan, these projections assume that the United States would use proportional distribution (based on population) in delivering the remaining Tamiflu® courses.

Discussing the U.S. Department of Health and Human Services Draft Pandemic Plan, Dr. Michael Osterholm, Director of the Center for Infectious Disease Research and Policy notes: "Beyond research and development, we need a public health approach that includes far more than drafting of

TABLE 9.1 **Potential Pandemic Influenza Deaths and Hospitalizations from a Midlevel Pandemic Flu***

State	Projected Dead	Projected Hospitalized	Number of Cases
Alabama	8,886	38,591	1,079,789
Alaska	886	4,558	152,328
Arizona	9,223	39,675	1,138,742
Arkansas	5,350	22,660	630,705
California	60,875	273,090	8,067,075
Colorado	7,192	32,978	973,161
Connecticut	7,054	29,932	817,465
Delaware	1,507	6,560	182,895
District of Columbia	1,155	4,974	132,241
Florida	35,737	142,386	3,663,486
Georgia	13,655	62,912	1,871,561
Hawaii	2,446	10,571	296,651
Idaho	2,279	10,157	302,558
Illinois	23,720	103,738	2,973,962
Indiana	11,817	51,711	1,466,027
Iowa	6,233	26,090	713,106
Kansas	5,373	22,946	654,335
Kentucky	7,930	34,748	977,031
Louisiana	8,334	37,148	1,087,942
Maine	2,651	11,333	310,513
Maryland	9,958	44,500	1,273,572
Massachusetts	13,136	56,038	1,529,313
Michigan	19,622	86,005	2,443,473
Minnesota	9,304	40,786	1,171,387
Mississippi	5,362	23,531	682,625
Missouri	11,274	48,240	1,350,515
Montana	1,804	7,787	219,703
Nebraska	3,441	14,697	414,218
Nevada	3,243	14,455	419,202
New Hampshire	2,333	10,301	293,177
New Jersey	16,980	72,791	2,013,212
New Mexico	3,244	14,504	432,438
New York	37,701	162,490	4,534,307
North Carolina	14,987	65,637	1,856,296
North Dakota	1,371	5,795	160,221
Ohio	23,197	99,979	2,796,583
Oklahoma	6,833	29,376	829,273
Oregon	6,724	29,047	810,872

Table 9.1 **Potential Pandemic Influenza Deaths and
Hospitalizations from a Midlevel Pandemic Flu*** *(Continued)*

State	Projected Dead	Projected Hospitalized	Number of Cases
Pennsylvania	27,185	112,658	3,004,915
Rhode Island	2,234	9,263	246,857
South Carolina	7,474	32,983	940,045
South Dakota	1,559	6,599	184,493
Tennessee	10,875	47,678	1,342,050
Texas	35,124	160,648	4,859,834
Utah	3,393	15,906	514,787
Vermont	1,185	5,213	147,245
Virginia	13,104	58,872	1,683,499
Washington	10,910	48,610	1,402,591
West Virginia	4,049	17,014	453,947
Wisconsin	10,620	45,842	1,292,419
Wyoming	915	4,086	119,936
U.S. Totals	541,433	2,358,089	66,914,573

* Projections are based on CDC's FluAid 2.0 program. The estimated deaths are for a pandemic strain three times more lethal than the 1968 pandemic, on which the default FluAid numbers are based. The hospitalization rate is three times the default 1968 rate. The Dead and Hospitalized numbers represent the most likely FluAid projection at a 25% rate of contraction. The Number of Cases is the projected number of residents contracting the flu, based on a 25% rate of contraction. State population numbers are from FluAid, using U.S. Census data gathered in 1999. Updated population data were not used to ensure consistency with estimated Dead and Hospitalized numbers.

general plans, as several countries and states have done. We need a detailed operational blueprint of the best way to get through 12 to 24 months of a pandemic."[24]

In August 2004, the U.S. Department of Health and Human Services (HHS) released a draft plan of U.S. strategy to deal with a flu pandemic. The plan, an updated version of a 1978 plan, outlines proposed collaboration among jurisdictions, as well as preparedness and response guidelines for federal, state, and local health officials. The goals of the plan are to (1) decrease the burden of disease; (2) minimize social disruption; and (3) reduce economic impact.[25] The draft plan can be accessed on the HHS Web site at http://www.dhhs.gov/nvpo/pandemics.[26] Related public comments on the plan were accessible at the site in March 2005.

TABLE 9.2   State-by-State Capacity to Treat Citizens with
Recommended Antiviral*

State	Number of Tamiflu Courses Available	Number of Cases	Number of Cases Without Tamiflu
Alabama	85,525	1,079,789	994,263
Alaska	12,065	152,328	140,263
Arizona	90,195	1,138,742	1,048,547
Arkansas	49,955	630,705	580,749
California	638,956	80,670,756	7,428,119
Colorado	77,080	973,161	896,081
Connecticut	64,748	817,465	752,717
Delaware	14,486	182,895	168,409
District of Columbia	10,474	132,241	121,767
Florida	290,168	3,663,486	3,373,318
Georgia	148,238	1,871,561	1,723,323
Hawaii	23,496	296,651	273,154
Idaho	23,964	302,558	278,594
Illinois	235,554	2,973,962	2,738,408
Indiana	116,117	1,466,027	1,349,910
Iowa	56,482	713,106	656,624
Kansas	51,827	654,335	602,508
Kentucky	77,386	977,031	889,645
Louisiana	86,171	1,087,942	1,001,771
Maine	24,594	310,513	285,918
Maryland	100,874	1,273,572	1,172,698
Massachusetts	121,130	1,529,313	1,408,183
Michigan	193,536	2,443,473	2,249,937
Minnesota	92,780	1,171,387	1,078,607
Mississippi	54,068	682,625	628,558
Missouri	106,968	1,350,515	1,243,546
Montana	17,402	219,703	202,301
Nebraska	32,808	414,281	381,409
Nevada	33,203	419,202	385,999
New Hampshire	23,221	293,177	269,956
New Jersey	159,457	2,013,212	1,853,755
New Mexico	34,251	432,438	398,186
New York	359,142	4,534,307	4,175,165
North Carolina	147,029	1,856,296	1,709,267
North Dakota	12,690	160,221	147,530
Ohio	221,505	2,796,583	2,575,078
Oklahoma	65,683	829,273	763,590
Oregon	64,225	810,872	746,646

TABLE **9.2** State-by-State Capacity to Treat Citizens with Recommended Antiviral*

State	Number of Tamiflu Courses Available	Number of Cases	Number of Cases Without Tamiflu
Pennsylvania	238,006	3,004,915	2,766,910
Rhode Island	19,552	246,857	227,305
South Carolina	74,457	940,045	865,589
South Dakota	14,613	184,493	169,880
Tennessee	106,298	1,342,050	1,235,752
Texas	384,925	4,859,834	4,474,909
Utah	40,774	514,787	474,013
Vermont	11,663	147,245	135,582
Virginia	133,342	1,683,499	1,550,157
Washington	111,093	1,402,591	1,291,498
West Virginia	35,955	453,947	417,992
Wisconsin	102,367	1,292,419	1,190,053
Wyoming	9,500	119,936	110,436
U.S. Totals	5,300,000	66,914,573	61,614,573

* Tamiflu® availability projections are based on state-by-state proportional distribution of the 5.3 million courses of Tamiflu® ordered or currently in U.S. federal government possession. For example, California, with approximately 12 percent of the U.S. population, receives 12% of the Tamiflu® in the above projection. The Number of Cases is the projected number of residents contracting the flu, based on a 25 % rate of contraction. State population numbers are from FluAid, using U.S. Census data gathered in 1999. Updated population data were not used to ensure consistency with estimated Dead and Hospitalized numbers. The Number of Cases Without Tamiflu® is the difference between the other two columns.

Questions for U.S. Pandemic Planning Efforts

■ Is there coordination among government, health, and economic infrastructures? State, federal, and international efforts must be coordinated, with instructions for specific implementation. Sufficient resources must be allocated to match what is needed to carry out the plan.

■ Does the National Strategic Stockpile include all necessary medical supplies that will be necessary to respond to a pandemic? In addition to stockpiling antivirals and vaccines, when they are available, the United States must also stockpile critical medical supplies such as masks, gloves, gowns, bed linens, and all other equipment needed to assure that hospitals and other health care providers are properly protected when the usual supply chain is disrupted either abroad or in the United States.

- Are there sufficient surge capacity capabilities? A pandemic or other mass-emergency scenario would overwhelm the normal operations of hospitals and the health care system. Readiness efforts must account for massive demand triggered by a pandemic. Local health officials and first responders must be included in planning efforts to maximize the on-the-ground ability to "scale up" capability in a rapid manner.

- Is there a prioritization of who would receive antivirals and vaccines based on a limited supply? Specific national guidance must be established on vaccine prioritization, including developing guidelines on the use of antiviral drugs and lists of priority groups for vaccine receipt and distribution, given that there are likely to be insufficient supplies during a pandemic.

- Is there a rapid response plan to develop, test, and produce a vaccine? It will take an estimated six to nine months after a pandemic emerges to develop a vaccine. Questions remain of how to rapidly review and test the vaccine once it is created, including concerns about speeding the approval process by the FDA, liability protection for vaccine manufacturers, and what type of preservative will be used in the vaccine. In addition, industry representatives have suggested that current production capacity is insufficient to meet the demand for a pandemic influenza vaccine, and that it could take twelve to eighteen months to meet appropriate production levels?

- Is there clear assignment of who in the government would control and distribute vaccine and treatments? Do plans exist to stockpile stopgap antiviral medications and vaccines, based on the small supplies of drugs that will be available versus the expected need and demand? As was evident in the 2004 flu season in the United States, when there was a shortage of available vaccine for the annual flu, there is no centralized infrastructure to control and monitor vaccine distribution.

- Are there clear plans to communicate and inform the public? Effective response to a pandemic would require a clear action plan for what information would be made available to the public and on what time frame.

- Are there coordinated plans for monitoring outbreak and managing containment? Coordinating containment efforts require sufficient surveillance and tracking systems to monitor and detect outbreaks, infected persons, and the vaccine supply, as well as the ability to examine the readiness of infected survivors to reenter the workplace. Survivors as a volunteer workforce would prove essential to helping combat the pandemic, because they will have developed immunity to the virus.[27]

In 2005 the U.S. Government Took Several Notable Steps to Begin to Prepare For a Potential Pandemic

- Congress has been increasingly concerned about the nation's readiness to respond to pandemic and annual influenza. Since convening in January 2005, the 109th Congress held a series of hearings on issues related to influenza, including:
- May 26, 2005, "The Threat of and Planning for Pandemic Flu," House Committee on Energy and Commerce, Health Subcommittee.
- May 4, 2005, "The State of Readiness for the 2005–2006 Flu Season," House Committee on Energy and Commerce, Subcommittee on Oversight and Investigations.
- April 12, 2005, "Pandemic Preparedness and Influenza Vaccine Supply — CDC, NIAID, and the Office of the Secretary of HHS," House Committee on Appropriations, Subcommittee on Labor, Health and Human Services, Education, and Related Agencies.
- February 10, 2005, "The Perplexing Shift from Shortage to Surplus: Managing This Season's Flu Shot Supply and Preparing for the Future," House Committee on Government Reform.
- In May 2005, Congress passed supplemental appropriations legislation that made available $25 million "for a coordinated program to prevent and control the spread of the avian influenza virus."[28] In addition, $58 million was appropriated for the purchase of influenza countermeasures for the Strategic National Stockpile. These funds are expected, in part, to be used to order an additional 3 million courses of Tamiflu®, to bring the U.S. stockpile order up to 5.3 million courses of treatment.
- In April 2005, U.S. Senator Barack Obama (D-IL) introduced the AVIAN Act of 2005, proposed legislation that includes a mandate for the federal government to stockpile the antiviral medication oseltamivir, commonly known as Tamiflu.[29]
- In April 2005, U.S. President George W. Bush approved use of quarantine in the event of an outbreak of "influenza caused by novel or reemergent influenza viruses that are causing, or have the potential to cause, a pandemic," which includes, but is not limited to the H5N1 strain of avian flu currently in Southeast Asia.[30]

- In April 2005, the U.S. Department of State issued an advisory statement about the avian flu and announced it is taking measures to support the World Health Organization's (WHO) efforts to contain the outbreak.[31]
- In March 2005, in the U.S. Department of State authorization bill (S.600), the U.S. Senate proposed including $25 million for International Famine and Disaster Assistance to prevent and respond to a possible outbreak of the avian flu and called for a task force to coordinate U.S. policy.[32]

"'We remain very vulnerable.' summarized Energy and Commerce Committee Chairman Joe L. Barton [R-Texas]. 'Think of it like this — a bad flu outbreak could kill more Americans than either or both of the last century's world wars.'"[33]

State Pandemic Readiness

Similar questions can be asked about the level of preparedness of state and local governments for a pandemic. America's public health system relies on a loosely affiliated network of approximately 3,000 federal, state, and local health agencies, often working with private sector and professional health organizations. State governments have primary responsibility for the health of their citizens under U.S. law. Therefore, a federal plan without state plans that are ready for implementation would be insufficient.

Most states have developed draft pandemic response plans, but they are in widely different phases of readiness. A recent examination found that only between 25 and 30 states have made their plans publicly available.[34] Making the plans publicly available is considered by many experts as an essential feature of pandemic readiness in order to improve integration with other jurisdictions as well as to add a level of accountability. In fact, in commenting on the draft U.S. pandemic influenza preparedness plan, the WHO stated, "We feel that in order to ensure broad commitment for the plan, it is essential to involve the community in the planning process."[35]

City and Local Planning: On the Ground and Facing Unique Problems

Pandemic planning efforts must incorporate local health departments and first responders in plan development. While states have legal jurisdiction to oversee much of a pandemic plan's contents, local responders will be responsible for the related operational, on-the-ground implementation. Surge capacity, antiviral prioritization, and outbreak tracking are among the areas especially critical to plan for in the local context. Additionally, a highly dense urban area poses a particular danger because of the possibility of massive virus transmission. Local areas, in coordination with state and federal officials, need to prioritize pandemic preparations to ensure that implementation and first response is as seamless and effective as possible. As Jean Taylor, head of Maryland pandemic-planning efforts, Maryland Department of Health and Mental Hygiene, said, "We have to plan for the worst-case event."[36]

Vaccine and Antiviral Medication Issues

National planners must focus on questions surrounding stockpiling antivirals and stabilizing vaccine development to protect people in the event of an outbreak. This is problematic given the limited production capacity for antivirals and vaccines in the United States and throughout the world.[37]

The Shrinking Vaccine Manufacturing Market

In 1976, 37 U.S. companies manufactured vaccines. In 2002, there were only three. Reasons given for the decline are mostly economic:

- Vaccine production can take decades of research and development and, according to industry estimates, costs about $800 million per licensed vaccine.
- Concerns about liability impact manufacturers' decisions to avoid vaccine production, especially after the significant compensation claims that followed the swine flu immunization program in the mid-1970s.
- Some companies also cite insufficient market size as reason to stay out of the vaccine market, due to the current low incidence of many diseases in the United States, such as tuberculosis. The flu vaccine demand is particularly seen as unstable due to the unpredictability of the size and scope of the market each year.
- There are only two manufacturers currently licensed to produce influenza vaccines in the United States, and a third overseas manufacturer who supplies vaccine to the the United States. One of the U.S. licensed manufacturers produces inactivated influenza vaccine and one manufacturer produces the live, attenuated vaccine administered through nasal spray.[38]

Flu Vaccine Crisis of 2004

The October 2004 announcement that approximately half of the expected flu vaccine for the United States would not be available heightened public awareness about the fragility of the public health system's vaccine development system and national readiness for a fast-moving influenza epidemic.

In early October, Chiron Corporation announced it would not be able to meet demand for its flu vaccine after problems at a British plant halted production of millions of doses. The dose shortage highlighted the fact that the United States relies on very few manufacturers to deliver the country's "projected need of 100 million doses."[39] As a result, CDC officials were forced to encourage changes in the nation's distribution and procedures for the flu vaccine supply, reserving doses only for the populations most in need. This illustrates the lack of coordination for the prioritization and distribution of vaccines, particularly in a crisis.

The shortage resulted in a focus of media and public attention on the issue, long lines at health clinics around the country, and calls for incentives, liability reforms, and other measures to encourage a broader range of vaccine producers.

Outdated Technology For Developing Flu Vaccine

Flu vaccines similar to those currently in use were first introduced during the 1940s. Since then, scientists have improved the standardization and purity of the process. However, the world remains dependent on the same basic technology that relies on inoculating the current influenza virus strains into embryonated hens' eggs for vaccine production.[40]

The influenza vaccine available each year is carefully engineered to respond to particular strains of the flu virus judged most likely to emerge, based on projections and the genetic composition of prior outbreaks. However, the creation of a vaccine to combat a particular flu season is an annual "best guess" by virologists. It takes approximately six to nine months to develop yearly seasonal flu vaccines or vaccines for new strains of the disease. Therefore, employing the currently used technology, there will be a lag time of at least six to nine months before a vaccine will be available after a pandemic outbreak strikes.

There are new technologies being developed to produce flu vaccine, which involve cell cultures, recombinant protein, and DNA-based approaches. They may help produce influenza vaccine more efficiently and provide more adaptability to unexpected problems or losses in production. The U.S. government invested approximately $150 million from 2003 to 2005 to stimulate development of these new technologies.[41]

However, this technology will not be available for use in the immediate future, including within the timeframe that most experts predict a pandemic outbreak will occur.

In the meantime, the federal government has invested $41 million to expand and maintain the chicken flocks used by Sanofi Pasteur, the only company that has a U.S.-based influenza vaccine production facility. The goal of the investment is to ensure that flocks can produce eggs year-round.

Stopgap Antiviral Measures

It is possible to protect people in the six to nine months or more it would take to develop a vaccine for a new, lethal flu virus strain. In February 2005, WHO released a report recommending that nations around the world stockpile antiviral medication to protect against the current lethal strain of the flu. The WHO has recommended this as a "stopgap" measure,because it would likely take a minimum of six months to develop a vaccine after a widespread outbreak. Therefore, antiviral stockpiling would be an essential interim step to have a way to protect people if a pandemic outbreak spread before a vaccine was developed. While health experts expect a pandemic will occur soon, there is no way to forecast the exact timing. The quantities of both the antiviral medication and a vaccine, once one becomes available, would be limited and countries around the world would all be seeking supplies.

An antiviral medication, oseltamivir (Tamiflu®), exists that could help alleviate symptoms of those who contract the pandemic flu and reduce mortality levels. It can also be used as a preventive treatment, to help protect emergency first responders and frontline health care workers.

The WHO estimates that a pandemic flu outbreak could impact approximately 25% of the population in nations worldwide. Unfortunately, there is limited production capacity for this vital medication.

The U.S. federal government reportedly has ordered 5.3 million courses of Tamiflu® for the Strategic National Stockpile; however, it would require approximately 70 million courses to cover 25% Several other countries have already ordered enough Tamiflu® to protect between 20 and 25% of their populations in case of an outbreak. The United States is already behind in the queue to place an order for the medication, for which there is a single manufacturer worldwide — Roche Pharmaceutical, which is located in Switzerland. In testimony before the U.S. House of Representatives Health Subcommittee of the Energy and Commerce Committee, the Roche company medical director for Tamiflu® explained that historically they have not produced the levels of Tamiflu® required for global stockpiling. To help accommodate the growing concerns and orders, they have increased production of the antiviral nearly eightfold since 2003.[43]

On March 1, 2005, the British government announced that it was taking steps to procure 14.6 million courses of Tamiflu®.[44] This procurement would cover 25% of the British population, the rate WHO has recommended.

Given the current and projected production capacity, if the United States did place a large order for Tamiflu®, Roche has testified before Congress that it could be the end of 2007 before they could deliver enough to the national stockpile for 25% of the population. Thus, antiviral treatment will be an effective part of the U.S. response only if a pandemic does not occur for several years and, of course, if the pandemic strain is responsive to antiviral medications. In testimony before the U.S. House of Representatives Health Subcommittee of the Energy and Commerce Committee on May 26, 2005, Dr. Dominick Iacuzio, Medical Director for Tamiflu® at Roche, said:

> Roche has received and is filing on schedule, a pandemic stockpile order for Tamiflu® from 25 countries worldwide. Discussions are underway for the U.S. government to purchase significantly greater amounts of Tamiflu. However, HHS stockpile purchases to date are sufficient to treat less than 1% of the U.S. population. We have also received a non-binding letter of intent for HHS to purchase additional treatments to cover under 2% of the population.

In contrast, countries such as the United Kingdom, France, Finland, Norway, Switzerland, and New Zealand are ordering enough Tamiflu to cover between 20 and 40% of their populations. Unfortunately, given the complexities I have described and the increasing global demand, any government that does not stockpile sufficient quantities of Tamiflu in advance cannot be assured of an adequate supply at the outbreak of an influenza pandernic.[45]

A Strategic National Stockpile and Ongoing Antiviral Concerns

Tamiflu® and other antiviral medications have shelf life considerations. The FDA has currently approved a five-year shelf life for Tamiflu®. In the event that a pandemic does not occur within the five-year window for use of the stockpiled reserve of the drug, the United States and other countries can still make use of the Tamiflu® they have ordered for use against annual regular flu concerns before it expires.

Shelf life concerns of antiviral medications, that are part of the Strategic National Stockpile, however, need to be taken into consideration. Budgeting to replace the reserve of medications, which have been determined to be essential to protecting Americans in the event of emergencies, needs to be factored into ongoing homeland and health security discussions. According to the European Commission, November 27, 2001,

> the next pandemic is imminent ... (and we) ... are not prepared. Vaccine availability is not secured, antiviral stocks do not exist and will not be under the current market forces.... in the event of a pandemic millions of people could die, economies could be affected and medical and civil services could collapse. members of the public will not excuse authorities, who will be held responsible for not having put in place up-to-date preparedness.[46]

New Concern: Birds to Pigs to Humans?

According to a May 2005 edition of Nature magazine, scientists are increasingly concerned that a "dangerous strain" of the avian flu virus may be growing in pigs in Indonesia.[47] Pigs are known to serve as a "mixing vessel" that incubates strains of disease that then become more easily transmissible to humans.[48]

Southeast Asia's Containment Capacity

Southeast Asia, the epicenter of the avian influenza outbreak, has a poor capacity to contain a pandemic if one should emerge. A pandemic would overwhelm the capabilities of local Asian health departments. With their insufficient capacity and technology, much of Southeast Asia's ability to detect and monitor the outbreak is severely limited. As a result, the United States must also decide how it will assist the world community in responding to the threat — and prepare for the worst.

The United States versus the World?

U.S. planning and preparedness for a pandemic lags behind a number of other countries. Table 9.3 below is a comparison of the United States versus the United Kingdom and Canadian efforts across a number of dimensions. While the preparations in the United Kingdom and Canada compare

Table 9.3   Relative Preparedness: The United States, United Kingdom, and Canada

	U.S.	U.K.	Canada
Leadership	No governmentwide coordination mechanism in place; Secretary of HHS receives daily briefings; within HHS divided authority between pandemic preparedness (Office of the Assistant Secretary for Health) and pandemic response (Office of the Assistant Secretary for Emergency Preparedness). Guidance provided to states for planning purposes; no release or systematic review of local plans.	Cabinet-level office coordinates governmentwide and "civil society" efforts. Department of Health leads public health and National Health Service response. Similar plans adopted and coordinated with Scotland, Wales, and Northern Ireland.	Pandemic Influenza Committee cochaired by federal government and provincial representative to coordinate national efforts.
Planning	Draft plan released August 2004 for public comment; no timeline for finalized plan.	Plan effective March 2005	Plan effective February 2004
Vaccine development	NIH funded research on pandemic vaccine; initial contracts for building potential vaccine stockpile.	Research being conducted on potential vaccines. Prototype "dossier" for regulatory approval of pandemic vaccine in development.	Canadian government has contracted for reserve production capacity for a pandemic vaccine with a Canadian manufacturing facility.

Vaccine Planning	Significant portion of U.S. vaccine supply manufactured abroad. HHS soliciting public comment on prioritization for receipt of limited supply of vaccine.	Discussions with manufacturers regarding optimizing capacity. Most UK vaccine production occurs in the UK. Priority groups for vaccination identified in plan.	Priority groups for vaccination identified in plan.
Antiviral Planning	5.3 million courses of antiviral ordered for stockpile (2% of population). HHS soliciting public comment on prioritization for distribution of limited supply.	Courses to cover 25% of the UK population have been ordered. Priority groups for antiviral treatment identified.	Courses order to cover 3% of the Canadian population. Priority groups for antiviral treatment identified.
Health care system surge capacity planning	Planning guidance provided for healthcare system.	UK has an integrated health care system through National Health Service and local Primary Care Trusts, under direct leadership of U.K. Department of Health.	Checklist of activities issued for surge capacity preparation.
Communications planning	Outline of steps to prepare communications tools and mechanisms.	Specific professional (provider), public and media communications messages and activities identified by stage of pandemic.	Checklist of activities by stage of pandemic issues.

favorably to the United States, some public health experts still have concerns about the degree of implementation-readiness in their efforts.

Recommendations

Congressman Jerry Lewis (R-CA), Chairman of the Appropriations Committee of the U.S. House of Representatives has said that: "U.S. pandemic readiness will depend on immediate and long-term strategies as well as dedicated, informed federal leadership." While U.S. Department of Health and Human Services Secretary Mike Leavitt said on May 16, 2005: "There is a time in the life of every problem when it is big enough to see and small enough to solve. For flu preparedness, that time is now."[49]

Overall, U.S. pandemic preparedness is inadequate. Both the federal pandemic plan and various state pandemic plans are insufficient blueprints for an effective national response to a pandemic influenza.

How prepared the United States and the rest of the world are to respond to and control a pandemic will be determined by how much time remains until an outbreak occurs. Preparations must he considered without knowing this exact time frame. Scientists predict it could happen as soon as this year, or it could take several years. Therefore, planning and policies must consider what would need to be done if an outbreak occurred very soon or with longer lead time to prepare.

- **Crucial immediate steps** that must be taken to minimize loss if a pandemic occurs in the near term include outbreak tracking, stockpiling medical supplies, and communications plans.
- **Intermediate steps** that must be considered if a pandemic occurs with several years to prepare, include stockpiling antivirals and developing additional surge capacity plans for hospitals and other medical providers.
- **Longer-range steps** that should be undertaken if there are a number of years to prepare include increasing vaccine production and the development of new technologies for vaccines.

Whether a pandemic emerges from the H5N1 virus or a different strain, the challenge remains constant. Responding quickly and effectively to a pandemic requires a comprehensive national plan integrated with state and local-based emergency planning efforts. Though wider national attention and a general acknowledgment of the virus's danger are important developments, specific U.S. pandemic planning efforts are in need of immediate attention. A review of both the federal pandemic plan and state pandemic plans found that many important planning topics remain underaddressed. Shoring up these weaknesses should be the highest of government priorities. In the interim, every effort should be made by the federal government to procure the antivirals as a stopgap measure.

To move toward operational plans capable of wider integration and implementation, and as a first step toward a strong, cohesive, and rapid-preparation U.S. pandemic flu strategy, Trust for America's Health (TFAH) recommends the following activities be addressed in federal, state, and local preparedness efforts:

Define Roles and Responsibilities

A clearly defined organizational structure and chain of command is essential for rapid and efficient control and response, both in the federal government and at the state level. At the federal level, the president should designate a single senior official whose primary responsibility is to assure Cabinet-level coordination of the federal government's response to a pandemic and also to ensure coordination between civil society (nongovernmental economic infrastructure) and government during a pandemic. Immediate planning should be occurring at the federal level to minimize disruption of the healthcare system and the overall economy. CDC must review and approve of state pandemic plans to ensure nation-wide preparedness standards and regional coordination. States must define and agree upon leadership roles and responsibilities with respect to who is in charge of a state's public health and healthcare decisions. Plans must also designate liaisons to work with other jurisdictions and federal officials.

Outbreak Tracking Plans

should ensure adequate laboratory surveillance of influenza, including the ability to isolate and subtype influenza viruses year-round. Following federal guidelines outlined by HHS, states should report all necessary data and information to federal and other health officials as soon as it becomes available. Congress should provide additional support for CDC's global surveillance activities, and the United States should support the WHO surveillance program to assure as early a warning as possible for U.S. preparedness purposes.

Vaccine Research, Development, and Production

The United States should continue to support and expand research into new technologies for influenza vaccine and clinical trials for potential avian flu and other pandemic vaccines. While the United States has issued limited contracts for stockpiling a potential pandemic vaccine, the federal government should also explore the Canadian approach of contracting for a reserve production capacity located in the United States. A vaccine stockpiling approach is successful if public health authorities have guessed correctly as to what the pandemic strain will be. A reserve production capacity can assure quick turnaround for production of a vaccine for the actual pandemic strain.

Prior to production, the FDA must approve a new vaccine. Other nations are putting protocols in place now with respect to creating a rapid review process for a pandemic flu vaccine. With clear advance notice of the scientific data that will be required for approval from regulatory agencies, vaccine manufactures can better anticipate how to comply. For example, regulators in the United Kingdom are already working with vaccine manufactures to develop a model application for approval of a pandemic vaccine, which they estimate could reduce production time by as much as two months. The FDA should adopt a similar strategy.

Procure Additional Antivirals for Treatment

Even during a pandemic, when efforts to contain transmission may seem futile, there exists a capacity to treat infected individuals. While the ultimate effectiveness of treatment depends on the particular strain, Tamiflu® may be an effective treatment option while scientists work on the development or a vaccine. Furthermore, Tamiflu® can be used prophylactically to protect hospital and health care workers on the front lines.

The recently enacted emergency supplemental appropriations legislation made available $58 million for the purchase of influenza countermeasures for the Strategic National Stockpile, including, but not limited to, antiviral medications and vaccines. These funds are most welcome, but TFAH believes that Congress should provide additional funds during the fiscal year 2006 appropriations cycle to continue to build the nation's antiviral stockpiles from the current level of 2% of the U.S. population.

Mass Vaccination and Treatment Systems

The federal government, in coordination with the states, must develop systems for tracking and distributing antiviral medication and vaccines. A national system is needed to assure targeted or equitable distribution of supply, so we do not have a repeat of the distribution problems we experienced in the 2004 to 2005 flu season. State-level systems also are needed to assure similar availability across a state. One of the best ways to improve vaccination preparations for a pandemic outbreak may be to enhance annual flu vaccination coverage for nontraditional high-risk groups (e.g., individuals with chronic diseases or compromised immune systems) to facilitate access to these populations during a pandemic.

Prioritize Who Would Receive Antivirals and Vaccines Based on Limited Supplies

It is important to determine a protocol for allocation among high priority populations, such as healthcare workers, prior to an outbreak. The federal government should provide specific guidance to states as to which sectors of the population should receive antiviral medications and vaccines, and in what order, particularly since the amount of available pharmaceuticals will be limited.

Public Information Campaigns and Materials

Communicating with the public in a clear and efficient manner is essential during a high-anxiety time. The federal government, in conjunction with the states, should develop coordinated messages for various audiences (media, public, providers, etc.) for each stage of a potential pandemic. States must identify and train spokespersons in multiple languages and educate public health officials, politicians, community leaders, partners, and the media about what information will and will not be available during a pandemic. States should ensure clear and consistent messaging by creating information templates in multiple languages ready for customization and distribution during a pandemic.

Stockpile Medical and Safety Equipment for Health Care Workers and First Responders

Efforts must be undertaken to ensure that basic medical and safety equipment will be available for healthcare workers and emergency responders in the event of a major outbreak. Currently, most health providers order and stock supplies on a "just-in-time" basis. This means they often have only a few days of reserve supplies, equipment, and medicines, including many basic protective items, such as masks, gloves, gowns, and clean hospital linens. In order to prepare for a mass event, steps must be taken immediately to stockpile additional supplies, particularly since during an outbreak, many production and delivery systems for supplies will likely be stalled or even stopped.

Surge Capacity Capabilities

Plans must account for the likelihood that hospitals will be quickly overwhelmed during a pandemic, by developing auxiliary sites such as shelters, schools, nursing homes, hotels, and daycare centers for surge capacity treatment and for treatment of the "walking well." States should be conducting surveys of potential sites and obtaining agreements. Cooperation and integration with local health officials and first responders is essential.

Secure a Backup Workforce

States should conduct and maintain an inventory of residents who are healthcare professionals, including current and retired doctors, nurses, veterinarians, emergency medical staff, and other potential volunteers. These workers could be an essential expanded workforce during a pandemic. Survivors of a pandemic are also a population of potential workers. States should plan for tracking and soliciting volunteer support from this population, which is presumably immune to the virus. Planning efforts should also incorporate private sector support whenever possible, especially in infrastructure and nonhealth service provider capacities.

Ensure Availability of Food, Water, and Other Supplies

States must account for high demand for food, water, and other basic supplies, and plan for distribution to general and hard-to-reach populations. Plans should factor in potential complications of infected food and delivery workers, possible infected store facilities, and limitations on public interaction both for those infected and the general population at risk of exposure. Planners must also weigh the issue of "just-in-time" manufacturing of food and supplies, since reserves of supplies will not be available. Additionally, planners must address the limitations of medical equipment manufacturing, much of which Asia exports to the world.

Quarantine Measures and Authority to Close Public Places

States must establish clear legal authority and emergency measures to effectively contain the spread of disease. States must have powers to prohibit public gatherings, close public facilities and schools, and restrict travel, if necessary.

Measures to Manage Mass Death

Planning for worst-case scenarios is a critical component of effective planning. States must conduct and maintain an inventory of facilities with sufficient refrigerated storage to serve as temporary morgues in the event of a pandemic.

As indicated, there are several concrete steps that the United States can take to better prepare against an influenza pandemic. Such policies and investments will help stabilize the nation's health and economy in the event of a pandemic while ensuring that pandemic readiness preparations are "commensurate with the scale of the threat we face."[50]

Table 9.4   State-by-State Range of Potential Pandemic Influenza Deaths and Hospitalizations*

State	Projected Dead		Projected Hospitalized	
	25% Contraction, Lower Severity Flu Deaths	25% Contraction, High Severity Flu Deaths	25% Contraction, Lower Severity Hospitalizations	25% Contraction, High Severity Hospitalizations
Alabama	2,962	17,771	12,863	77,178
Alaska	295	1,771	1,519	9,114
Arizona	3,074	18,446	13,225	79,350
Arkansas	1,783	10,700	7,553	45,318
California	20,292	121,750	91,030	546,180
Colorado	2,397	14,383	10,993	65,958
Connecticut	2,351	14,107	9,978	59,868
Delaware	502	3,014	2,187	13,122
District of Columbia	385	2,310	1,658	9,948
Florida	11,912	71,474	47,462	284,772
Georgia	4,552	27,309	20,970	125,820
Hawaii	815	4,892	3,524	21,144
Idaho	760	4,558	3,385	20,310
Illinois	7,907	47,439	34,579	207,474
Indiana	3,939	23,634	17,237	103,422
Iowa	2,078	12,465	8,697	52,182
Kansas	1,791	10,746	7,648	45,888
Kentucky	2,643	15,859	11,583	69,498
Louisiana	2,778	16,668	12,383	74,298
Maine	884	5,302	3,778	22,668
Maryland	3,319	19,916	14,833	88,998
Massachusetts	4,379	26,271	18,679	112,074
Michigan	6,541	39,244	28,668	172,008
Minnesota	3,101	18,608	13,596	81,576
Mississippi	1,787	10,723	7,844	47,064
Missouri	3,758	22,548	16,080	96,480
Montana	601	3,608	2,595	15,570
Nebraska	1,147	6,882	4,899	29,394
Nevada	1,081	6,486	4,819	28,914
New Hampshire	778	4,665	3,434	20,604
New Jersey	5,660	33,960	24,264	145,584
New Mexico	1,081	6,488	4,835	29,010
New York	12,567	75,401	54,163	324,978

Table 9.4 State-by-State Range of Potential Pandemic Influenza Deaths and Hospitalizations*

State	Projected Dead		Projected Hospitalized	
	25% Contraction, Lower Severity Flu Deaths	25% Contraction, High Severity Flu Deaths	25% Contraction, Lower Severity Hospitalizations	25% Contraction, High Severity Hospitalizations
North Carolina	4,996	29,973	21,880	131,280
North Dakota	457	2,742	1,931	11,586
Ohio	7,732	46,393	33,326	199,956
Oklahoma	2,278	13,666	9,792	58,752
Oregon	2,241	13,447	9,682	58,092
Pennsylvania	9,062	54,369	37,553	225,318
Rhode Island	745	4,467	3,087	18,522
South Carolina	2,491	14,947	10,995	65,970
South Dakota	520	3,118	2,199	13,194
Tennessee	3,625	21,750	15,893	95,358
Texas	11,708	70,247	53,550	321,300
Utah	1,131	6,786	5,302	31,812
Vermont	395	2,369	1,738	10,428
Virginia	4,368	26,207	19,624	117,744
Washington	3,637	21,820	16,204	97,224
West Virginia	1,350	8,097	5,671	34,026
Wisconsin	3,540	21,240	15,281	91,686
Wyoming	305	1,830	1,363	8,178
U.S. Totals	180,478	1,082,866	786,032	4,716,192

Note: Projections based on CDC's FluAid 2.0 program. The estimated deaths and hospitalizations assume the following: The projections range from the most likely number of deaths and hospitalization at a 25% rate of contraction for a relatively mild pandemic, similar to the 1968 pandemic, to the most likely number of deaths and hospitalizations for a more severe pandemic, similar to the 1918 pandemic.

Chapter 10

Influenza: Biology, Transmission, Course, Complications, Prevention, and Treatment

Hillel W. Cohen, M.P.H. Dr. P.H. and
Christina M. Coyle, M.D., M.S.

CONTENTS

"I've got the flu." On a typical winter morning, thousands of workers in the United States will call their jobs and cough, wheeze, or sniffle these words as they prepare to suffer in bed for a day or two or as long as a week. They'll be joined by thousands of children staying home from school and thousands more grandparents too ill to do their daily chores. For most, it will be a temporary inconvenience. But for as many as an estimated 36,000 people in the United States alone each year, that case of flu will lead to death.[1]

Worldwide, influenza is estimated to be the underlying cause of 250,000 to 500,000 deaths each year.[2] This estimate is for a typical year and for typical strains of influenza virus. At least three times in the last century, worldwide pandemics of particularly virulent and deadly strains of influenza virus claimed the lives of millions more. The worst of these pandemics occurred from 1918 to 1919. Although record-keeping was poor, estimates of those who died in the pandemic range from 20 million to as many as 100 million, or one-twentieth of the world's population at that time.[3] In the last 200 years, the longest time between recognized flu pandemics has been about 40 years. With the last documented flu pandemic in 1968, many public health officials are concerned that another pandemic may be imminent.

In this article, we will review the biology of influenza, characteristics of transmission, the course of illness, complications, and the current methods of prevention and treatment. We will then present the public health issues concerning flu, including a discussion of what is being done and not being done with regard to the growing fear that an avian flu (H5N1) pandemic may emerge at any moment.

Epidemiology of Influenza

Most people with the flu never see a doctor, and are more likely to treat their own symptoms with bed rest, home remedies, and over-the-counter preparations. Some states ask doctors to report tallies of numbers of flu cases seen to local health departments, but such reporting is not uniform nationally and the extent of adherence to reporting is also not known. Thus, estimates of national and international incidence extrapolated from limited reports and samples must be considered very approximate.

Estimates of mortality rates from flu derived from death-certificate data are more reliable, but here too, there is considerable room for uncertainty. Patients who enter a hospital with a diagnosis of flu and later die from its complications will likely have influenza listed as an underlying cause of death. But for the substantial majority, hospital admission and cause of death may be listed only as unspecific pneumonia or upper respiratory infection, even if a case of influenza may have been what gave rise to the pneumonia. Mortality from influenza takes place primarily among the elderly, and for those who die at home, the death may be recorded as a sudden death or a heart attack when flu symptoms may have precipitated the event.

Nonetheless, the CDC estimates that in the United States anywhere from 5 to 20% of the population will contract flu in a given year; that about 200,000 of these will be hospitalized for flu or complications, and that about 36,000 people will die.[4] WHO estimates that the corresponding figures for the world are that 5 to 15% who will contract the disease; about 3 to 5 million will have severe illness, and from 250,000 to 300,000 will die.[5]

A substantial number of these deaths may be preventable. For many years, flu vaccines have been available. As described above, flu vaccinations have to be administered each year since the prevalent strains of flu change sufficiently to limit the effectiveness of earlier vaccinations. Similarly, it is believed that unlike some viral infections, surviving a case of a particular strain of flu does not provide sufficient protection from infections by new strains in subsequent flu seasons.

In the 2004 to 2005 season (through January 31, 2005), approximately 62.7% of those 65 years old and over, 35.7% of healthcare workers with patient contacts, and 25.5% of adults aged 18 to 64 with high risk conditions received a flu vaccination.[6] This compares to national goals for 2010 of 90% coverage of those 65 and over and 60% of those 18 to 64 years with risk factors.[7] International data for vaccination coverage are not readily available, but it is likely that comparable proportions are vaccinated in the industrialized countries and much smaller proportions in most developing countries, either due to lack of resources or problems with the healthcare delivery system.

For the most part, individuals in the United States have to take the initiative in seeking out vaccination, whether by inoculation, or more recently with a nasal mist. Individual primary care doctors provide most vaccinations. Institutions such as hospitals and nursing homes also offer inoculations for their staffs and vulnerable patients. Some health departments at the state and county levels will also provide or facilitate inoculations, but these are not universally available.

Limited public awareness that flu can have serious consequences is a major factor limiting more widespread use of vaccine. Government sponsored announcements are sporadic as are media reports. When a disease is hot news, broadcast and print media provide running tallies of cases and fatalities; for example, human cases of West Nile fever will be reported, and any fatalities will make headlines. Some years ago, Lyme fever was "hot" and outbreaks were featured stories. During the anthrax outbreak in 2001, every reported and even suspected case would be the top news story for days. On the other hand, flu and other major causes of death are viewed as too commonplace to be newsworthy. There are no running totals of cases and fatalities and few if any special reports.

One aspect of flu, however, has made the news — repeated shortages of flu vaccine. Most recently there have been shortages due to disruptions in either manufacturing or distribution of vaccine in the 2000 to 2001, 2001 to 2002, and 2004 to 2005 flu seasons.[8] Each time, increased supplies have become available later toward the end of the season, but the earlier shortfalls likely impact both vaccination and infection rates.

The shortfall in the 2004 to 2005 season was particularly serious. The Chiron company had contracted with the U.S. government to be one of the largest suppliers. Although a U.S. based company, the primary flu vaccine manufacturing facility was in the United Kingdom, and the plant there had been warned several times by British health officials that safety/quality control conditions were below acceptable standards.[9] The company failed to make the necessary corrections and British health officials closed the plant. This plant was to be a major source of vaccine for Britain as well as the United States, but when the early warnings were announced, British authorities made alternate plans to get vaccine supplies. The CDC and other U.S. health officials however accepted company assurances that the plant would satisfactorily meet the minimum safety standards and avoid a shutdown, so no alternate plans were made. Almost half of the anticipated supply was not available and health officials had to scramble to direct vaccine supplies to the higher risk populations. When more supplies became available later in the season, others were allowed to get the vaccine.

To make up for the shortfall, those in high-risk categories were given priority in getting the vaccine. As a result, coverage for the over-65s was only slightly lower than in 2003 (62.7 vs. 65.6%), as was that of health care workers (35.7 vs. 40.1%).[10] Adults under 65 with other risk facts had substantially less coverage in 2004 vs. 2003 (25.5 vs. 34.2%), while coverage among adults 18 to 64 without other risk factors fell by about half (8.8 vs. 17.8%). At the time of this writing, data are not available to assess whether the shortage resulted in a higher than expected mortality, hospitalizations, or days of work lost in the 2004 to 2005 season. Even when mortality figures become available, it will be difficult to ascertain to what extent any differences can be attributable to the supply. Factors such as the relative virulence of the flu strain, the relative effectiveness of the vaccine, weather conditions, and other variables that are unknown or difficult to measure can conceivably confound comparisons from one year to the next.

Difficulties in assessing the impact of vaccinations on mortality and morbidity may also provide a barrier to achieving higher coverage or the goals set for 2010. Because new formulas for vaccine must be developed each season, it is questionable as to what extent data from previous vaccines can be generalized for each new formulation. Even so, the lack of systematic, active monitoring of the results of vaccinations, and comparisons with nonvaccinated individuals make assessments even more difficult. Implementation of a Behavior Risk Factor Surveillance System (BRFSS) has improved estimation of coverage, but does not address outcomes. Skepticism among substantial sections of the population about effectiveness will continue to be hard to dispel without better evidence of the benefits.

Pandemic Flu

The ongoing problems with regular flu vaccination will be compounded in the event of a pandemic outbreak — especially if the mortality or serious morbidity from the pandemic virus is greater than usual. This is why so many public health officials are alarmed about the possibility of so-called avian or bird flu, and in particular the H5N1 strain. Avian flu virus is prevalent among waterfowl, including ducks, that can carry the virus without getting symptoms. H5N1 can also infect commercially grown chickens. In some East Asian countries the preponderance of chickens may be infected with H5N1. In the last few years, several cases of avian flu virus in humans have been identified. So far, human cases are believed to have been limited to farms in China, Thailand, Vietnam, Cambodia, Indonesia, and elsewhere in East Asia. At least 50 people have died from the disease. In small-farm production where people come in very close proximity with the animals they tend, the potential for animal to human transmission increases substantially. It is unclear in these cases whether the virus jumped directly from chicken to human or whether there were intermediary animals such as pigs that are also routinely kept on farms and whose genetic makeup may provide a bridge from birds to humans.

Such jumps between animals are not rare, but usually the virus that jumps does not have the genetic capability of being transmittable from human to human. Human to human transmission is a prerequisite for any large-scale outbreak and in order to have that ability, the virus must exchange genetic material with its new host or with other viruses in the host. In a couple of instances of reported avian flu in humans, several people in a household succumbed to the virus. This raised the specter that a mutation or genetic exchange had already occurred. On the other hand, the clustering of cases in a household could also have been due to common exposure to the same infected animals.

In the human cases that have been observed so far, the mortality rate has been between 35 and 50% [11] — much higher than the 1 or 2 per 1,000 usually observed with ordinary flu. No one can predict what the mortality rate will be if the H5N1 makes the genetic changes necessary to acquire the ability of human to human transmission. Also, the very high mortality rate observed so far in the animal to human transmission of H5N1 may be due to a heavy selection bias. In these remote rural areas, it is possible that only those who are already very sick come to the hospital for treatment or are given the special tests necessary to identify H5N1. Nonetheless, just the possibility of a mortality rate substantially higher than 0.1% would be enough to lead to serious concern as to what might happen if the H5N1 virus mutates or otherwise transforms sufficiently to achieve an efficient human to human transmission.

There is no way of knowing when influenza first appeared in human populations, and reports of illnesses with symptoms resembling flu can be found in ancient texts. Flu pandemics have probably been taking place for at least 300 years, although even these estimates from earlier centuries have to be construed from scattered sources and accounts of symptoms.[12] Since 1900 there have been three major pandemics recorded, starting respectively in 1918, 1957, and 1968.

Pandemics Past and Future

By far the most deadly was the "Spanish" flu pandemic of 1918/19. The name arose because Spain seemed like the epicenter of the illness when it first came to public awareness.[13] Later examinations of the pattern of the disease suggest that the earliest cases were in military camps in the United States, and that it might have traveled to Europe with U.S. troops sent there to fight in World War I. But, wherever it started, it spread quickly around the globe and resulted in huge numbers of deaths. Estimates range from 20 to 100 million victims with about 500,000 deaths in the United States.[14] However, international record-keeping was too sporadic in the midst of the world war, the Russian civil war, and the worldwide pandemic for any estimate to be reliable.

It is also very hard to estimate the case fatality rate that varied markedly according to age. Older people, who tend to be more frail and infants without much natural immunity, are often more vulnerable to infection and death from flu and other infectious diseases. But the 1918/19 pandemic was unusual in that death rates were especially high for people in their twenties as well as the very young and very old. This made it particularly devastating in the military camps where young men were crowded together in large groups. In some of these camps where it was easier to count, death rates as high as one in ten of those showing symptoms were observed. In other populations the rate was probably much lower.

At that time, it was only possible to tell if people were infected if they showed the usual symptoms. When antibody and other tests became available decades later, it was possible to show that many more people are exposed to a prevalent flu virus than succumb to symptoms, and it is likely that many of these will have sufficient amounts of virus to transmit the disease even if they themselves had a strong enough immune response to avoid illness. If the "Spanish" flu virus was transmitted in that fashion, then one can surmise that a very large proportion of the world's population was infected and it is possible that very few indeed were not exposed. In recent years, some partial samples of the 1918/19 virus have been recovered, including from lung tissue in a body exhumed from the frozen ground of a remote Inuit fishing village in Alaska where 85% of the adult population had died from the disease.[15]

Even though virtually the full genome of the 1918/19 virus genome has been sequenced, based on RNA samples extracted from several sources, the full origins of this type A H1N1 virus are still not clear. Some features similar to swine flu and other features similar to avian flu gave rise to a hypothesis that the virus had adapted from bird to swine to human. Despite the similarities, however, there are enough differences to deter characterizing it as directly related to either. More recently a hypothesis has been raised that the virus stems from an isolated avian virus that spread to both humans and swine.[16] Type A H1N1 strains still circulate in human populations, although none of these has shown the deadly force of the original "Spanish" flu. [17]

The other two pandemics in the twentieth century took place in 1957/58 and 1968/69. The 1957/58 type A H2N2 strain was first identified in China and dubbed "Asian" flu. This pandemic is believed to have caused about 70,000 deaths in the United States and as many 10 million deaths worldwide. In 1968/69, a type A H3N3 virus was first identified in Hong Kong and dubbed "Hong Kong" flu. Death estimates were about half of that of the 1957/58 pandemic. Whether the lower number of deaths was due to a higher level of palliative care or a less virulent virus or other factors is difficult to establish. Both the H2N2 virus of 1957/58 and the

H3N3 virus of 1968 /69 are considered to be of avian flu origin that combined genetic material with human influenza viruses and gained the ability to be transmitted from human to human. [18]

With a pandemic appearing in each of the two preceding decades, concern about an imminent pandemic was high in the 1970s. In 1976, a 19-year-old U.S. soldier stationed in Fort Dix, New Jersey, died within twenty-four hours of reporting severe flu symptoms. Four other soldiers on the same base were hospitalized and doctors determined the virus to be swine flu. At that time, some virologists believed that the 1918/19 pandemic had also been a version of swine flu and were aware that the earlier pandemic had started in a U.S. military base. Fear that the Fort Dix outbreak might be the index cases of a new 1918/19 pandemic led the CDC, the Department of Health, Education and Welfare (predecessor of today's HHS), and the administration of President Gerald Ford to organize an emergency response. In the context of a presidential election campaign, the Ford administration called for a massive program to rush a new vaccine into production and vaccinate the entire U.S. population.[19] Although many other soldiers on the base came down with flu, and at least 500 of these had antibodies suggesting exposure to swine flu, no other illnesses or deaths from the swine flu virus were recorded. Nonetheless, within a year over 40 million people in the United States received vaccinations.

The mass vaccination program ran into considerable trouble. There were disruptions in vaccine supply that would be expected in such a huge and hurried campaign. More importantly there were disturbing reports of sudden deaths taking place in close proximity to vaccine administration. Most of these were considered coincidental heart attacks but a connection could not be decisively ruled out. Public awareness that the vaccine might not be safe combined with the absence of any confirmed illnesses from swine flu after the Fort Dix cases led to an abrupt halt of the vaccination program. In the months and years that followed, there were numerous reports of neurological disorders, especially Guillain-Barre syndrome, which might have been connected to the vaccine.

In retrospect, the episode was viewed as a debacle for the Ford administration and for U.S. government health officials. The lack of any additional cases suggests that it is likely that the soldiers infected had gotten it from the same source (which was never identified) and that the virus was not transmitted from human to human. It is unlikely that the vaccinations prevented the spread: It took several months before a vaccine was widely available and one would expect many cases to have been detected during that time. Public attitudes toward massive vaccination for flu were influenced for years after, although memories of the episode seem to have little impact today. Instead, public concern about a pandemic is again growing.

Can the Next Pandemic Be Stopped?

Representatives from the World Health Organization (WHO), Centers for Disease Control and Prevention (CDC), and other national and regional health organizations have been meeting to map strategies to forestall a worldwide pandemic of the kind that claimed tens of millions of lives in 1918/19. One strategy that has been put forward is to try to set up an early-warning system in areas where H5N1 has been observed to catch any early outbreaks and to concentrate all available vaccines and antiviral medication in that region in an effort to contain the outbreak before it spreads widely.

Two scientific teams have already published computer generated simulation models to try to estimate how quickly a pandemic might spread from a point source in Southeast Asia.[20, 21] The teams used somewhat different methods and assumptions, which included the basic assumption that an H5N1 avian virus reassorted with an H3N2 human virus that allowed sustained human to human transmission. Assumptions for transmission rate (virulence), the limited effectiveness of existing vaccines (not reformulated for the specific new virus), effectiveness of antiviral medications, and other factors were based on experience with current prevalent H3N2 human viruses. Sensitivity analyses to assess how changes in the assumptions would affect the results were also performed. Both reached similar conclusions, namely that containment was possible provided that a massive intervention could be mounted quickly and provided that the natural transmission rate of the newly emerging virus was not extraordinarily greater than what is now prevalent.

Nonetheless, several big obstacles will have to be overcome in order for there to be any reasonable chance of employing the containment strategy. First, surveillance has to be adequate to identify the index cases early enough to mobilize the human and pharmaceutical resources into the infected area to begin an intervention within about two weeks. However, surveillance in the rural areas of underdeveloped countries will be a challenge. If the first human transmission cases take place in a rural area, under usual conditions, one would expect a substantial delay before illnesses and deaths are identified as resulting from a new and potentially pandemic virus. Will the WHO be able to put into place increased surveillance measures? Will the communication and education infrastructure be adequate to identify the index cases and then inform and mobilize the community for the intervention? Will there be cooperation from those countries affected and the communities within them that are deemed likely to host an originating outbreak?

Second, the models presume that an effective antiviral medication will be available. To date only oseltamivir (Tamiflu®) is believed to be effective against H5N1. It is likely to be effective against a reassorted or mutated strain, but how effective it will be is unknown. Also unless a huge increase in production is begun soon, there may not be adequate supplies of the drug. Will the wealthier nations that have ordered supplies make those supplies available for a containment effort in another country or will they hoard their stocks? Once word of an outbreak spreads, political pressures may mount to make the drug available for their own populations.

Third, the models to a degree presume at least a partly effective vaccine will be available. A prototype vaccine for H5N1 has already been developed but at this writing, larger safety and efficacy tests still remain to be done. Even if the virus proves efficacious in early trials, it will still be unknown whether the efficacy of the vaccine that is measured in the early trials will be altered by the mutations and genetic rearrangements that must occur before such a virus can be transmitted between humans. The above-mentioned models assume that any vaccine on hand will be only of limited effectiveness, but just how limited they are could have a major impact.

Assuming the vaccine is relatively effective, will there be sufficient stocks available for a containment strategy? As with the oseltamivir antiviral, large-scale production will be needed and this is most likely to be done in one or several of the industrialized countries. The same problem of whether vaccines in short supply will be shared and allocated for a containment effort or horded for the citizens of the country possessing the stocks could make or break an international containment intervention.

An additional problem is that if a putative outbreak is identified, might containment measures be initiated prematurely? That possibility seems to be one lesson that can be drawn from the swine flu experience in 1975/76. The cluster of cases on a New Jersey military base set in motion a national vaccination campaign. But no other cases ever presented. Was the entire episode an overreaction to a false alarm or was it an example of prudent, preventative countermeasures? Thirty years later this question still remains subject to debate.

One could argue that a massive, preemptive containment effort is a small price to pay to avoid the consequences of inordinate delay. However, a potentially premature reaction also has consequences. The substantial supplies of vaccine and antiviral medication that will be mobilized could exhaust or at least seriously decrease the stocks available for a subsequent outbreak. Flu pandemics have been rare enough for us to make a reasonable estimate that more than one in a particular year is highly unlikely. But if an early warning proves to be a false alarm, the possibility of a real outbreak soon after cannot be excluded. Fire departments are well aware that false alarms cost lives by diverting resources and delaying response time to real fires. The question then arises how early can one distinguish the next pandemic from a small, self-limiting outbreak?

Current preparations for a pandemic of H5N1 flu present still another problem. Much of the preparations are based on an assumption that the danger of pandemic is primarily from H5N1 or a related avian flu virus. What will happen if it is yet another strain? Will we be building the biological equivalent of a Maginot Line — a formidable defense that might be readily circumvented if the chance appearance of a new strain of virus doesn't conform to current assumptions? It is noteworthy that the single largest recorded flu pandemic, that of 1918/19, is now believed to be not directly related to the H5N1. Nature has a propensity for coming up with surprises.

Reviving Trouble

It is still unclear what factors led to the extraordinary transmission rate and the high case fatality of the 1918 Spanish flu. At least some of the factors may not be related to the virus itself. The pandemic took place at the height of World War I. The close quarters of the troop barracks must have greatly facilitated transmission. Soldiers carrying the infection while on leave or on weekend furloughs could easily have spread it to neighboring communities at home and abroad. At the same time, the war devastation in Europe and elsewhere had created conditions that disrupted sanitation, water and food supplies, rodent control, and all the other public health consequences of war. In some areas these conditions would have led to the deterioration of the general health of the affected communities, suppressing the ability of individuals to ward off or survive infection and thus giving the flu pandemic a virulence that may have transcended the flu itself.

However, other areas, not directly affected by the war were also hit hard. Some have hypothesized that many may have died not from the flu itself but from an exaggerated immune response that had been primed by exposure to a flu epidemic some years earlier.[22] Still others believe that some special characteristics of the virus itself, some special adaptation in its genetic makeup made it the consummate killer. Several teams of researchers sought to retrieve samples of the virus to study those characteristics more closely.

One of these research teams has described their efforts to essentially revive the virus, so that its potentially unique genetic characteristics could be studied.[23] The authors maintain that their purpose is to promote understanding that will help in the detection, prevention, and possible treatment of particularly virulent virus strains. They suggest that even if a form of the virus itself does not reappear, genes that may have made it so virulent might appear in related form in other flu viruses in the future.

It seems reasonable, one can argue, that vaccines or antiviral medication developed specifically to target particularly virulent proteins might be an important addition to protection against another pandemic.

Notwithstanding these arguments, the research into the 1918/19 "Spanish" flu virus raises a number of serious concerns. The lead research team to isolate the 1918/19 virus is affiliated with the Armed Forces Institute of Pathology, a branch of the Department of Defense. Other Pentagon scientific organizations such as the Army Medical Research Institute of Infectious Diseases (AMRIID) were involved in developing and manufacturing biological weapons. In years past, scientific agencies controlled by the military supervised the production and storage of tons of weaponized anthrax spores and smallpox virus. Biological weapons production was supposed to have been halted in 1973 with the signing of the Biological Weapons Convention (BWC). Much of the stocks of anthrax and smallpox virus were reported destroyed but AMRIID never discontinued its research activities at Fort Detrick, Maryland or at other secret laboratories and facilities around the country. Secret biological weapons research was substantially increased in the mid-1990s under the rubric of bioterrorism preparedness.

The BWC has a loophole that allows for research into potential biological weapons agents for "peaceful" and "defensive" purposes. However, it does specifically ban production of more than research quantities and devices or vehicles that could be used to "deliver" biological weapons. In other words, bombs, aerosols, and other devices for unleashing biological agents on targets are in violation of the treaty. Because the United States signed and ratified the treaty, it is, according to the U.S. Constitution, also part of U.S. law.

The BWC, however, has never had meaningful enforcement or inspection provisions. The U.S. government has long maintained that U.S. military laboratories and even commercial and academic laboratories researching biological and chemical weapons will not be open to international inspections. Both the Clinton and Bush administrations refused to support an international convention that had been negotiated by many countries to implement inspections and enforcement of the BWC. Pentagon sponsored research into biological bombs, that was publicly exposed in the *New York Times*,[24] genetic engineering of biological weapons agents, and similar research may be violations of the BWC.[25] Federal investigators into the anthrax outbreak of 2001 believe that the anthrax spores that were released into the mail that year came from an AMRIID related laboratory.[26] Ironically, the five deaths, dozens of injuries, and widespread panic that followed the release of the anthrax spores was used as justification to increase spending for bioterrorism research and has resulted in a proliferation of laboratories studying anthrax and other potential biological weapons agents. The additional laboratories and increased number of personnel working in them necessarily increases the opportunity and risk for another release of these agents whether intentionally or accidentally.[27]

It is in this context that reviving the Spanish flu virus and studying it in laboratories under the auspices of the Pentagon should raise alarm. Not withstanding the intent of individual researchers, the possibility cannot be excluded that some among the military sponsors of the research may see a modified Spanish flu virus as a potent biological warfare agent, especially if an effective vaccine is also developed that can protect those who are not targets. Even if this is not now on anyone's agenda, just the possibility could be reason enough to spark the militaries in other countries to embark on similar research as countermeasures or deterrents. The prospect of a renewed arms race in biological weapons is precisely what the BWC was supposed to avert, and the apparent violations of the BWC and active opposition to enforcement and inspection provisions are a very worrisome indication.[28,29]

Just reviving the virus itself, or using recombination techniques to transpose pieces of the virus genes onto other organisms, raises great dangers, even with the best of intentions. The anthrax episode made clear that supposedly fail-safe containment measures sometimes fail. In addition to the purposeful diversion there have been several examples of accidental release of agents or of individuals leaving the facilities infected with an agent. In one of these cases, a researcher was walking around the community after unknowingly being infected with glanders.[30] Glanders is a disease that primarily affects horses and livestock but that has been considered a potential biological weapons agent. It cannot be transmitted from human to human, and the accidental infection could not spread. What if, instead of glanders, a researcher leaves a containment compound unknowingly infected with a virulent flu agent such as a version of the Spanish flu virus? In another incident, live anthrax samples were accidentally sent by Federal Express from an institute with military connections to a laboratory in a children's hospital.[31] And in yet another unrelated incident, Meridian Bioscience Inc., a biologics company that provides what are called proficiency samples of virus to laboratories for the purpose of testing vaccines, accidentally distributed samples of H2N2 virus to as many as 5,000 laboratories in 18 countries.[32] H2N2 is the avian flu-related type A virus that was responsible for the 1957 pandemic that killed millions of people worldwide. Fortunately the error was discovered in time, and the WHO was able to direct the 5,000 laboratories to destroy the samples before they were used in actual testing. What if, however, an accident occurs that is not caught in time; and what if it was with a virulent form or parts of the recovered 1918/19 pandemic virus? Perhaps the chances are unlikely. But they seem to be much more likely than the implausible scenarios that have been used to justify huge expenditures of bioterrorism research.[33] Is it not possible that research purportedly undertaken to provide defense against a pandemic could become the starting point of one? Is it not reasonable to ask what purported benefits can be expected from reviving and experimenting with the 1918 virus that could justify such a risk?

What Can Be Done?

As described above, there are inherent problems developing and producing adequate supplies of effective vaccines in time to contain or mount an effective defense against a pandemic outbreak. But even within current state of vaccine technology, a great deal more could be done to prevent the catastrophic emergence of a pandemic that could claim tens of millions of lives and have a huge and devastating impact on the global economy that could severely affect the lives of the survivors as well. While there undoubtedly are many possible measures, we see three main approaches that could be undertaken.

Greatly Expanding the Flu Vaccine Production Infrastructure

Because it cannot be determined in advance with any certainty what strain of virus might trigger a pandemic, what is needed is the capability of producing huge amounts of vaccine quickly, once the threat has been identified and a vaccine developed. This means having the factories, the skilled labor, and the resources (such as huge quantities of fresh eggs for growing the vaccine cultures) ready to go. But assembling all of this productive capacity and leaving it idle would be a tremendous waste of resources if it had no other use than being ready for something that may not appear for years, if at all. Further, there are unavoidable problems of scaling up production and distribution, and it cannot be left to chance whether such problems can be overcome in time. Instead, why not build a productive capacity and use it each year to produce and distribute adequate supplies of ordinary flu vaccine for the world's population? As the techniques of developing safe and effective vaccine improve, and awareness that widespread use of even ordinary flu vaccine can save millions of lives worldwide, acceptance of routine vaccination for flu could gain the regularity that had existed previously for smallpox and polio vaccinations. Clearly this would need international cooperation and would have to be organized by a consortium of governments (perhaps through WHO), and could not be left to the vagaries of for-profit pharmaceutical production. It would also need a greatly expanded commitment of resources and couldn't be done with the paltry budgets now allocated to public health. But resources can be found. For example, it was reported in August 2005 that the United States had ordered an additional billion dollars worth of smallpox vaccine,[34] even though substantial stocks already exist and there has been no hard evidence of any chance of an outbreak of what has been an eradicated disease for over 25 years. Flu takes place every year, and all public health experts agree that the chance of a flu pandemic far, far exceeds the chance that smallpox virus will escape the freezer and reenter the human population. In addition, if just a portion of the huge expenditures on new weapons procurements were diverted to defense against this real and present danger, it would make a big difference.

As this article is being written, the Senate has just passed a $4 billion allocation for defense against avian flu, of which $3 billion is supposed to increase the government stockpile of Tamiflu® from 2 million doses to enough to cover about 50% of the U.S. population.[35] Presuming this bill becomes law, it will be a substantial step forward but still leaves unanswered many questions. Will there be sufficient productive capacity to produce the medications rapidly and with adequate safety and quality controls? If the public knows that there is a supply for just half the population, in the event of an outbreak will there be frenzied panic among those fearful that they will be in the uncovered half? In the 2001 anthrax outbreak, when the actual danger of exposure to anthrax by any one individual was miniscule, there were still huge lines of people trying to procure the Cipro antibiotic. Production and stockpiling of supplies are necessary preconditions, but not at all sufficient in the event of an outbreak. How will distribution of the drugs be carried out? Will it be limited to those who can afford to pay for a doctor's visit and for the drug itself? Will there be priorities for those who are most vulnerable or will it follow the pattern of the current health system where those with the most resources get the most access? For decades, the public health infrastructure has been allowed to erode in favor of privatization and for-profit medicine. Few communities around the country still have active, well-staffed neighborhood public health stations. Public hospitals in many communities have been closed, and in the event of an outbreak emergency rooms will quickly be swamped with both the ill and the worried-well. If a serious outbreak does occur, many of the usual transportation/shipping pathways will be hit with disruptions as the transportation workers themselves fall ill. Flu is transmitted primarily by droplet infection, so it may not be wise to encourage large concentrations of people to gather at centralized distribution centers to get medications, because such congregations of people might of themselves accelerate transmission.

What about coverage for people in other countries, some of whom may be in areas harder hit by an outbreak? Will U.S. purchases monopolize the available production and will other countries enter into an international competition for supplies, and will those countries that have within their borders the supplies or factories producing Tamiflu® try to block exports and requisition the supplies for their own populations?

Expanding Public Health Awareness and Surveillance Infrastructure

Provided the resources for mass rapid response vaccination were available, early warning of outbreaks of new flu virus strains would be needed to mount an effective containment effort. Trainings for hospital and clinic personnel worldwide on what to look for, the availability of laboratories to test for new strains, and international coordinating centers that can centralize and process the information would be needed. This is now being done to some extent by WHO, CDC, and the health officials of some other countries, but it could be greatly expanded. In addition, early warning would be greatly enhanced by international support for establishing indigenous, community-based clinics, and training local community health workers to monitor the potential of outbreaks in their neighborhoods. This would be even more important for the rural farming areas where many believe an outbreak of human to human avian flu could start. Of course, the expansion of local clinics, preventive health information, disease surveillance, and a community health worker system could improve disease control for a wide variety of infectious and parasitic diseases (such as HIV and malaria) that still plague much of the world.

Improving international public awareness of the nature of infectious disease could help. Simple educational measures could have a large impact. Even in advanced industrial countries like the United States, handwashing habits among children and adults leave much to be desired. Covering one's nose and mouth when sneezing and coughing seems like simple good manners, but it could be an important public health measure in the event of an epidemic. In some countries it is not unusual for those with a cold or flu to wear a facial mask when traveling in public. Providing masks and encouraging their use, could slow transmission rates. Most workers will avoid staying home with the flu because they get no sick leave and will not be paid, and may even be fired if they stay home. Mandating compensation for sick leave would be an important measure in the event of a serious epidemic.

Improving Basic Conditions of Life and Health

It is quite likely that the Spanish flu pandemic of 1918/19 spread so rapidly and took such a large toll because of the devastating conditions brought on by World War I. It is well known that communities that are malnourished, or without potable water or organized sanitation, or otherwise lacking in the basic necessities of life have compromised immune systems and are much more susceptible to infection, and much more likely to spread disease and succumb to it. Thus, international efforts to provide the resources to improve these conditions would have the double benefit of slowing down the international spread of infection as well as improving the lives of billions of people.

The era when geographic and national boundaries could limit the spread of disease passed decades, if not centuries ago. But as the global economy ever more rapidly integrates individuals from even the remotest areas, the ability for the rapid spread of disease has increased exponentially. To date, this potential danger has been counterbalanced in part by development of infrastructures for safe water, sanitation, food supply, and pest control as well as technological advances in disease control and prevention. Nonetheless, the counterbalancing forces are not distributed evenly and are clustered in a manner consistent with the unequal socioeconomic levels between and within regions, countries and communities. The uneven development inevitably creates conditions that are amenable to the emergence of new pathogens such as a new pandemic flu virus and the reemergence of old ones like malaria and tuberculosis. Rather than raising all boats, the flood tide of the globalized economy has shown a tendency to raise the best-equipped yachts to new heights while drowning huge numbers of the ill-equipped in an economic tsunami. Unless checked and consciously reversed, the economic dislocations may provide fertile ground for a new flu pandemic as much as the disruption of World War I did for the 1918/19 pandemic.

Lessons from Katrina

The category 5 Hurricane Katrina that devastated New Orleans and the U.S. Gulf Coast in 2005 provided an example of the consequences that may ensue from a flu pandemic. At the time of this writing, it is still unknown to what extent hurricane survivors will be hit with outbreaks of infectious disease, including flu as well as illnesses from environmental toxins such as heavy metals, petroleum, and other chemical spills, and human and animal waste that were released into the flood waters, and that were drained, without treatment, into the coastal waters. The lack of preparedness for the hurricane and the inept, inadequate response in the days that followed hold many lessons relevant to the danger of a flu pandemic and other potential natural disasters and to important issues in public health.

Race and Class

Perhaps the most important lesson is how social conditions, including race and class, will impact public health. Those hardest hit by Katrina were Black and poor. These were the people who did not have the transportation and other resources to evacuate after it was clear that Katrina was going to hit land. Many perished in the flood, but even the survivors suffered greatly because the government had made no preparations or provisions to assist evacuation or to adequately supply evacuation shelters. For infectious diseases like influenza, those who are poor, with substandard diets, housing, and access to health care will also have higher rates of impaired natural immunity, will face the greatest barriers to vaccination, and will live in crowded conditions where the potential for transmission will be greatest.

There are many ways that race and class will adversely affect the impact of a pandemic flu outbreak. Neighborhoods with concentrations of people of color have the lowest concentrations of healthcare providers and facilities. This will hamper the distribution of antiviral medications and vaccine if one becomes available. Unless an alternate distribution network is built and tested in advance, it is likely that the existing disparity in facilities and services will be the automatic path that will lead to even great disparities during an outbreak. Is it unreasonable to expect that government and health officials will prioritize their own communities and business districts ahead of the oppressed neighborhoods? One need only recall how the trapped residents at the expensive hotels in New Orleans during the flood got evacuated long before those stranded in the Black communities.

Poor communities are characterized by overcrowded housing. This will necessarily increase transmission rates in those areas. People of color, and poor working people in general are most likely to have jobs without adequate health insurance, are least likely to have savings, and are least likely to maintain extra supplies for food and necessities in their homes. What will happen to the millions of people who live from paycheck-to-paycheck if an outbreak disrupts their precarious access to money for every day expenses? During Katrina, people without ready cash or credit did not have transportation money to evacuate. Check-cashing offices quickly ran out of cash as did ATMs, so that even those with a paycheck or a small account could not get the cash to leave. When Hurricane Rita threatened only weeks after Katrina, highways were packed with motorists who soon ran out of gas and drinking water while stranded in miles-long traffic jams. While a flu outbreak won't lead to evacuations, the same problems will manifest for those without ready cash, or pantries stocked with reserves of food and drink. If a neighborhood is put under quarantine (as President Bush has just proposed), how will those residents get food and other necessities of life? Even if stores are open, will the stores get adequate supplies? And where will the people get the money to pay?

These problems will exacerbate the already severe disparities in health and well-being. Ordinary flu hits the young and the elderly the hardest because they are the ones with the weakest immune systems. Those who have been chronically malnourished, or without regular healthcare or stricken by the numerous diseases and conditions that afflict those with the least resources, will also be the most likely to succumb to a pandemic flu (that the 1918/19 pandemic severely hit young adults was a notable exception).

Limitations of Preparedness

It is a great irony that just prior to the Katrina disaster, the Department of Homeland Security (DHS) and the American Red Cross had declared September 2005 as "National Preparedness Month."[36] While the DHS and the Federal Emergency Management Agency (FEMA) which had been placed under its jurisdiction were prepared for a public relations exercise, they were woefully unprepared for a hurricane, despite the fact that weather forecasters had days earlier warned that Katrina might hit New Orleans and that a similarly sized hurricane had just missed New Orleans the year before. But since 9/11 and the creation of DHS, emergency preparedness has been geared almost exclusively to the highly unlikely threat of a nuclear, chemical, or biological attack. Such an orientation has been driven by a political agenda rather than scientific concerns. Real public health needs have been subordinated to hypothetical threats that have been used to justify the Iraq war and militarization in general.[37]

New Orleans was one of many large cities that had undergone an emergency drill to practice a response to a simulated biological weapons attack, even though the likelihood of such an attack approaches zero. On the other hand, instead of a real preparedness drill for a hurricane or flood, there was only a table-top exercise the year before. Nonetheless the table-top exercise had estimated that as many as 100,000 residents of New Orleans would not have the wherewithal to respond to an evacuation order — a number chillingly close to the actual number stranded by Katrina.[38] The best theoretical planning and preparation will fail if the priorities are skewed, if the identified threats are not realistic, or the resources are not available to implement plans. In contrast, Cuba with much less material resources than the United States is able to evacuate all the people and animals threatened by an encroaching hurricane and has survived direct hits from category 5 hurricanes with minimal loss of life.[39]

When Hurricane Rita was approaching Galveston and Houston, Texas, and the western part of the Gulf, officials tried to avoid the fatal mistakes made with Katrina and more vigorously promoted evacuation. They did not foresee, however, that instead of overcrowded, undersupplied shelters, they would be faced with overcrowded, undersupplied highways. A 40-mile drive out of Houston took over 12 hours. Cars overheated and broke down while stuck in the many miles long traffic with air conditioners unable to overcome the heat. Those who were used to driving without a full tank, ran out first and with no access to a gas station blocked traffic even more. When the government sent tanker trucks to provide gas to stranded motorists they found that the truck hose nozzles were too large to accommodate ordinary auto gas tanks. Emergency response plans developed on table-tops in committee rooms are prone to collide with real-life conditions, especially in a crisis.

The Bush administration has drawn exactly the wrong lessons from the Katrina disaster. In a major speech, President Bush announced that in the event of a pandemic outbreak he would bring the military into the nation's city streets to enforce quarantines in affected neighborhoods.[40] Prior to the flooding, military vehicles would have facilitated evacuation and military-style encampments could have been set up to house evacuees. In the early hours and days of the Katrina flooding, military helicopters, boats, and rescue vehicles could have prevented loss of life. But neither of these measures was undertaken. Instead, troops were sent to protect stores from "looting" when in fact people were for the most part only desperately trying to procure the necessities of food and drink. In the meantime the oil companies were allowed to loot the whole country by doubling oil prices, and politically connected companies like Halliburton began the process of looting the public treasury by overcharging for inadequate cleanup and supply services.

The idea that armed troops enforcing quarantines will resolve the problems of a pandemic is both ludicrous and dangerous. It is ludicrous because the nature of flu epidemics is that the contagion spreads rapidly and widely and will not be confined to a particular house or street. If people are confined to their homes, who will do the jobs that all the workers in those neighborhoods generally do? Who will provide food and drink or resupply those with diabetes or heart disease who depend on regular medications? A flu season can last for months. Will the quarantine be for everyone or only those who are sick? If the latter, how will they be identified? If everyone, how long can it be sustained? How will those who are sick get any palliative care and who will remove the corpses of those who die?

The proposal is dangerous because it is based on overturning the long-standing practice and law — in particular the Posse Comitatus Act of 1878 — that prohibits the federal army from assuming police duties. This law was originally part of the reactionary dissolution of Reconstruction after the Civil War. The white aristocracy in the South was reasserting itself and federal troops were protecting the civil rights of the newly emancipated Black population. But since that time, the U.S. military has long ceased to be an agent of emancipation, and the law has been seen as a protection against the abuses of martial law. And it is precisely martial law that Bush is proposing. Since September 11th, 2001 a number of state legislatures have passed laws allowing forced quarantines in the event of biological emergencies. Bush's proposal seeks to give himself the unilateral authority to impose martial law wherever and whenever he chooses, under the rubric of national emergency. Critics have pointed out that forced quarantine without adequate protection of civil liberties could be a disaster of its own. Will whole cities be "locked down" for weeks and months like a huge prison? And will shoot-to-kill orders be issued? And if so, is it hard to guess in which neighborhoods such orders will be carried out?

Ounce of Prevention or Ton of Cure

An analysis conducted almost 10 years before Katrina identified that the levees protecting New Orleans from flood were vulnerable to a hurricane of category 4 or 5 strength. A large scale program undertaken to shore up the levees was virtually halted in 2003 when funds for the Army Corps of Engineers project were diverted to the Iraq war.

Over decades, the wetlands and barrier islands, which provide natural protection from hurricanes and floods, had been allowed to erode or were swallowed up by developments. The changes from year to year seemed small, but the accumulation of erosion and developer inroads left the whole region more vulnerable. In general, public health is much more cost-effective than acute medical care because public health tends to focus on prevention rather than cure. But the healthcare system in the United States is heavily skewed toward acute care and not prevention. Thus, as the advanced technologies of diagnosis and treatment have driven up the cost of health care, funding for public health projects and infrastructure have been in steady decline. Some believed that the influx of billions of dollars for so-called bioterrorism preparedness would help restore the public health infrastructure but in reality the reverse has been true.

In sum, the possibility of an influenza pandemic in the near future underscores the need for addressing the underlying social determinants of health and reclaiming public health priorities and leadership away from the militarized and politicized agendas of "homeland security." There are prevention and preparedness measures that can be undertaken, but the most effective measures will take resources that are orders of magnitude greater than currently allocated to public health. Tens of millions of lives and the worldwide economy may be at risk from something seemingly as simple as pandemic flu. Whether realization of the extent of the danger will motivate the necessary preventative measures in time remains to be seen.

Appendix A: The Influenza Virus

Influenza, a virus that belongs to the family *Orthomyxoviridae*, is classified into three distinct types: influenza A, B, and C. The types are based on major antigenic differences. An antigen is any substance that causes the immune system to produce antibodies, which are produced by the cells of the immune system and which destroy foreign antigens. The three major influenza types have significant differences in host range, epidemiology, and clinical characteristics.

Influenza virus has the ability to rapidly evolve changes in two surface glycoproteins, hemagglutinin activity (HA), and neuraminidase activity (NA). The changes can be either small or large. HA impacts the antibody effect on red blood cells and NA is a key enzyme that the virus uses. Influenza A viruses are further classified by their combination of HA and NA subtypes such as H1N1 or H3N2.[42] Minor antigenic changes (antigenic drift), occur frequently (every year or every few years). The antibodies that an individual develops to a particular strain of virus will not be as effective against a new strain that has undergone even these minor antigenic changes. The new strain that emerges then can become the predominant virus in a new epidemic. This phenomenon explains why humans do not develop long-term immunity to influenza and why the population at risk requires yearly vaccination.[43]

An outbreak of influenza in one location, such as a city, town, or country is referred to as an epidemic. Epidemics are associated with morbidity and mortality greater than what is usual for that region in a comparable time. Epidemics in the U.S. usually peak in the winter months.

Attack rates (the number of people at risk who develop an illness/the number of people at risk) are usually highest in the very young. Influenza is emerging as an important health problem in healthy children less than 2 years of age and increased rates of influenza-related hospitalizations in this age group have been noted.[44] On the other hand, morbidity is usually highest in older adults. Outbreaks of influenza in nursing homes can have devastating results. For example, influenza-related death rates in nursing home residents with comorbid conditions are as high as 2.8% In general, serious influenza complications, including mortality, are highest in older and debilitated individuals, but the majority of hospitalizations are in healthy ambulatory individuals. The influenza epidemic curve with respect to age typically reflects a U-shape.[45]

Pandemics — unusually large, worldwide outbreaks — are due to viral strains which have undergone major antigenic shifts. Populations have no acquired immunity to the newly emerged virus, leading to rapid transmission throughout the globe. Characteristics of pandemics include lack of association with a season, attack rates in all age groups, and high levels of mortality seen in healthy young adults. The interval between historic pandemics has varied and the time, place, and source of the next outbreak is unpredictable. Nonetheless most experts agree that there will be future pandemics and there is considerable concern that one may be imminent or that the virus that will cause one has already emerged.

Transmission and Clinical Manifestations

Influenza virus infection is acquired by a susceptible person exposed to respiratory secretions containing virus from an infected person. Respiratory secretions of an individual ill with the flu contain large amounts of virus. When the individual coughs, sneezes, or even simply talks, the virus can be transmitted by aerosolized droplets. The droplets can come directly in contact with mucous membranes through the nose and mouth or indirectly via the hands. A single infected individual can transmit the virus to a large number of susceptible individuals. A hospitalized patient can spread influenza to individuals at highest risk for morbidity and mortality; therefore early recognition and isolation of the patient is important. In hospital settings, healthcare workers should wear masks when caring for an individual with influenza and hand hygiene is important in this setting. Transmission of the virus is especially efficient where people are in close contact such as on public transportation, in schools and childcare centers, and at workplaces. Effectively covering the mouth with tissues during sneezes and coughs and frequent hand-washing in schools and workplaces could substantially reduce infection rates.

Uncomplicated influenza has an abrupt onset with fever, chills, headache, myalgias (muscle pain), and anorexia (loss of appetite). The incubation period is usually one to two days. Fever is a universal finding with peaks of 104° F that are usually continuous, but can be intermittent. Myalgias can be quite severe as can the headache. Patients may experience respiratory symptoms such as dry cough, sore throat, and nasal discharge. Prostration can occur in severe cases. It is the systemic symptoms that help distinguish influenza from the common cold and other upper respiratory tract infections.

Pulmonary complications, including primary influenza, viral pneumonia, and secondary bacterial infection are well described. Primary influenza viral pneumonia presents initially as typical influenza but patients will rapidly deteriorate, complaining of difficulty breathing, and become cyanotic from lack of oxygen. Many of these patients will require ventilation. Patients with secondary infection describe a classic influenza syndrome with improvement, but within two weeks they experience renewed fever associated with respiratory symptoms. A chest x-ray will reveal pneumonia. Pathogens such as *Streptococcal pneumoniae* and *Staphylococcus aureus* are commonly isolated. Patients with underlying asthma may also experience exacerbations.

Nonpulmonary complications include myositis, myocarditis, and pericarditis, which are inflammation of the heart and its lining, respectively. Very rarely Guillain-Barre syndrome, an immune system disorder, has been reported to occur after influenza A infection.

Diagnosis, Treatment and Prevention

Diagnosis of influenza can be made by virus isolation or by detection of viral antigens in respiratory secretions using a rapid test. In each rapid test, a sample of respiratory secretions is treated with a mucolytic agent and then tested for a color change indicating the presence of antigen. There are a variety of rapid tests on the market. All of them are designed to detect the presence of influenza A and B, and take about 30 minutes to perform. The reported sensitivities of each test in comparison to cell culture have ranged between 40 and 80%. Rapid tests can help with early diagnosis and institution of therapy. In the hospital setting, isolating patients early is important to prevent further spread. In particular, the rapid flu tests are used to distinguish influenza from other upper respiratory virus. Serologic tests are more accurate in sensitivity and specificity, but they do not provide results in time to affect clinical decisions.

Four drugs are currently available for the prevention and treatment of influenza. The greatest benefit is seen when therapy is instituted within the first forty-eight hours after symptoms appear. Amantadine and rimantadine are administered intravenously. Treatment with amantadine results in significantly more rapid improvement in small airways dysfunction in healthy adults. Drug resistance has been a factor in limiting the more widespread use of these antiviral agents.[46,47] Another class of antiviral drugs recently released are the neuraminidase inhibitors. These agents act by inhibiting the functioning of the critical influenza virus enzyme neurmanidase. Influenza B viruses are approximately tenfold less sensitive to such treatment than influenza A viruses, but they are still considered sensitive. Two neuraminidase inhibitors are commercially available. Oseltamivir carboxylate is administered orally and zanamivir is a dry powder for oral inhalation. Both have shown similar results in clinical trials. Early treatment of uncomplicated influenza with oseltamivir in ambulatory adults resulted in 30 to 40% reductions in the duration of symptoms and severity of illness. An earlier return to work and reduction of complications was also observed.[48]

The most effective measure available for the control of influenza is the annual administration of inactivated influenza vaccines. The current vaccine is generally formulated as a trivalent preparation, containing one example each of influenza A (H1N1) virus, A (H3N2) virus, and influenza B virus, thought to be most likely to cause disease in the upcoming season on the basis of epidemiologic and antigenic analysis of currently circulating strains. The vaccine is generally well tolerated in adults with local side effects being the most common complaint. People with hypersensitivity to egg products should not get the vaccine because it is grown in hens' eggs. Ninety percent of healthy adults will have an appropriate immune response. The response takes about two to four months to develop after vaccination, so that vaccination prior to or early in the flu season is most effective. Patients require yearly vaccination because an adequate response is not generally sustained beyond a single season and variations in the virus can also attenuate effectiveness.

Inactivated influenza vaccine has been shown to be effective in the prevention of influenza A in controlled studies conducted in young adults, with levels of protection of 70 to 90%, when there is a good correlation between the vaccine and the current predominant influenza strain. Vaccination results in decreased absenteeism from work or school and has been shown to be cost effective when the higher levels of protection are achieved. There have been few studies looking at protective efficacy in high-risk populations. In one placebo-controlled prospective trial in an older adult population, inactivated vaccine was approximately 58% effective in preventing laboratory-documented influenza.

Recently, the first live-attenuated influenza vaccine for use in humans, the cold-adapted influenza vaccine trivalent, was licensed for use in individuals five to forty-nine years of age. This vaccine has a nasal administration route and induces a mucosal immune response that mimics the response of infection with natural influenza virus. The vaccine is well tolerated, but shedding of virus does occur in vaccinated adults and children. In other words, it is possible that susceptible individuals may contract influenza if they are exposed to vaccinated adults or children. Administration of the nasal vaccine to health care workers is not recommended due to the potential of exposing patients. Early studies in children suggest that the vaccine will be effective in this age group, but studies in older adults have yet to be done.

In general, three main groups of individuals have the highest priority for influenza vaccination: individuals at increased risk for complication, individuals who can transmit the virus to those at high risk such as healthcare workers and people over the age of fifty, although vaccination is offered and encouraged for the whole population. In the United States, about a third of the population receives a flu vaccination in any given year, even though health officials would like to see a much higher proportion. Some of the barriers to more widespread use include lack of access to and information about the vaccine, concern that it is not effective or concern about side effects. Periodic disruptions in vaccine supply have also hampered more widespread coverage.

INFECTIOUS DISEASES AND THE BUREAUCRACIES THAT ARE ACCOUNTABLE FOR PUBLIC SAFETY

VI

Chapter 11

Hospital Cleaners and Housekeepers: The Frontline Workers in Emerging Diseases

Bernadette Stringer, Ph.D., R.N., and
Ted Haines, M.D., M.Sc.

CONTENTS

Background

Because hospitals and other healthcare institutions are environments where people with a wide variety of illnesses are brought together, infection control is of paramount concern. Florence Nightingale, the founder of the modern hospital and the nursing profession, was one of the first infection control practitioners. Through the experience of caring for injured soldiers in the Crimean War[1] and applying her keen mind and common sense, she learned the critical importance of cleanliness and infection control in determining patient outcomes.[2] By implementing hand washing, cyclical cleaning, segregating infectious patients, and implementing other infection control practices still in use today, under her supervision death rates of hospitalized soldiers decreased from 48 to 2%.

Doing a good job of cleaning and hand washing has never been more important than it is today. North Americans are aging, and therefore suffering more chronic disease, and the availability of invasive medical technologies grows by leaps and bounds. At the same time, the prevalence of antibiotic-resistant organisms grows and is further compounded by new viruses easily able to spread around the world. In this context, cleaning in hospitals and other healthcare facilities is more important and more complex than ever before. And especially at this time of a looming pandemic, cleaning of healthcare facilities will be crucial to controlling cross-contamination.

This chapter reviews aspects of the published research literature on healthcare acquired infections, also called nosocomial infections (NI): their prevalence and cost; the role of cleaning as an infection control measure used to prevent them; and some current, "accepted," best practices to ensure hospital cleanliness.

Introduction

Hospital-acquired or healthcare-acquired infections, known as nosocomial infections (NI), are those that patients develop while in a hospital or healthcare facility or are due to treatments received in such facilities, or they are infections acquired by healthcare workers as a result of work-related exposure.[3,4]

Patient NI may occur for numerous reasons; for example, organisms normally carried on one part of the human body or in the intestinal tract can relocate to an open wound. Most often, NI develop in patients with lowered immunity, such as those suffering from certain diseases or receiving treatments such as chemotherapy, or at stages of life when they are more vulnerable; for example, in infancy or in the later years of life.[5] They also occur because of "opportunity,"[6] such as when the skin, normally a very effective barrier, is perforated during surgery.

While the U.S. Centers for Disease Control (CDC)[7] estimate that about 30% of NI could be eliminated through the use of appropriate hand washing, judicious prophylactic antibiotic and antiseptic use, and implementation of cleaning regimens as often and thoroughly as recommended, along with other relatively simple measures, these claims may be exaggerated.[8] Nevertheless, reducing even a much smaller proportion with interventions, especially cost effective ones, would be a substantial health benefit. After all, the many NI that can be considered medical errors because of not employing procedures able to prevent them, are not only costly to the healthcare system but also blatantly contradict the basic tenet to do no harm.[9]

It should be noted that in the past, actual cleaning practices around the world seem to have been as much a function of national and local culture than based on more substantial scientific evidence. The perceived importance of cleanliness has seemed to be more associated with the perception of cleanliness in the society as a whole or to the perception that clean hospitals were markers for the quality of patient care,[10,11] not because of clear examples that cleanliness and healthcare acquired infection were linked. So if the societal value for cleanliness was strong then hospitals were clean, especially because this was recognized as important for patients and healthcare providers' psychological comfort.[12] The following quote from a document published by the United Kingdom's Department of Health, summarizes the problem:

> Cleanliness and infection control are closely linked in the public mind, but there are important distinctions to be made. Cleanliness contributes to infection control, but preventing infections requires more than simple cleanliness. Cleanliness produces a pleasant, tidy, safe environment that makes us feel better; however, the scientific evidence that the environment is an important contributor to infection rates is not always clear cut.[13]

In fact it is well recognized that visual cleanliness does not necessarily mean that the environment is not contaminated.[14,15]

Nosocomial Infections

NI are believed to be the sixth leading cause of death in the United States.[16] Between 5 and 10% of patients admitted to acute care U.S. hospitals acquire one or more NI,[17] and since NI rates have been more reliably measured they are believed to have increased steadily, even though patient hospital admissions in the U.S. have decreased. For example, in 1975 there were 38 million admissions but in 1995 admissions had decreased to 35 million with the average length of stay declining from 7.9 days to 5.3 days, yet the CDC estimated that in 1972 there were 7.2 NI/1000 patient days and that this had increased to 9.8 NI/1000 patient days by 1995.[18] This rate is expected to increase even more because of our aging populations and the increased factors that increase patient vulnerability.[19]

About 25% of NI occur in intensive care unit patients and about 80% originate in urinary tracts, surgical incisions, the blood, or the lungs, and organisms causing about 70% of them are resistant to one or more antibiotics.[20] Mortality from nosocomial pneumonia and bloodstream infection is most common, and in the United States it is believed that incidence of NI of the blood tripled from 1975 to 1995.[21] Canadian trends appear to be similar. Approximately 200,000 NI are thought to occur in Canadian patients per year and between twenty-eight and forty deaths result per 100,000 per year, translating into between 8,500 to 12,000 deaths Canada wide.[22]

Although the CDC has estimated that the cost of NI exceeded $4.5 billion in 1992,[23] which adjusted for inflation factors increased to $5.7 billion in 2001, based on Stone, Larson, and Kawar (because CDC estimates have relied on data collected in the 1970s), the true estimates of costs related to NI in the United States remains unknown,[24] although based on a systematic review of the literature, they have estimated individual costs of types of NI and NI due to MRSA. For example, they estimate that overall, each case of NI costs $U.S.13,973 and each MRSA infection (all types) costs $U.S.35,367.

Infection Control

For healthcare facilities to be recognized (accredited) as capable of providing quality patient care, they should meet certain infection control standards, but this is a voluntary process not one that is legally required. Nevertheless, having functioning infection control committees with qualified infection control personnel managing recommended types of programs such as ongoing, targeted surveillance and NI rate estimation is highly recommended. Surveillance, defined by Benenson is, "the scrutiny of all aspects of occurrence and spread of disease that are pertinent to effective control."[25] Surveillance is to ensure that elevated NI rate estimates (especially those at epidemic or outbreak levels), are recognized and addressed with appropriate interventions. Unfortunately, infection control personnel cannot ensure compliance because in most countries they lack legal authority and their role is consultative in nature.[26] Nevertheless, recommended practices specify the type of monitoring that should be in place, such as laboratory monitoring of surgical wound infection rates, for example, or the number of trained personnel required to oversee programs, train, and interact on a one to one level in response to potential and real NI problems. And while it can be argued that legal authority should have been given to infection control experts long ago, there are many factors determining NI rates that will always make it hard to develop prescriptive guidelines.[27] That is why it is extremely important that principles underlying NI are well understood by professional personnel and by nonprofessional personnel alike, including cleaners.

Outbreaks

Outbreaks, which are defined as, "the occurrence of disease at a rate greater than that expected within a specific geographical area over a defined period of time,"[28] can occur as a result of most organisms causing NI, although some of the most common are *Pseudomonas aeruginosa*, MRSA, Vancomycin resistant enterococci, and *Acinetobacter baumanii*.[29–31]

The main reason for routine surveillance is the rapid identification of outbreaks,[32] although there are many examples of outbreaks that have gone undetected for long periods, resulting in considerable expense, as well as suffering and mortality.[33]

Whether or not outbreaks are suspected at the laboratory or ward level, it is important to immediately involve infection control personnel. It is their role to gather necessary ward and laboratory data required for confirmation while simultaneously providing directives to nurses, cleaners, and other personnel about appropriate control measures; for example, placement of patients in isolation and implementing enhanced cleaning practices. Usually, in outbreak investigations, the custom is to err on the side of safety as data required for confirmation are gathered. If outbreaks are confirmed, infection control personnel will develop explanatory hypotheses and as a result may implement additional control measures. Regardless, controls should be assessed over time, to ensure that they are effective.

A scientific publication on a nosocomial *Pseudomonas aeruginosa* outbreak illustrates well why cleaners in healthcare institutions require good training.[34] This outbreak occurred over two months on an oncology ward in a German hospital, resulting in five patients developing bloodstream NI and one developing a surgical site NI; two were admitted to the ICU, and two died. Based on the investigation, which included patients' blood and wound secretion cultures and 209 environmental cultures of tap water, cleaning cloths and solutions, and washbasin drains, the most likely reason found for the outbreak was cleaning items in close proximity with patients with a solution of soap and water and not a solution of soap and water and a disinfectant. This was a decision that cleaners were required to make on their own after the nurse supervising them had retired and was not replaced. Although the way that cleaning practices are organized in Germany appears somewhat different compared to North American practices, cleaners with a solid understanding of the principles underlying cleaning methods would have been less likely to make such a mistake. There is little controversy, for example, about cleaning all areas near patients who are immunocompromised due to chemotherapy. Expert bodies clearly recommend that solutions containing a disinfectant rather than soap and water alone, be used.[35,36]

CDC Recommendations

Based on the "Study of the Efficacy of Nosocomial Infection Control (SENIC)" in the 1970s,[37] the Centers for Disease Control recommended that there should be at least one infection control practitioner per 250 beds. In addition, they recommended that there should be direct involvement of physicians or PhD trained practitioners in all infection control programs. Since the 1980s, they have recommended individual reporting of surgical wound infection rates to surgeons.[38]

More recently, the CDC has also made recommendations that it claims could reduce NI rates by about 30% based on the National Nosocomial Infection Surveillance (NNIS) system that consists of more than 300 participating hospitals.[39] The CDC has been using reports from hospitals included in the NNIS system to justify specific types of infection control activities and staffing levels.[40] For example, it has claimed that when staff can routinely measure NI rates,[41] and then compare them to baselines and benchmarks derived from similar institutions, they will implement interventions that have been shown to decrease rates. Although considered to have been of value and likely to have motivated certain declines in NI rates, CDC claims are believed by several experts to be exaggerated.[42,43] NNIS hospitals volunteer to participate, and are therefore not necessarily representative of U.S. hospitals as a whole, and it remains difficult to identify which interventions have been effective, including cleaning interventions. But most importantly, as highlighted by Burke, "this system has not yet addressed many important safety issues, such as clinical errors or omissions leading to failures to diagnose infections or delays in the diagnosis of infections."[44]

Resistant Organisms

Many experts around the world consider resistant organisms an emerging public health crisis.[45,46] In fact, in the United Kingdom (UK) the strains (phage types 15 and 16) of methicillin-resistant *Staphylococcus aureus* (MRSA) showing up at unusually high levels are now referred to as Epidemic MRSA or EMRSA.[47]

Evidence is mounting of a link between environmental contamination, increased levels of colonization with organisms resistant to one or more antibiotics, and evidence that NI related to resistant organisms is mounting.[48,49]

In one study, personnel without direct patient contact, but only contact with objects or surfaces in MRSA infected patients' rooms, were found to have gloves contaminated with the same MRSA strains as the patients.[50] In another study, 30 cases who were patients colonized with Vancomycin resistant enterococci (VRE) were matched to 60 randomly selected controls. Just being placed in a "high-risk" room where VRE patients had spent time, was found to be an independent risk factor for acquisition of VRE, after adjusting for other potential risk factors.[51] Although it should be noted that it appears that gloves do not as easily become contaminated with VRE as they do with MRSA, so cleaning to remove VRE may not be as important a means of control as cleaning to remove MRSA.[52]

In a study by Rampling et al., for example, it was reported that the reduction of MRSA cases from 30 in the six months prior, to three in the six months postintervention was likely due to doubling the time spent cleaning. But as is the case in most cleaning studies, this study suffers from a methodological problem; other potentially responsible factors such as improved staffing levels, were not accounted for.[53]

It is noteworthy that in the United States, where for more than three decades infection-control programs have not successfully managed to control resistant organisms, the Society for Healthcare Epidemiology of America (SHEA), recommends much more aggressively trying to identify, culture, and isolate patients at high risk of carriage or colonization with multiresistant staphylococcus and enterococcus strains, than the CDC recommends. The SHEA recommendations were made after they conducted an independent systematic literature review to reexamine the best evidence and found that such policies were in place in countries that were successfully controlling MRSA, such as Denmark and The Netherlands.[54]

Norwalk Virus

Viruses, bacteria, and protozoa that can cause intestinal infections are usually transmitted via the fecal–oral mode, and a link between these types of NI and contaminated healthcare environments also exists. The most prominent of these is the Norwalk virus. Transmission to two plumbers 12 days after the end of an outbreak on a ship[55] is reported in one of several studies indicating its viability and infectivity. Strict measures need to be instituted to control an outbreak, including prohibiting health-care workers from moving between wards, specifically cleaners, as was done in a Toronto outbreak that occurred in an emergency department.[56] Unlike other enteric organisms, the Norwalk virus can be transmitted via the airborne route, when it remains suspended in the air as a result of projectile vomiting.[57] Speculation about respiratory transmission results from drastic measures that have been necessary to control several outbreaks, including the closure of wards and entire hospitals. And in a 2004 study, implementing strict hand washing alone was not able to control a Norwalk epidemic. Only when hand washing was implemented in combination with strict environmental cleaning was the outbreak brought under control.[58] In one outbreak, 300 patients and staff at a Scottish hospital contracted Norwalk virus and the hospital had to shut its doors for several weeks. The union representing the facility's cleaners reported that in 2002, there was one cleaner for every 360 patients, compared to one cleaner for every 60 patients in 1985, when outbreaks were unheard of.[59] Although not frequently lethal, because of its infectiousness, potential for airborne spread, and relatively short incubation period, Norwalk virus can cause serious havoc.

Clostridium Difficile

Another disease of particular concern in Canada among nosocomial diseases is Clostridium difficile diarrhea (CDAD), thought to be becoming more common and more virulent.[60] Data collected between January and April 1997, from 19 Canadian hospitals, already estimated that healthcare acquired CDAD cost $128,200 per year per hospital.[61] In 1997, the incidence rate of CDAD was 3.4 to 8.4 cases per 100,000 in Canada,[62] much lower that the incidence rate in Quebec in 1991, which was 35.6 cases per 100,000. The 2003 incidence rate in Quebec, which was suffering an ongoing outbreak, has been reported to have increased to 156.3 cases per 100,000. Not only more cases but much sicker cases have been seen in Quebec, requiring more ICU care and surgical intervention than expected, and resulting in more deaths than expected.[63,64] In Sherbrooke, one of the cities where the outbreak took place, the proportion of people who died within 30 days of diagnosis went to 13.8% in 2003 from 4.7% in 1991.[65] In hospitals attached to McGill University, 84% of cases are considered to be NI, and along with patients, a doctor, a volunteer, and several healthcare workers, as well as members of the community, have also contracted CDAD.[66]

Antibiotic use resulting in disruption of the gut's normal organisms is considered most responsible for CDAD, but environmental contamination, having older hospitals with cramped spaces that are difficult to properly clean, fewer single rooms, and sinks that do not facilitate hand washing, have also been highlighted as important.

It may be even more likely that environmental contamination increases CDAD risk because Clostridium Difficile spores survive in the environment for months and they are resistant to most disinfectants,[67] although bleach solutions have been found to be effective.[68,69] As well, it has been demonstrated that the more heavily contaminated the environment is, the more likely that healthcare workers will be found to have contaminated hands.[70]

Whether not having a cleaner on the night shift increased risk, as proposed by infection control personnel in one of the Montreal outbreak hospitals,[71] remains unclear.

Evidence that Cleaning Reduces Nosocomial Infection

Can it be said that the high rates of NI are partly due to the lack of cleaning occurring in healthcare institutions? The answer to this question is more complicated that simply saying yes or no. There is some evidence implicating a direct link between contaminated hospital environments and NI, but not enough good evidence. That is because too few studies have been done investigating the link and because many of the studies that have been carried out are of poor quality.

On the other hand, there is good evidence that hospital workers' contaminated hands are responsible for many NI. More specifically, ample evidence exists that appropriate hand washing by personnel results in decreased rates of NI,[72,73] and that increased environmental contamination results in much higher levels of contamination on healthcare workers' hands.[74-76] That is why there has been so much emphasis placed on appropriate hand washing. And while we would not disagree that there is a need for improved hand washing among healthcare providers, it is also obvious that if environments were cleaner, healthcare workers' hands would not be as contaminated. The important point that should be made is that clean environments are intricately linked directly and indirectly to decreasing NI. It may be that cleaner environments alone will result in fewer NI, but it is definitely the case that cleaner environments will lead to less contamination of healthcare workers' hands and in that way result in fewer NI.

Scientific Literature

Many weakness have been identified in studies assessing the association between cleaning and NI, including the absence of longitudinal studies, studies that have inadequate sample sizes, or that have not measured the level of environmental contamination. For example, there are studies that have taken too few swabs, or have only taken swabs in small areas of the environment, or studies that have not selected appropriate outcomes when evaluating the link.[77] For example, some studies of short duration have selected a decrease in NI rates when they should have selected a more appropriate outcome. The reason for this is because even high NI rates in the short term consist of few events and therefore NI rates cannot be used as the outcome in a study unless the study follow-up time is long enough that a true fall in NI rates is likely to be seen, if it has occurred. Therefore when a study does not find a link between increased or improved cleaning and decreased NI rates, one of the first questions that must be asked is, "was the improved cleaning (the intervention) implemented long enough, or, were NI rates followed long enough to determine whether or not improved cleaning led to decreased NI rates?"[78] Or another, more appropriate type of outcome should have been selected; for example, determining whether fewer contaminated swabs were found in the environment after cleaning practices were enhanced. Another issue with studies reporting on cleaning effectiveness is that generally enhanced cleaning regimens were implemented along with other intervention measures, making it difficult to interpret the specific effects of cleaning regimens.

When epidemiologic evidence is inadequate, the use of other scientific perspectives to aid in making decisions about cleaning routines is warranted. Two that apply well to our topic are: biologic plausibility and the precautionary principle.

It is biologically plausible that a more heavily contaminated hospital environment is linked to increased rates of infection,[79] and there is some supportive evidence for this: a study on salmonella food contamination found that homes with good hygiene practices had fewer organisms recovered in the kitchen and fewer family members who developed salmonella infections.[80]

The precautionary principle, a risk management approach supported by Canadians over time, is believed reasonable to put in place when there is "the need for a decision, a risk of serious or irreversible harm and a lack of scientific certainty."[81]

While there is agreement that cleaning is important in infection control,[82] the state of scientific knowledge is strongest when it comes to the ways of dealing with particular organisms in particular circumstances, especially during outbreaks.

The growing evidence of the link between environmental cleaning and decreased NI is reflected more in United Kingdom practices than in other industrialized countries, including North America. And this is because in 2001, a group of researchers in the United Kingdom conducted a systematic review of the published scientific literature to develop guidelines to prevent NI. They considered all scientific publications to do with cleaning, hand washing, the appropriate use of antibiotics before patients undergo certain types of surgery, and other measures considered to be associated with the growing NI rates. These were called the "Evidenced Based Guidelines for Preventing Health Care Associated Infections," or the EPIC guidelines.[83]

When the researchers developing these guidelines first looked at the literature on cleaning and NI, they determined that it was of such poor quality that only three identified studies had used methods considered rigorous enough to be included in their review, and that based on these studies, they could not develop recommendations for cleaning routines. There just wasn't sufficient evidence.[84,85] But three years later, in 2004, when the same researchers looked at the scientific literature again, in order to update the previous EPIC guidelines, they found that ten studies could be included, and that while they could not give specific directives, they stated firmly that

> New evidence supporting the maintenance of hospital environmental hygiene is focused on the importance of ensuring that the physical environment is free of microbial contamination,... recommendations for hospital environmental hygiene need to include regular assessment and regular monitoring of cleaning and hygiene standards within all clinical areas, particularly ICUs and environments where patients with multi-resistant organisms have been placed.[86]

Again, this is just a small taste of the back and forth debate in the literature. It is clear that the state of evidence-based infection control, as it relates to cleaning, is best characterized as uncertain. One of the fundamental problems, of course, is what does "clean" mean? Most argue that the answer is relative to the environment. In other words, a "clean" floor in a medical ward hallway is something different from a "clean" toilet in a bathroom used by many patients. But even here disagreement arises. Is an element of "clean" even for a hallway floor always situational? When, if ever, are bacteria counts on hallway floors part of the definition of clean? Always or only when a facility is suffering an outbreak that it is trying to control?

At least two researchers are arguing for more objective standards, similar to those in the food industry, such as the use of "marker" organisms, like the total aerobic colony count (ACC) which counts organisms from an area with specific dimensions,[87] or the use of ATP bioluminescence, a rapid hygiene test which has been shown to be highly correlated with surface microbiological counts. Microbiological counts cannot be used to routinely monitor contamination since they require inoculated culture medium to be observed for 24 to 48 hours; too long a period for regular monitoring purposes.[88] In one study, for example, when surface contamination in bathrooms, kitchens, patient rooms, and other high risk areas in four hospitals were assessed only visually, or using ATP bioluminescence or with microbiological samples, it was found that over 90% were determined to be clean visually while none of them were determined to be clean based on ATP bioluminescence, and only 10% based on microbiological culturing.[89] Combining visual assessment with a more objective method to quantify bacterial load is used to assess cleanliness in the food industry so there is no reason why similar methods could not be used to assess cleanliness in healthcare facilities. Much research remains to be done comparing actual cleaning methods and outcomes such as overall levels of microorganisms compared to levels of specific organisms.

This sort of research must be carried out, as does research to determine the effectiveness of (including perhaps cost-benefit analyses) of particular overall cleaning programs, even though this would likely consist of carrying out long-term interventions, in which comparisons in multiple equivalent facilities would be done that required training scores of people working in complex social environments, over significant periods of time.

It is therefore fair to say that defining the relationship between cleaning and preventing NI is likely to be a work in progress. As healthcare technology changes, as the populations of the United States, Canada, and other industrialized countries ages, as old facilities are torn down and new ones built, and infectious organisms mutate and evolve, so should cleaning standards and practices.

Cleaners' Working Conditions

Workers in jobs with low job status and little chance of skill development,[90,91] with high job demands and low decision latitude (job control),[92,93] sometimes in combination with poor support from managers or coworkers,[94,95] and in institutions that have been downsized, are at risk of suffering greater sickness absence and work related injury.[96] There is also some evidence that female workers in the same jobs as male workers have worse outcomes.[97] Based on this, cleaners could be expected to manifest similar outcomes, and in fact there is research to support that.[98–101]

For example, in a Finnish survey study of 7,375 hospital workers in 10 hospitals, about 10% of them cleaners,[102] found that after adjusting for other potential risk factors, workers reporting that they worked in a monotonous job, or had low control, had 26 and 27% more occupational injuries, respectively, and 40% more injuries if they also reported that they had conflicts (poor support) at work. In fact in this study, men with low control jobs were almost three times more likely to have a work related accident when compared to workers with jobs in which they had high control.

In another study, 5,342 workers from seven Finnish hospitals, of whom 17% were cleaners or working in maintenance (150 men and 546 women), were asked about procedural and relational justice at work.[103] The questions included: your supervisor considered your viewpoint; your supervisor was able to suppress personal biases; your supervisor took steps to deal with you in a truthful manner; procedures (in your hospital) are designed to collect accurate information necessary for making decisions; and, procedures are designed to provide opportunities to appeal or challenge the decision. The study found that workers with poor scores were up to twice as likely to have medically certified sick leaves, compared to workers with higher scores, even after adjusting for individual levels of job control, workload, and support from coworkers.

As well, in an interview study of 225 cleaners,[104] cleaners reported that they felt a "lack of respect" (for their jobs and themselves). In another interview study by the same researchers,[105] cleaners referred to themselves as "hospital trash," rating themselves at the bottom of the occupational hierarchy. Exclusion from processes such as ordering furniture or redesign of the workplace, with implications on cleaners' work, and being excluded from social functions, were examples they provided. In the same Canadian province where these studies were conducted, women cleaners suffered the highest level of mental illness among all women workers.[106]

As part of the debate on preventing medical errors, Berwick stresses the need for a shift in attitude in order to accelerate healthcare improvements and highlighting the need for a "bias toward teamwork."[107] Based on the above, it appears that a very large shift in attitude will be required.

Close Relationship Between Infection Control and Cleaning Personnel

Consistent with the work environment and behavioral literature, is the need for a close and collegial relationship between infection control personnel and cleaning supervisors and cleaners themselves, when considering the implementation of infection control practices. The need for such a relationship is also acknowledged in the infection control literature; for example, the need for ongoing collaboration between cleaners and staff in higher risk areas such as adult and pediatric intensive-care units,[108,109] the OR,[110] and renal units.[111] A quote in one publication illustrates this well: "domestic staff who work in these areas require a high level of training in disinfectant use and dilution to ensure optimal effectiveness in reducing the risk of infection."[112] A close and good quality relationship between infection control personnel and cleaning staff is important in "low risk" areas as well, as is indicated by reports that a weak or nonexistent relationship can lead to inappropriate or overuse of disinfectants, for example.[113,114]

Knowledge of the principles of infection control assists cleaning personnel in making the dozens of decisions facing them on every shift. Given the fact that appearing to be clean does not necessarily correlate well with being properly cleaned,[115,116] a "professional" attitude by cleaning personnel should be cultivated. If cleaning personnel are treated as part of the hospital team,[117] if their jobs are acknowledged as necessary and important, as is now being emphasized in UK guideline and summary documents particularly,[118] cleaners are much more likely to act in a professional manner, as is the case for nurses and physicians. This type of treatment of cleaners is of even greater importance in an environment where best practices are evolving and feedback about changes is critical.[119]

Training

Hospital environments frequently act as reservoirs, or sites where conditions are ideal for the growth and replication of the organisms that cause NI. Conditions such as the right temperature, moisture, light, or darkness are frequently found in bathrooms, in the apparatus used for patient treatment, or on surfaces of equipment at the bedside.[120]

Cleaners need a sound understanding of the conditions that encourage the risk of the replication of microorganisms, such as moisture, and those that decrease the risk, such as drying, and they also need to know that certain organisms survive for days[121] and that dusty environments pose a risk because some very small, light organisms can travel on dust.[122] Moreover, it is critical that cleaning staff understand that while environments contain microorganisms that can cause disease in any human given the right circumstances, those who are immunocompromised such as the very young or cancer patients, are much more vulnerable and therefore more at risk.[123,124] Cleaners must also understand the principles that guide the preparation of disinfectant solutions and the application of disinfectants so that they can kill organisms. For example, if organic material like blood and feces are not removed from an item the action of disinfectants will be impeded,[125] and as well, that clean water and that disinfectant solutions themselves can be good mediums for certain organisms to grow and why disinfectants should not be diluted with recommended amounts of water and not more than recommended by manufacturers.[126] They should understand which items pose greater or lesser risk based on the chance for contact — for example, bedrails compared to blinds or curtains — because hospital environments vary in the risk that they could cause. For example, bathrooms where patients go to the toilet are usually expected to have higher levels of microorganisms compared to patients' bedside areas on most wards (not ICU bedsides though), and why there has been considerable debate about whether or not all areas in healthcare facilities should be cleaned with soap and water as well as disinfectants. Cleaners should also understand differences between routine and targeted cleaning[127] and terminal cleaning,[128] and should be aware of the research demonstrating that while good routine cleaning reduces contamination for short periods, contamination levels return to precleaning levels within hours.[129,130]

It is hard to determine exactly how much initial and ongoing training and supervision cleaners require, although there seems to be agreement that they should be trained both in classes and on the job. Healthcare cleaners cannot be considered unskilled workers. To be effective cleaners in the healthcare environment, it is essential that the training is long enough and incorporates educational techniques and materials that are able to transmit essential information, and that there is the type of supervision that will pick up breaches if they occur. The training given cleaning personnel must lead to a solid understanding of the important principles that underlie the removal of contaminants and include refresher training at regular intervals so that cleaners are reminded of priorities and that there is the opportunity for the clarification of directives, when there is the need. This will ensure that the knowledge base of the cleaning staff grows and keeps up with ongoing research findings.

Best Practices

It is understood that cleaning solutions should always contain water and a detergent whether or not a disinfectant is added. Cleaning guidelines that appear to be noncontroversial are summarized below; however, generally, as indicated earlier, these are not based on well-designed intervention studies, but rather are supported by certain experimental, clinical, or epidemiologic studies or a theoretic rationale.

- Cleaning is needed to remove organic material before disinfection
- There are low and high risk cleaning environments. Low risk environments consist of those where patient contact is unlikely, while high risk environments are those where patient contact is likely.[131]
- Generalized routine cleaning of environments and items such as furniture is necessary, and cleaning requires easy access to items and floor spaces.[132]
- Floors should be routinely (daily) cleaned with soap and water,[133] unless there has been a blood and body fluid spill, in which case a bleach solution should be immediately applied after the blood or body fluid is removed.[134]
- Proper cleaning with water requires that it be hot with detergent added, and that water is changed frequently.[135]
- Clinical areas such as ICUs, and environments that are known to have been contaminated by multiresistant organisms, should be considered priority areas, and in addition to water and detergent, to clean these areas a disinfectant should be added to the solution.
- ICUs (adult and pediatric), burn units, and operating rooms should be cleaned with disinfectant solutions.[136]
- Proper cleaning with disinfectants requires that they should be diluted according to directions and disinfectant solutions also frequently changed.
- Mop heads should be not be used for more than 24 hours; after that they should be either laundered, or disposed of.[137]
- Routine, thorough dusting that prevents the dispersal of dust should be used.[138,139]
- After use, cleaning equipment should be cleaned and dried and then stored.[140]

But it remains unclear how much time should be allocated to cleaning, as this appears to depend on the unit being targeted, its size, age of the hospital, and the organisms of particular concern, and as well, the use of disinfectants to clean the general healthcare environment remains controversial.[141,142] As discussed further below, some organizations recommend disinfectant use for what are considered low risk areas, those less likely to come into contact with the patient, while others recommend water and detergent only. In the UK, it is recommended that toilets, sinks, and baths be cleaned every day and that color coding be used[143] to distinguish cleaning materials used to clean toilets from those used to clean at the patient's bedside.

For low risk areas, recommendations by William Rutala and David Weber, two well-known disinfection researchers, and other U.S. agencies such as the CDC and APIC, and some European countries, favor disinfectant solution use, such as water and quaternary ammonium compounds (QACs), phenolics, and sodium hypochlorite (bleach), while the Canadian and UK government agencies do not; the latter recommend cleaning with hot water and detergent only.

One reason Rutala and Weber recommend routine use of disinfectants on bedside tables, bed rails, and other such items, is that appropriate disinfectant use has been shown to reduce the number of microbes most often responsible for NI that may contaminate low risk items and because it remains unclear how important it is that contaminated environments result in healthcare personnel hand contamination. Routine screening for MRSA (and VRE) occurs very little in the United States, even though the prevalence of MRSA (and VRE) contamination among hospitalized patients is relatively high. In fact, VRE prevalence rates between 1989 and 1995 increased by twenty times from 0.3 to almost 10%, and are much higher than in Canada.[144,145] Second, Rutala and Weber state that routine use simplifies the training required by cleaners.[146]

Since 1996, the CDC has recommended that only the highest risk patients should be cultured for MRSA and placed on strict barrier precautions.[147] But it is important to remember, as mentioned previously, that even among U.S. experts there has been disagreement that the CDC has not taken a more aggressive approach. Which is why the infectious disease experts in SHEA have publicly disagreed with the CDC since undertaking the organization's own systematic literature review in 2000. They argue that even though CDC recommendations to control multiresistant organisms have been in place for more than three decades, NI have not decreased but increased. After reviewing the literature, SHEA experts believe that controlling MRSA and VRE will only occur when more targeted surveillance to identify those colonized and infected with resistant organisms, takes place.[148] To date, the CDC has not changed its practice.[149]

Cleaning Routines

Supervision of cleaners and feedback, and cleaning times will depend to some extent on such issues as, how critically ill patients are, when specific cleaning and disinfectant dilution procedures must be followed, and the age of the hospital.

Terminal cleaning, which is much more labor intensive, is recommended after a patient on isolation leaves the room. Routine cleaning recommendations differ. In one recent infection control manual from the UK,[150] daily wet cleaning of low risk surfaces and ad hoc cleaning if body fluid spills occur is suggested, while in another manual daily cleaning of low risk surfaces, was not recommended.[151]

There is very little specific research on how much time should be spent cleaning or how often cleaning should be carried out, although it is generally recommended that ICU rooms of patients with MRSA should be disinfected three times per day and once per day on the wards.[152]

In the study by Rampling et al.,[153] cleaning time was doubled for six months to 123.5 hours/week instead of 66.5 hours/week, and there was a statistically significant reduction in MRSA cases and approximately $U.S.60,000 was saved after subtracting cleaning costs from costs that would have been expended to treat MRSA cases.

It should be noted that in one report on time spent cleaning by a health and social affairs reporter at the *Guardian*,[154] who worked as a cleaner in a large London hospital, contracting out led to a loss of hospital management's ability to time cleaning jobs. More specifically, the reporter was given three hours to do a cleaning job that he could have been done in one hour, and throughout the time he worked as a cleaner, it continued to be scheduled for that duration because the contractors claimed that three hours were required.

In audits conducted in 74 hospitals in Scotland,[155] in which four wards and a number of public areas were inspected to assess cleanliness levels, the number of hours spent cleaning, the cost of cleaning, and other factors, it was found that 93% of cleaners were assigned to a specific ward, that external contractors were providing services in 20% of healthcare facilities, and that only 70% of facilities were determined to be adequately clean when compared to benchmarks. As well, 40% did not have cleaners cleaning for the minimum number of prescribed hours because of scheduling problems and absence due to sickness.

The Standards for Environmental Cleanliness (SEC) tool,[156] was developed by the NHS for use by managers, clinical staff, and infection control practitioners in the UK, to ensure that cleaning procedures are being followed. But because cleanliness is based on visual assessment alone, the SEC has been criticized as insufficiently sensitive.[157] When compared to an assessment incorporating the SEC with rapid hygiene testing, a method used regularly in the food industry, it was found to be inaccurate in three out of four hospitals. In other words, visual assessments were not able to detect certain types of contamination of concern. Considering that the Scottish hospital audit, which relied on visual inspection alone, found 30% of facilities insufficiently clean, it may be that using a second method to assess cleanliness, such as rapid hygiene appraisal, would have identified a higher percentage insufficiently clean.

Cleaning regimens are not standardized, and even when comprehensive cleaning regimens are put in place they may not be met because of increased sickness absence and inability to recruit. No state of the art cleaning guidelines exist. This means that infection control recommended cleaning practices must be communicated in a comprehensive, systematic, ongoing manner to those who clean and disinfect at all levels. It means that cleaning is complicated. It means that choices are faced, and decisions made on an ongoing basis. It means that infection control personnel must be in constant close contact with cleaning staff, supervisors and cleaners, in order to make certain that cleaning is part of the infection control solution rather than one more problem.

Pandemic Hypothesis

During a pandemic, cleaning and infection control will become central issues. Good practices as described in this chapter are relevant, but how long and what disinfectant solutions will have to be used for a virulent pathogen may require changes to some of the parameters. Training housekeeping staff to wash their hands, apply appropriate barrier protection including respiratory protection, may pose new challenges in a system that is already stressed. And the potential for pandemics is one of the most important reasons housekeeping services should not be contracted out. It is important to highlight that Taiwan's Center for Disease Control singled out contracted out housekeeping, laundry and nursing aid services as one of the important contributors to the SARs outbreak in that country. And in Canada, during SARs, housekeepers in one hospital did not initially know how to put on and take off gloves without cross-contamination.

Now is the time for healthcare facilities to make sure that their cleaning staff are properly trained and to consider how much cleaning there should be in patient rooms to err on the side of safety.

Conclusion

Infection control is growing in importance as the human and financial costs of healthcare-associated infections rise. While research on the links between cleaning and infection control is evolving, it is clear that cleaning personnel are the "instruments" necessary to carry out many recommendations from the infection control program. It is also clear that the relationship between infection control and cleaning personnel must be close and ongoing, as it should be between cleaners and healthcare providers on the units they work on. A good quality relationship is one that goes both ways, with both sides respecting and learning from the other. This is especially important in an environment where change is constant, where best practices and the organisms they are designed to control are evolving, and where ongoing evidence-based research is critical. Simply put, it is a mistake to treat cleaning personnel as machines that can be programmed to perform necessary tasks. Rather, they must be treated as an important part of the team that delivers quality healthcare, capable of reflection and problem solving.

While there would never be a good time to demoralize and increase stress among a necessary part of the healthcare team, an era of increasing healthcare-associated infections is an especially bad time for cutting wages and benefits and contracting out of cleaning personnel, a practice becoming more and more widespread in our countries. Contracting out will almost certainly lead to difficulties in communication between infection control and cleaning personnel. Cutting wages and benefits will inevitably weaken morale among cleaning staff and therefore do harm to the infection control system. Trying to save money by cutting wages and benefits of cleaners by any means or by contracting out cleaning services, may end up costing more in the long run because of the costs associated with outbreaks of NI.

In conclusion this is just a small taste of the back and forth debate in the literature related to NI and environmental cleaning in healthcare facilities. It is clear that the state of evidence-based infection control, as it relates to cleaning, is best characterized as uncertain.

Much research remains to be done, especially in comparing actual cleaning methods and overall programs. Of course, this sort of research is difficult because it necessarily involves interventions. For example, the best way to determine the infection control effectiveness (including perhaps cost-benefit analyses) of a particular overall cleaning program would be to carry out a long-term intervention in which it was compared to another type of cleaning program at equivalent facilities. This would involve looking at the work practices and educating scores of people in complex social environments over a significant period of time. But even if difficult, such research must be carried out now and in the future.

Chapter 12

Establishing Cooperative Synergy: Which Agencies, Which Departments?

Nora Maher, M.Sc./Occupational Hygiene

CONTENTS

Occupational health has long been considered to be a natural subspecies of public health. If public health practice is charged with the responsibility of protecting and improving the health of the community, then occupational health practice can be identified as protecting and improving the health of the working community. There are two significant distinctions; the population being served and the magnitude and inventory of the hazardous agents. Contagious disease control represents a point of convergence for public health, infection control, and occupational health. In the 1980s the convergence was realized with the emergence of a new threat, HIV. The human immunodeficiency virus was primarily spread in

325

the community, and yet changed virtually every aspect of healthcare delivery. In the early twenty-first century, the global community is facing viral threats previously unknown, which have the potential to change the occupational as well as the public health landscape.

Services are offered by different agencies, and depending on the jurisdiction involved, it can become quite complex and unclear. In the U.S. model, oversight for public health is the domain of the Centers for Disease Control (CDC), under the direction of the Department of Health and Human Services. One of the departments governed by the CDC is the National Institute for Occupational Safety and Health (NIOSH). The NIOSH mandate includes research, guidance, information, and service in issues relating to workplace health and safety. Separate from both of these is the Occupational Safety and Health Administration (OSHA), whose mission it is to assure the safety of America's workers by setting and enforcing standards, providing training, and establishing partnerships. OSHA references CDC in many of its directives.

In Canada, the federal department responsible for helping Canadians maintain and improve their health is Health Canada. Each province within Canada is responsible for administering the healthcare system and providing services. Among the array of services offered, occupational health is notably absent.

A new body recently arose in Canada, the Public Health Agency of Canada (PHAC), reportedly "out of concerns about Canada's public health system's capacity to anticipate and respond effectively to public health threats." When asked about the role of PHAC in the occupational health of healthcare providers, the director responded that the health and safety of these workers is of paramount importance to the agency. He went on to say that this responsibility is shared with the provinces and territories, other federal departments and agencies, including Human Resources and Skills Development, Canada's Labour Program, Health Canada's Workplace Strategies Bureau, and the Canadian Centre for Occupational Health and Safety.[1] Clearly, in Canada at least, accountability is hard to see in the murky pond of bureaucracy.

Layering and division of responsibilities becomes a navigational challenge at times, and never more intensely than during an emergency. SARS was such an emergency. In each jurisdiction in which SARS was found, complexity was encountered in determining leadership and partnership arrangements. And just in case domestic lines of accountability, authority, and responsibility weren't confusing enough, this was an emergency that was experienced, and managed, across international boundaries. Signatories to the United Nations become members of the World Health Organization (WHO) by accepting its constitution: these 192 nations share governance of the WHO. Membership implies a choice to follow the directives of the WHO, which sends teams of experts and specialized protective equipment for infection control to countries requesting such assistance. The WHO, in a status report issued on May 20, 2003[2] stops short of identifying jurisdictions which chose not to avail themselves of these resources, or where territorial issues prevented optimum cooperation. What is clearly stated, though, is that failures were seen and lessons learned about the importance of international collaboration, privileged access to all countries, and strong but politically neutral global leadership. Each of the contributing partners has to be willing to work laterally with all other partners. And leadership must be identified, agreed upon, and then respected. In general this does not happen. It is more common for silos to be established and protective barriers to be erected. Like so many bull moose during the rut, territories are established and not willingly shared. Reasons for this tend to be rooted in political policy. Bureaucracies have been encouraged to compete for increasingly scarce resources; this competition may result in electing not to collaborate if the threat of collaboration may be the sharing of resources. Synergy implies a whole that is greater than the sum of its parts, and when all partners are given equal voice and equal ear, this tends to be the outcome. Clearly the isolationist behavior seen during the SARS outbreak in Ontario is in direct contradiction to the principles of cooperative synergy. Public health specialists, infection control specialists, and occupational health specialists are equally important in the management of an outbreak of contagious disease, but certainly did not earn equal voice in the SARS outbreak. The voice of the public health sector drowned out the statement of concern from the workers. This resulted in illness and death that could have been avoided.

The population served under the public health mandate includes all socioeconomic classes, all levels of education, and ages from conception to death. The full spectrum of health as well as physical, mental, and psychosocial wellness must be taken into account. Conversely, occupationally, there will be a narrower age range and a generally healthy population with some commonality in language, literacy, and education. It is possible therefore to make occupational health more focused to the needs of the community at risk; public health needs to be broader but does not require the same depth. The hazards may be similar but the intensity will often differ, sometimes by many orders of magnitude.

This point can be illustrated by considering the experience with a well-known hazardous substance, asbestos. Asbestos is certainly in the community. It is in the public buildings, in many homes, in water pipes and other manufactured items that are part of daily life. A certain%age of asbestos fibers in all of those environments have the potential to become airborne, and vigilance and control are necessary for the protection of the public. But far and away it is the occupationally exposed who are suffering most of the health impacts. Pipe fitters, insulators, miners, electricians, brake-repair technicians, and their spouses and offspring, have been the populations in which mesothelioma and other asbestos-related diseases have been most evident. Over time it was recognized that workers comprise a particular risk group, and the community movement to ban the mining and use of asbestos is because of *their* risk, not the risk of exposure among the population at large. Different populations require risk assessment that is specific to the probability, frequency, and intensity of their exposure; assessment in one population does not cover the needs of another. In occupational hygiene these are referred to as similar exposure groups.

Flexibility and sensitivity, described by Tyler and Last[3] as pivotal to public health surveillance, will define whether public health practice is reliable to meet occupational health needs in the event of emerging transmissible disease. More than any other factor it will come down to recognizing the population at risk and having the program capacity (flexibility) and the specialized knowledge (sensitivity) to address the needs of the critically exposed population, the healthcare providers.

Inevitably the lead on the management of a communicable disease will fall to public health services, so public health practitioners will need to recognize healthcare workers as a unique exposure group, different from the general public. Included here are professional caregivers and first responders, allied workers (dietary, laundry, laboratory service providers) as well as unpaid caregivers, usually family members, in homes as well as institutions., caregivers are at particularly high risk for viruses that are airborne or transmitted by droplet. And these risks are best understood, and best addressed, by the specialists in occupational health and safety, including the workers at risk. The obvious vehicle for occupational hazard identification, risk assessment, and hazard control is the labor–management joint health and safety committee, which in many workplaces already exists. The members of these committees have rights and responsibilities under the law, and a defined mandate within the Health and Safety Act. The right to know, the right to participate, and the right to be protected are entrenched in occupational health and safety law.

Several worthy investigations have been undertaken regarding the strengths and weaknesses of joint health and safety committees, among them a review by John O'Grady.[4] He concludes that joint health and safety committees can play an important part in improving workplace health and safety if appropriately resourced. These resources include access to information and training of committee members. Participation in the occupational health and safety management of an organization requires that a set of tools be supplied to practitioners. Members bring to the table expertise in their work; they may require additional information on hazards and controls as well as the rights, responsibilities, and liabilities which attach to their role on the health and safety team. Support for the joint health and safety committee in communicable disease management needs to come from both the public health and the occupational health specialties, not one or the other. Communication is vital. All stakeholders are required to participate if a robust risk assessment is to result. The risk assessment is the process by which identification and characterization of the hazard, identification of the exposed populations and segregation into similar exposure groups, and the development of control strategies appropriate to the relative risk of each of the exposure groups will be achieved. It is a scientific discipline used to evaluate whether an event has the potential to occur, and if so, with what frequency, with what range of outcomes, and under what conditions, or in the words of the U.S. National Research Council, to "characterize potential adverse effects resulting from human exposure to hazardous agents or situations."[5]

What follows is a case study of the events which unfolded in Toronto, Ontario, Canada in the spring of 2003. It was an extreme test of the city's emergency response preparedness. It generated a great deal of discussion about the gaps in the public health system, and this is important. There has been much less discussion of the gaps in the *occupational* health system. Though this represents only one jurisdiction, it may be recognizable in many.

Case Study 1: Severe Acute Respiratory Syndrome (SARS) in Ontario

It may be tempting to think that if public health is delivered well, occupational health is naturally protected. In the case of SARS, the reverse may be closer to the truth. Jurisdictions that were successful in managing the hospital and caregiver control had little to worry about in the broader community. Information emerging from Asia early in the WHO alert about the atypical pneumonia was that large numbers of healthcare workers were succumbing to this disease. This was known before the first case was identified in the West, before the virus had been characterized, before the alarm was raised among the public, which virtually shut down cities and decimated economies.

According to the report published in the *Morbidity and Mortality Weekly Report* [6] in March 2003, the Chinese Ministry of Health notified the WHO of the occurrence of 305 cases of SARS; the disease was *"characterized by transmission to healthcare workers and household contacts."* This is a substantial difference from the pattern seen in, for instance, influenza outbreaks. Generally influenza outbreaks are mapped by tracing absence from schools, workplaces, and an increased rate of visits to family doctors. Often family doctors are recruited as sentinels to keep public health services informed of a rise in rates of patient visits. What this atypical pattern of spread was telling us, we now know, is that this was a severe illness that drove sufferers to emergency health care, that it was infectious, and that it was most communicable when sufferers were at their sickest levels, therefore when they were most probably receiving invasive and intensive health care. On March 11th, 2003 in Hong Kong it was recognized that this outbreak appeared to be confined to the hospital setting and healthcare workers seemed to be at the highest risk. The WHO issued a global alert about the mounting number of SARS cases among hospital staff in Hanoi and Hong Kong. Svoboda et al.[7] note that "SARS in Toronto was primarily a nosocomial illness, largely restricted to persons who were exposed in affected hospitals and their household contacts. The actual risk to individuals in the community was always very low." Should the pattern repeat itself, this is a critical branch of the risk assessment decision tree about whether to emphasize public health or occupational health control measures.

Critical also is the ability to communicate with one another. The World Health Organization recognized the atypical pneumonialike illness in China, and issued a communication to the rest of the world. Communication problems were a common thread running through the SARS crisis. Finding an empirical measurement tool that will identify the adequacy of information distribution is not likely to happen. It is often quite obvious, though, when communication is clearly inadequate. China came under much criticism globally for dealing quietly with the health crisis and not alerting the international community sooner. And at the other extreme the WHO came under criticism for posting a travel advisory warning against visiting Toronto. Mr. Justice Campbell[8] addressed the issue of communication, or failure to communicate, in his Inquiry report. In his first interim report Mr. Campbell writes that "successful public communication provides everyone with vital information, helps them make an informed assessment of the situation and the attendant risks, bolsters trust between the public and those solving the crisis, and strengthens the community bonds." Conversely, "A failed effort can breed confusion and antagonism, disrupt an orderly response, poison relations with public authorities and sow mistrust." Though Justice Campbell has not yet issued his report dealing specifically with the occupational experience, the same can certainly be said of the workplace. It is obvious to stakeholders that the area of communication needs to be given high priority in planning for the next communicable disease threat.

When SARS hit Canada it hit hardest in the healthcare provider population. Forty-three percent of the recognized cases of SARS in Canada were in healthcare workers; nurses, physicians, laboratory and radiology personnel, and those providing dietary and cleaning services in hospitals. In Taiwan it was 20%, in Hong Kong 22%, in mainland China 19%. In these four jurisdictions, the ones with the greatest numbers of SARS cases, 1565 health care workers contracted the disease. Healthcare workers clearly constituted an identifiable exposure group. Occupational hygienists tend to look at workers in groupings in order to target control measures toward those with the highest risks and interventions, which yield the greatest bang for the buck.

Workers in what we now recognize as a distinct exposure group were frequently not involved in developing a plan for their own protection while providing care to others. There were limitations to the role of joint health and safety committees in planning for the infectious disease outbreak hazard prior to SARS, up to the complete exclusion in some institutions, and problems resulted. Personal protective equipment was provided on a one size fits all basis. In some cases there was insufficient personal protective equipment to go around. Workers were asked to "make do." Training about the use, care, and limitations of respirators, gloves, and other items of personal protective equipment has long been missing from the orientation or annual training review in most healthcare institutions. In their joint submission to the SARS Commission of Inquiry, the Ontario Public Service Employees' Union and the Ontario Nurses' Association[9] wrote that:

> If the hospital sector had a properly functioning health and safety system with safety conscious and responsive employers, competent and active supervisors, and active joint health and safety committees made up of well-trained members, both unions believe that a number of problems could have been avoided and perhaps fewer workers would have become ill with SARS.

Their experience was that the joint health and safety committees even prior to the outbreak of SARS in many healthcare workplaces were "weak and ineffective." The submission went on to state that:

> Effective joint health and safety committees would have been able to quickly assess where the risks of exposure to SARS were the greatest and would have worked to ensure that workers understood the directives [being issued by the public health leaders].

The Ontario Occupational Health and Safety Act, and parallel legislation in all other jurisdictions, requires correct fitting of safety equipment and training in respect to the selection, use, and care of such equipment.

It was by random chance that certain cities became targets of the SARS virus. While there may not have been enough information to prevent that, we have the knowledge to protect against transmission of airborne and droplet-borne viral disease. Even while all the discussion about bioterrorism has been about inhalable viral spread, healthcare workers apparently did not have the facilities to routinely isolate fevers of unknown origin, were not fit tested and equipped with appropriate respiratory or dermal protection, were far too mobile within the healthcare community, and in large part, were left out of the health and safety management and planning activities.

In the workplace context, while the precautionary principle endorses a philosophy of extreme caution until the hazard is well understood, often the opposite approach is taken. Asbestos, lead, vinyl chloride, benzene, and a host of other entities, now universally recognized to pose severe health hazards, were used copiously and handled cavalierly until the epidemiology could not be ignored. One of the questions to ask with respect to the transmission of communicable disease is, "Has the safety of workers been placed ahead of all other considerations?"

Those who fail to learn from history are doomed to repeat it. Incident review and risk assessment revision are essential. Not only is it important and valuable to revisit a risk assessment, it is important to predetermine the frequency which will be needed. And each risk assessment should be viewed in the context of any incidents or accidents it was designed to prevent. The risk assessment for the transmission of new hazards in the future, like SARS, needs to be a dynamic process, keeping pace with the daily information release, and reviewed in the context of each new case. SARS demonstrated how quickly the body of knowledge could grow. The use of case series reporting to map the movement of the virus through the healthcare worker and patient populations should be used as a source of information. Checkoway[10] makes reference to *case series* when a disease cluster is identified and reported. Characteristic of case series is the contribution to rapid response which may be missing in other epidemiologic investigation techniques. These reports are considered to be particularly valuable in rare conditions for which there are few if any established causal factors, or where there are early indications of a single, as yet unidentified, causal factor. Park[11] believes that a focus on case series could make a significant contribution to the prevention of workplace injuries. While not all investigations of apparent clusters will result in significant findings, case series reporting makes an important contribution to the science of epidemiology. Like all science, epidemiology is dynamic, continuing to evolve and progress.

The International Labour Organization (ILO) in their guidance document[12,13] advises close cooperation between public authorities and employers' and workers' organizations. The ILO also advises a national policy to ensure effective occupational safety and health, and national laws and regulations to support the policies. In Canada, it appears, based upon the review document of the Public Health Agency of Canada,[14] that there was and perhaps remains a gap in awareness of the occupational health resources that are in place to serve the needs of working populations both day to day and in times of crisis. The report makes numerous references to the U.S. Centers for Disease Control as a model for addressing many of the concerns raised. What neither this report, nor the report of the Standing Senate Committee on Social Affairs, Science and Technology[15] raise in their discussion of it, is contemplation of a Canadian national body charged with the task of occupational health and safety oversight. Within the CDC, NIOSH serves this critical function; in the United Kingdom, it is the Health and Safety Executive. In the Canadian response to SARS, failure to include expertise in occupational health and safety resulted in an elevated threat to public as well as worker health. Federal, provincial, and territorial public health experts convened the first of their daily information-sharing teleconferences on March 13, 2003. This was three weeks after the index case arrived in Canada and two weeks after her death. It was known by March 12th that healthcare workers were hugely overrepresented in the numbers of cases abroad, and yet the ministries of Labour, federally and provincially, were not at the table for the daily briefings. WHO issued a global alert on March 12th that a mystery illness was occurring, primarily among healthcare workers. From an occupational health perspective, several findings of the Public Health Agency of Canada invite further discussion. The Scientific Advisory Agency, described as an "ad-hoc group of experts," included experts in infection control, administrators, and physicians, but no members knowledgeable in occupational hazard and risk management. Questions were raised about the exclusion of specialists in anesthesia, pediatrics, and respiratory therapy, but not about workers and their workplaces. A status report posted on the WHO website dated May 20th, 2003 reports that the government of Canada advised the WHO on March 14th that they had taken steps to alert hospital workers, ambulance services, and public health units across provinces that four cases and two deaths had been attributed to SARS in Toronto. While it was clear that workers were at risk, there was no evidence of recognition by the committee that there exist specialists in occupational health and safety. The report refers to the requirement for healthcare workers to use fit-tested N-95 respirators as *controversial,* and the fit-testing process as complex. The Scientific Advisory Committee deemed fit testing to be *operationally impossible.* The Ontario Public Service Employees' Union health alerts and recommendations were seen as "confusing matters" rather than embraced as a partnering voice. It is observed in the report, without editorial comment, that the Ontario Nurses' Association was compelled to launch grievances regarding the lack of fit testing and noncompliance with provincial directives, in an attempt to protect front-line nurses. Awareness of professional organizations such as the American Industrial Hygiene Association (AIHA) and in the case of Toronto, the Occupational Hygiene Association of Ontario (OHAO) were overlooked; they can provide sources of expertise in fit testing as well as other aspects of workplace risk reduction that can be brought on line rapidly in an emergency.

Were workers optimally prepared for the SARS outbreak? How could a different approach to risk assessment have better protected this high-risk population? There is an opportunity now that the dust has settled to review the state of readiness, and revise the plan as necessary. This is the time to evaluate and retool. There is no way to predict when or where the next transmissible organism will strike, nor is it safe to assume that it will behave like SARS. Healthcare workers have been recognizing for a long time that understaffing, overcrowding, early discharge, and noninclusive management have had a significant negative effect on the ability to provide the highest level of health care. Public health, acute care, continuing and long-term care providers have all voiced their concerns that the cuts have been too deep, that they have been left vulnerable. In the report of the Public Health Agency of Canada, tribute is paid to the healthcare workers who died in Toronto, and those who cared for SARS patients and survived. That the committee "salute[s] each and every one of them for their courage and commitment" is frightening, because it fails to recognize or appreciate that their hazardous exposure was completely preventable. An administrator compared the nurses' post-SARS grievances to "having your own soldiers shooting at you." Rather than a defensive response, though, great gains could have been made by hearing and understanding what the occupational community had to offer in the post-SARS debriefing period. I use this jurisdiction as an example of how the interests of public health can be better served by the recognition of exposure groups, including the occupationally exposed. Toronto had the unfortunate experience of being a guinea pig for the testing of emergency response to a biological hazard.

There were missed opportunities to recognize both the special vulnerability of the workplace exposure group, and the benefit of focusing resources within this group. It would be hopelessly naïve to imagine that SARS was a one-of-a-kind event. Whether as a deliberately released weapon of terror, or an illness carried by an innocent traveler, we will meet this crisis again. There is a synergy to be gained by combining strengths, skills, and access to expertise.

Where a joint occupational health and safety committee exists, decision makers and workers can be encouraged to utilize it to its full capacity. Where there is none, it is essential that a committee be appointed. Federal oversight is important, as has been pointed out by the ILO.

SARS has had a devastating effect on the well-being of each community it hit. In the healthcare community, the effects are still being felt. One way to promote the recovery of a sense of trust among healthcare workers is to take meaningful steps to ensure that there is adequate preparation for future threats. In health care as in all industries, the first step toward the protection of health and safety is to anticipate and identify the risks.

Taiwan and Singapore took an occupationally aware approach. These jurisdictions took seriously the risk to healthcare workers, and mounted an occupational response. In a letter posted to the CDC website, Koh et al.[16] report that audits were performed by occupational health professionals in hospitals in Singapore, at the height of the crisis. These specialists were important in modifying the ventilation system for effective infection control, in reviewing work processes, and identifying issues of staff protection and medical surveillance. Occupational groups other than healthcare workers were similarly scrutinized to ensure maximum protection.

Case Study 2: Emergency Response to Hurricane Katrina, New Orleans, Louisiana

In August of 2005 a hurricane struck Louisiana. Hurricane Katrina was as bad as it gets: category 4. Unlike the tsunami of Christmas 2004 in Southeast Asia, everyone knew Katrina was coming. Unlike the victims of the tsunami, U.S. citizens live in a country where there are resources: a transportation infrastructure, communications, and great wealth. And yet emergency responders again were compelled to assume excess risk to their lives and health. Even absent a legal obligation, humans experience a compelling social commitment to one another that does not allow for leaving others in need to fend for themselves. The U.S. government responded with inadequate supplies, with inadequate personal protective equipment, and with inadequate access. Resources that might have been utilized in reducing human distress were dedicated, under the emergency response plan, to protecting property against citizens in dire need.

That flooding in New Orleans was a possibility was optimistic; even the probability was optimistic. The truth is that it was an absolute certainty. In a city where undertakers bury their clients in above ground crypts out of respect for the surrounding water hazards, the expectation of an incursion of the Mississippi and Lake Pontchartrain into the New Orleans basin ought to have caught no one off guard. This reclaimed land mass sits below the level of the surrounding waters. A series of dams, called levees, hold the water back, and a series of pumps move groundwater from under the city to canals and ultimately to Lake Pontchartrain and the Mississippi River. Engineers achieved a remarkable feat. For the responders to this emergency, however, this is a workplace. Control of hazards requires recognition based on risk analysis. The Louisiana Superdome was designated as an evacuation refuge site. However, if the 9000 residents had been taken to safe refuge on higher ground, fewer National Guard, police, fire, healthcare providers, and relief workers would have been needed in the city center where hazards were at their greatest. Interestingly, it had already been identified that the Superdome was a poor choice. In 1998, 14,000 people sheltered there had difficulty accessing supplies, and looting was rampant. The arena flooded during Hurricane Katrina; there was damage to the roof, which may have completely compromised the structural integrity. The plumbing didn't work, there was no power, supplies were slow to arrive, and it took five days to complete the evacuation of the residents. During this time, relief workers were required to try to provide medical, nutritional, and psychosocial care. Many relief workers in turn found themselves in need of the same services.

Again, there was poor coordination among responding bodies. OSHA responded after two days with a public service announcement that workers faced dangers from falls, downed power lines, and chain saws. Recovery and cleanup workers were advised to take proper safety and health precautions. NIOSH, by September 9th, posted on their website recommendations for personal protective equipment for physical hazards associated with relief work; puncture wounds, blood and body fluids, floodwater exposure and electrical hazards. (No mention is made of body armor, although the prevalence of physical violence had been well documented.) NIOSH has undertaken to survey workers regarding their injuries resulting from participation in relief efforts, and the occupational health and safety community will look forward to the opportunity to critically review those results.

Each of these case studies illustrates the common failure to anticipate and control worker exposure to preventable hazards. Could SARS be prevented? Not so far as the world has been able to tell. Could Hurricane Katrina have been prevented? Not at this time in our history. But exposure could have been reduced and engineering controls could have been designed and implemented in anticipation of need. Experiences with both the SARS crisis and Katrina will require an adjustment to the way occupational hazards are regarded and managed. Failure to anticipate is based not upon lack of knowledge but upon a value system that incorporates the principle of acceptable risk. Mathematical formulae that assign a numerical probability to worst-case outcome have no credibility among vulnerable populations. Where death, disability, permanent injury, loss of quality of life, and permanent psychological trauma are the potential outcomes, there is no acceptable risk.

There is a great deal of work to be done in order to bring together the expertise needed to prevent repetition of these terrible events. It will begin with a recognition that workers can and must have safe workplaces. In OSHA's own words, "employers are responsible for providing a safe and healthful workplace for their employees."[17]

Chapter 13

A Rural Hospital's Preparedness for an Emerging Infectious Disease Epidemic

Jeanette Harris, R.N.

CONTENTS

Overview

In 2001, a letter containing anthrax spores was mailed to NBC one week after the September 11th terrorist attacks on the Pentagon and World Trade Center. Then, in November of 2002, SARS appeared. The focus for hospital preparedness had changed dramatically in just one year from preparing for natural disasters to preparing for terrorism and new and emerging infections. This was true, not just for large urban hospitals, but also for small, underfunded, rural hospitals. The world of preparedness changed for all of us, we just did not know it yet. In November 2002, SARS was beginning to spread. There were no headlines yet, but in Guangdong Province, an "atypical pneumonia" was noted in a few humans. Later we found that this virus had crossed the species barrier, attacking a new host, humans. Not only was it capable of infecting humans, it was able to spread from human to human like the common cold or flu. The majority of those afflicted were healthcare workers.

Small rural hospitals today are not isolated from the rest of the world. A hundred years ago they were isolated. Rural people lived and died within an area of 50 miles. Travel was expensive and took considerable time. Local populations were spread thin and it took considerable effort to gather in large groups. Rural small town populations were thin compared to cities.

Starting in the 1950s and 1960s with the building of better roads and the interstate highways, the population became more mobile. It became common to go for a drive in the country for weekend family outings. During these drives, families were able to experience the clean air, open spaces, and to view relatively inexpensive property. Word spread and families began to move to the country. People could commute to their jobs in the urban setting using the improved roads and freeways to drive home at night to enjoy the "farm."

Most rural towns within a two-hour drive of a city have become "bedroom" communities. Farms have given way to planned developments, weekend farms, and a population density that far exceeds the infrastructure of the original farming community. In addition, there is very little industrial tax base to upgrade the community health infrastructure.

All these conditions are very familiar to today's rural based hospital. The population has swollen and the numbers of patients have grown, but since the mid-1980s the small, local country hospital has struggled to keep up. Rural areas are home to 65 million Americans, along with farms, power generation facilities, and the nation's storage facilities for weapons.

As the world became smaller, easier travel times for both goods and humans from far reaches of the globe meant the potential to spread new emerging diseases increased exponentially. Rural hospitals experience the same risks of infection as large urban hospitals. Today, rural populations *live* in the country but most *commute* to the "Big City" every day. With this commute, they are exposed to public transportation, crowded offices, packed elevators, urban homeless populations, worldwide travel, and exotic foods and pets. With the exposures, they get sick, and where do they go? Not to the urban hospital, but to their local clinic and the local, rural emergency room.

Adequate hospital preparedness in rural communities depends on public health departments, emergency medical services, existing medical personnel, and hospital involvement in education and planning. Primarily, all emergency-planning personnel in the community must have awareness that *the rural hospital is a part of, and not separate from, the rest of the world*. It is all too common to believe still, that a rural community "Won't get that out here." Alternatively, there is the mindset, "Who would want to attack us? There's nothing important here." Rural hospitals are the nucleus of health planning, activity, and resources in the community. In the past, national policy changes have forced hospitals to downsize bed capacity in an effort to contain costs, resulting in a lack of surge capacity for personnel and patients. In addition, rural emergency services rely on volunteers and usually lack funding for education and equipment. Equipment and personnel needs in the rural setting are adequate for normal operation. However, the new "as needed" supply chain of materials will be a hindrance in the event of a widespread epidemic or pandemic. During the SARS outbreak, masks were in short supply, with manufacturers scrambling to fill the enormous demand worldwide, even though there were only four to five outbreak areas. Most small hospitals have two- to three-day supplies of most items, from pharmaceuticals to disposable paper products. In the event of a worldwide pandemic, the rural healthcare community will most likely be on their own and at the end of the line when it comes to support from centrally located suppliers, public health, security, reference laboratories, and even morgue storage.

With the enormous needs listed above, it is important to begin planning, education, and support *as soon as possible* and not wait until the next SARS event or pandemic is at the door of the emergency department of the rural hospital.

How to Begin

The need for effective preparedness effort in the rural setting led to U. S. Health and Human Services Secretary Tommy Thompson's (2001) inclusion of the State Offices of Rural Health (SORH) in the HHS Bioterrorism Hospital Preparedness Program. This bioterrorism funding, provided by the states, is perfect for preparing a rural hospital for new emerging infections. The plan is to improve the capacity of the nation's hospitals, their emergency departments, and all associated healthcare services to respond in the case of any surge event, be it bioterrorism or epidemic/pandemic.

The List for Rural Hospitals

1. Your state has developed a plan to help rural hospitals revise their Emergency Preparedness Plan. Have you contacted your state Public Health Office for help with this plan?
2. Find funding. The office of Homeland Security has funding resources available. http://www.dhs.gov/dhspublic. The main barrier to preparedness is lack of funding.
3. Is there an office specifically for rural health planning in your state? If yes, who are they, and what process must be developed to work with them?
4. Is your local health department conducting adequate surveillance for diseases?
5. Designate a department or individual to research outbreaks worldwide via the Internet. Many organizations track disease outbreaks continually. Do not rely on a phone call, *do your own research*. By the time your hospital gets a phone call, it is too late. Some sites with continuing surveillance information:
6. The World Health Organization: http://www.who.org
7. Centers for Disease and Control: http://www.cdc.gov
8. The National Institutes of Health: http://www.nih.gov
9. Most public state health offices have a website. Find it.
10. Emerging infections publication/web site from the CDC: http://www.cdc.gov/ncidod/EID/index.htm
11. The Department of Agriculture: http://www.usda.gov/wps/portal/usdahome
12. The Association for Professionals in Infection Control: http://www.apic.org
13. How would your hospital respond to mass casualties, decontamination, patient placement, security needs, and loss of communications? Make a plan. Use the worst case scenario. In a real emergency, it will probably be even worse than you could have imagined.
14. What are the training needs for all employees? Make sure you include local physicians, emergency personnel, transport systems, police, and do not forget the hospital administration.

Implementation

Assuming that the rural hospital recognizes the importance of preparation, and the hospital administration is in full support, a leader/chief/director needs to be in charge to coordinate all activities. Nothing will be accomplished if all those involved work on the same aspect of preparedness, such as emergency communication, and neglect surveillance measures or equipment acquisition. A regular meeting time of all participants is essential for an active Preparedness Committee, initially on a monthly basis, where assignments, duties, and progress is assessed.

Education

Train the trainer! Send one individual for each part of the preparedness pie to an outside facility to train. For example, most small, rural microbiology laboratories can run routine cultures on urine, sputum, and superficial wounds. The routine is easy and it is easy to look for the regular bacteria that always show up. However, will the working microbiologist recognize when the unusual, different, abnormal, and dangerous organism appears and not just call it a "contaminant"? The microbiologist who questioned the growth of gram-positive rods in blood culture and sputum culture of a patient with "pneumonia" in Florida is a hero. Instead of calling this "just a contaminant," a light bulb went off over his head and he called the local health department. This was not normal, nor was it a contaminant. It was the first anthrax case in 2001. It set off alarms all over the country and a search was on for the source. Many more people would have died if this microbiologist had not recognized the unique properties of these cultures.

Many state and county laboratories conduct training, informational sessions, and meetings on a regular basis. Participate in them! Not everyone in the department needs to go. Send just one or two people and make sure they come back and conduct training for the rest of the department's personnel.

Every time new equipment is ordered, or a new procedure added, everyone who has a remote possibility of using the new equipment or procedure needs training on the proper way to use it. It is often difficult to include training for personnel on all shifts. Accommodations for evening and night shifts need to take place. Offer training sessions when those employees are at work. No excuses allowed for missed training sessions. Murphy's law applies: if is going to happen, it will happen at 2:00 a.m. on a Saturday. If training is scheduled, it should be listed on proficiency or skills update for documentation purposes.

Figure 13.1 Examples of N95 masks.

Equipment

Rarely will a rural hospital have all the necessary equipment to handle emerging infections or bioterrorism. A process of researching necessary new equipment, evaluating its effectiveness, training personnel, and implementation should be adopted for uniform introduction. For instance, many hospitals use N95 masks for personnel while treating TB patients. N95 masks are half-faced air-purifying respirators that fit over the nose and mouth. These same N95 masks are used for many respiratory infections other than TB. SARS and chickenpox are two examples.

The problem with N95 masks is that OSHA has determined that all personnel that wear N95 masks be fit tested on a yearly basis. It can take up to an hour to properly fit test just one employee. In addition, the N95 mask is not a "one size fits all" mask. There are small, medium, and large sizes for small, medium, and large faces, necessitating storage of all sizes. Moreover, facial hair excludes anyone from wearing an N95 mask because there has to be an airtight fit around the nose and mouth. They are "one use only" and must be discarded after each use.

Many hospitals, large and small, have opted for powered air-purifying respirators (PAPRs) that uses a blower to force ambient air through HEPA filters into a loose fitting face piece or hood that covers the entire face or head. They are for use outside a sterile environment and can be used by all personnel; even those who cannot be properly fit tested or have facial hair that interferes with the respirator-to-skin seal. Moreover, PAPRs can be reused if proper cleaning, disinfecting, inspection, storage, and maintenance are addressed. Each facility must evaluate the pros and cons for each equipment decision. Overall, healthcare workers who must wear multiple layers of personal protective garments find the PAPRs easier to wear and more comfortable over long periods of time while caring for patients.

Figure 13.2 Example of a powered air-purifying respirator (PAPR).

TABLE 13.1   Advantages and Disadvantages of PAPRs

Advantages of PAPRs	Disadvantages of PAPR's
Does not require fit testing if head cover, hoods, or helmets are used	Equipment must be: Cleaned Stored Maintained Disinfected Inspected
May be an option for individuals who have facial hair or are unable to fit N95 respirator models	Batteries must be stocked and kept charged
Provides a higher level of respiratory protection than N95 masks	Special protocols need to be instituted for removing, handling, and decontaminating between patients
Cooler for healthcare workers who need to have multiple layers of protective clothing to care for patients	May increase the complexity involved in removal and decontamination, increasing the potential for self-contamination
Provides built-in eye protection	

High-cost decontamination equipment such as biohazard suits, decontamination tents, water heaters, hoses, and patient processing units were part of the HRSA grants to rural hospitals. If your facility has not received or is unaware of the process of receiving this equipment, see to it now! Contact your public health department as soon as possible to get in line for this equipment. Not only is it to be used for any biohazard or chemical hazard event, with the proliferation of rural meth labs, all of which use toxic and flammable chemicals, all hospitals are required to decontaminate potential meth lab patients before admission into the hospital.

Figure 13.3

Isolation Procedures

With the emergence of SARS in 2003, it became apparent that early identification and isolation of potential patients was very important. The vast majority of infections started in hospital emergency department waiting areas, followed by hospital interiors due to improper isolation precautions.

There should be basic and enhanced isolation precautions in place. The foundation of the proposed approach is a set of fundamental elements on which rural hospitals might base their preparedness and response activities.

Examples of these basic response elements are:

- Surveillance for cases of new, emerging diseases or suspicious clusters of pneumonia, with appropriate diagnostic testing.
- Rapid isolation and appropriate management of potential cases of disease.
- Rapid and efficient identification, evaluation, and monitoring of contacts.
- The issuance of travel alerts/advisories, screening of ill travelers at airports, and implementation of other border control measures to prevent international spread of a disease is an indication of which the rural hospital should be aware. Increased possibility exists of infection being introduced into the facility.
- Timely dissemination of communication messages to the local community and healthcare clinics.

Communities can supplement these basic elements with increased control measures that might be needed to address an escalating outbreak, changing transmission patterns or characteristics, variations in compliance, uncertainties about the effectiveness of basic control measures, feasibility and acceptability of specific interventions, or political pressures.

Possible enhanced activities might include:

- Establishment of designated sites for evaluation of possibly infected patients.
- Screening of incoming patients for travel history at airports, ports, and land border crossings.
- Quarantine of close contacts of cases or of persons potentially exposed to infected patients by their presence at a particular function, setting, or institution.
- The rural hospital facility may be asked for information concerning closing schools, canceling large gatherings, or implementing other "snow day" measures for increasing social distance as temporary measures to slow transmission in an affected community.

As the level of transmission during an outbreak changes from day to day, even hour to hour, response activities must also be dynamic. The key to understanding transmission dynamics and knowing when to escalate the response at the local level is a surveillance system. This system provides ready access to timely information on the number of new cases, the likely source of exposure for cases, the number of cases not previously identified as contacts, and the number of contacts (prospective cases) with high-risk exposures to known cases. Access to this information can be found above.

Although rural hospitals will need to adjust the types and level of response measures to local conditions and resources, they will also need to coordinate with adjacent jurisdictions to ensure consistency among responses and minimize confusion or mistrust that may derive from differences in outbreak control.

Isolation is based on capability. The need for rural hospitals to have increased numbers of negative pressure isolation rooms is apparent. Therefore, the ability to have portable negative pressure equipment and air scrubbers with HEPA or ULPA filtration is very important to containment of transmission. The engineers should purchase and be able to install these units to convert either positive pressure or equal pressure rooms to at least 0.01" of negative pressure to assure containment.

Laboratory Safety

All healthcare workers should be empowered to initiate effective measures to protect themselves and others from potentially contagious diseases, in a "safety first" approach. This means that laboratory healthcare workers should be aware of potential clinical and epidemiological risk factors;

have ready access to the biohazard safety cabinets, and knowledge needed to protect themselves and others; and be allowed to initiate the most appropriate infection-control measures immediately — with validation and approval from senior members of staff later.

Assess Hospital and Community Capacity to Respond to Emerging Diseases/Bioterrorism

Use simulations ("table top" or other exercises) to test the facility's response capacities. Use observers to determine criteria and methods for measuring compliance with response measures (e.g., infection control practices, case reporting, and patient placement). Constant surveillance of the rural hospital setting is needed to upgrade and continually train new employees and retrain employees.

Tabletop exercises are drills in miniature. Instead of large numbers of people, smaller groups are used in one large room. The room is much like a "situation room" in a war movie. Table 1 is incident command; table 2 is clinical services; table 3 is epidemiology and outbreak detection/control; and table 4 is hospital support services. First, read the script for each module, and then participants at each table talk about the scenario in the module and come up with answers to a set of questions. Halfway through each module, the tables get their own "update" that either adds some new information to the scenario or creates a bigger challenge.

Endnotes

Prologue

1. Institute of Medicine, *The Future of Public Health* (Washington, D.C.: Institute of Medicine, 1988).
2. Institute of Medicine, *Who Will Keep the Public Healthy: Educating Public Health Professionals for the 21st Century* (Washington, D.C.: Institute of Medicine, 2002).
3. Rand Corporation, *Protecting Emergency Responders,* vols. 1–3 (Santa Monica, CA: The RAND Corporation, 2003).
4. Institute of Medicine, *Public Health Risks of Disasters: Communication, Infrastructure, and Preparedness* (Washington, D.C.: Institute of Medicine, 2005).
5. Elizabeth Fee and Nancy Krieger, "Understanding AIDS: Historical Interpretations and the Limits of Biomedical Individualism," *American Journal of Public Health,* 83 (1993): 1477–86.

Preface

1. These data (and much of this preface) are adapted from E. Frank, "Funding the Public Health Response to Terrorism," *British Medical Journal* (and *BMJ USA*) 331: 526–27.
2. M. R. Bloomberg, "Mayor Michael R. Bloomberg's Statement on First Anniversary of September 11th Attacks," PR-244-02. September 11, 2002. Available from: http://www.nyc.gov/html/om/html/2002b/pr244-02.html
3. "Patterns of Global Terrorism. Appendix A: Chronology of Significant Terrorist Incidents, 2001." Released by the Office of the Coordinator for Counterterrorism. May 21st, 2002, http://www.state.gov/s/ct/rls/pgtrpt/2001/html/10250.htm

4. A. H. Mokdad, J. S. Marks, D. F. Stroup, and J. L. Gerberdingm, "Actual Causes of Death in the United States, 2000," *Journal of the American Medical Association* 291 (2004):1238–45.

5. National Center for Health Statistics, Annual Mortality Report for 2001, http://www.cdc.gov/nchs/fastats/deaths.htm

6. *Washington Post*, December 24th, 2002.

7. Associated Press Worldstream, International News, January 29th, 2003.

8, *Morbidity and Mortality Weekly Reports*, "Terrorism Preparedness in State Health Departments — United States, 2001–2003," *Journal of the American Medical Association* 290 (2003):3190.

9. Ibid.

10. Governor, "$1.3 Million in Funds for Heart Disease Pr evention," http://www.state.ny.us/governor/press/year02/sept27_02.htm

11. U.S. Department of Health and Human Services Press Office. Bioterrorism Preparedness Grants. http://www.hhs.gov/news/press/2002pres/20020606b.html

12. North Dakota Receives Funding for Prevention of Heart Disease and Strokes.

Introduction

1. "The Global Infectious Disease Threat and Its Implications for the U.S.," U.S. government report, www.cia.gov/cia/reports/nie/report/nie99-17dhtml

2. Ibid.

3. P. Bierbaum and M. Lippman, eds., *Proceedings of the Workshop on Engineering Controls for Preventing Airborne Infections in Workers in Healthcare* (Atlanta, GA: Centers for Disease Control and Prevention, 1993).

4. M. Davis, *The Nation*, July 18th, 2005.

5. Ibid.

6. Ibid.

7. B. McCaughey, "Coming Clean," *New York Times*, June 6th, 2005.

8. P. S. Gardner et al., "Virus Cross-Infection in Pediatric Wards, *British Medical Journal* 2, no. 5866(1973): 571–75

9. P. Farquharson, and K. Baguley, "Responding to SARs," *Journal of Emergency Nursing*; 29, no. 3 (): 222–28.

10. R. Ulrich and C. Zimring, "Role of the Physical Environment in the Hospital of the 21st Century," cited in Report to the Center of Health Design, 2004.

11. R.L. Riley, Nardell, E.A., "State of the Art," *Am. Rev. Respir. Dis.*, 1989, 1786–1794.

12. W. Charney, "HEPA Surge Tests for Surrogate Penetration for TB, in *Handbook of Modern Hospital Safety,* ed. William Charney (Boca Raton, FL: CRC Press, 1999),221–46.

13. R Ulrich et al., *Role of Physical Environment in the Hospital of 21st Century* (:Center of Health Design, 2004).

14. E. Bunja and L. McCaskell, "Presentation to the Commission to Investigate the Introduction and Spread of SARs.Toronto, Ontario, November 17th, 2003," http://www.ona.org

15. P. Bierbaum and Lippman, *Proceedings of the Workshop on Engineering Control.*

16. W. Valenti, "Selected Viruses of Nosocomial Importance,"in *Hospital Infections,* 3rd ed. (Boston: Little, Brown, 1992),789–821.

17. R. W. Chanock et al., "Respiratory Synctial Virus," *Journal of the American Medical Association* 176 (1961): 647–53.

18. P. C. Hoffman and R. E. Dixon, "Control of Influenza in Hospitals," *Annals of Internal Medicine* 87 (1977): 725–28.

19. M. R. Moser, "An Outbreak of Influenza Aboard a Commercial Airliner," *American Journal of Epidemiology* 110 (1979): 1–6.

20. I.T. Yu, Y. Li, T.W. Wong et al. "Evidence of Airborne Transmission of SARS. *News J. Med.* 2004; 350(17): 1731–1739.

21. J. Kool, "Risk of Person to Person Transmission with Pneumonic Plague," *Healthcare Epidemiology CID* 40 (2005):

22. L. A. Sawyer, "25-30μm Virus Particles Associated with a Hospital Outbreak of Acute Gastroenteritis with Evidence of Airborne Transmission," *American Journal of Epidemiology* 127 (1988): 1261–71.

23. Barack Obama and Richard Lugar, "Grounding a Pandemic," *New York Times,* June 8th, 2005.

24. U.S. Department of, "The Global Infectious Disease Threat."

25. Ibid.

26. Ibid.

27. Government Accountability Office,

28. Centers for Disease Control, *Guidelines* (Atlanta, GA: Centers for Disease Control and Prevention, www.CDC.gov).

29. Province of Ontario, Minister's Report

30. Sawyer, "25-30μm Virus Particles Association with a Hospital Outbreak of Acute Gastroenteritis."

31. add a note re MJA 2004 181(2):62-63

32. National Intelligence Council, January 2000

33. Ibid.

Chapter 3

1. T. V. Inglesby et al., "Plague as a Biological Weapon," in *Bioterrorism: Guidelines for Medical and Public Health Management,* ed. D. A. Henderson et al. (Chicago: AMA Press,), 121–40.

2. J. H. Lange, "Respiratory Protection and Emerging Infectious Diseases: Lessons From Severe Acute Respiratory Syndrome (SARS). *Chinese Medical Journal,* 118 (2005a): 62–68.

3. A. S. Monto, "The Threat of an Avian Influenza Pandemic, " *New England Journal of Medicine* 352 (2005): 323–25.

4. CNN, "Special Report: Severe Acute Respiratory Syndrome," 2003, http://www.cnn.com?SPECIALS/2003/sars/ (accessed July 18th, 2005).

5. Ibid.

6. Ibid.

7. Reuters. "China Reports Suspected New SARS Case," December 27th, 2003, http://www.wireservice.wired.com/wired/headlines.asp (accessed December 28, 2003).

8. M. Varia et al., "Investigation of a Nosocomial Outbreak of Severe Acute Respiratory Syndrome (SARS) in Toronto, Canada," *Canadian Medical Association Journal* 169 (2003): 285–92.

9. J. H. Lange "Respiratory Protection," 2005a.

10. A. Yassi et al., "Research Gaps in Protecting Healthcare Workers from SARS and Other Respiratory Pathogens: An Interdisciplinary, Multistakeholder Evidence-Based Approach," *Journal of Occupational Environmental Medicine* 46 (2005): 613–22.

11. Z. Shen et al., "Superspreading SARS Events, Beijing, 2003," *Emerging Infectious Diseases* 10 (2004): 256–60; http://www.cdc.gov/ncidod/EID/vol10no2/03-0732.htm (accessed May 30th, 2005).

12. I. T. Yu, et al, "Evidence of Airborne Transmission of the Severe Acute Respiratory Syndrome Virus," *New England Journal of Medicine* 350 (2004): 1731–39.

13. T. K. Koley, "Severe Acute Respiratory Syndrome: A Preliminary Review," *Journal of the Indian Medical Association* 101 (2003): 308–310.

14. M. Nicas, et al., "Respiratory Protection and Severe Acute Respiratory Syndrome," *Journal of Occupational and Environmental Medicine* 46 (2004a):196–97; Lange, "Respiratory Protection," 2005a; Lange, "SARS, Emerging Diseases, Healthcare Workers and Respirators," *Journal of Hospital Infection* 60 (2005b): 293 (letter).

15. V. M. S. Oh and T. K. Linn, "Singapore Experience of SARS," *Clinical Medicine* 3 (2003): 448–51.

16. A. S. Ho et al., "An Outbreak of Severe Acute Respiratory Syndrome Among Hospital Workers in a Community Hospital in Hong Kong," *Annals of Internal Medicine* 139 (2003): 564–67.

17. L.Y. Hsu, C.C. Lee, J.A. Green, B. Ang, N.I. Paton, L. Lee, J.S. Villacian, P.L. Lim, A. Earnest, Y.S. Leo, "Severe acute respiratory (SARS) in Singapore: clinical features of index patient and initial contacts," *Infection control* 9 (2003): 713–717.

18. T.S. Li, T.A. Buckley, F.H. Yap, J.J. Sung, G.M. Joynt, "Severe acute respiratory syndrome (SARS): infection control," *Lancet* 361 (2003): 1386.

19. D.C. Scales, K. Green, A.K. Chan, S.M. Poutanen, D. Foster, K. Nowak, J.M. Raboud, R. Saskin, S.E. Lapinsky, T.E. Stewart, "Illness in intensive care staff after brief exposure to severe acute respiratory syndrome," *Emerging Infect. dis.* 9 (2003): 1205–1210.

20. P.T. Tsui, M.L. Kwok, H. Yuen, S.T. Lai, "Severe acute respiratory syndrome: clinical outcome and prognostic correlates," *Emerging Infect. Dis.* 9 (2003): 1064–1069.

21. W. H. Seto et al., "Effectiveness of Precautions Against Droplets and Contact in Prevention of Nosocomial Transmission of Severe Acute Respiratory Syndrome (SARS)," *Lancet* 361 (2003): 1519–20.

22. Lange, "Respiratory Protection," 2005a.

23. Yassi et al., "Research Gaps"; C. Drosten et al., "Identification of a Novel Corona Virus in Patients with Severe Acute Respiratory Syndrome," *New England Journal of Medicine,* 348 (2003): 1967–76; Ho et al., "An Outbreak."

24. Yassi et al., ibid.

25. World Health Organization, "Update: Outbreak of Severe Acute Respiratory Syndrome — Worldwide," *Journal of the American Medical Association* 289 (2003): 1918–20.

26. J. H. Lange, "The Best Protection," *Canadian Medical Association Journal* 168 (2003a): 1524 (letter).

27. I. H. G. Escudero et al., "Surveillance of Severe Acute Respiratory Syndrome (SARS) in the Post-Outbreak Period," *Singapore Medical Journal* 46 (2005): 265–71.

28. D. T. Wong, "Protection Protocol in Intubation of Suspect SARS Patients," *Canadian Journal of Anesthesiology* 50 (2003): 747–48.

29. Koley, "Severe Acute Respiratory Syndrome," 2003.

30. Drosten et al., "Identification of a Novel Corona Virus," 2003.

31. W. Nyka, "Studies on the Infective Particle in Airborne Tuberculosis. I. Observations in Mice Infected with Bovine Strain *M. tuberculosis,*" *American Review of Respiratory Disease* 85 (1962): 33–39; M. Nicas et al., "The Infectious Dose of Variola (Smallpox) Virus," *Applied Biosafety* 9 (2004b): 118–27; M. Nicas, W. W. Nazaroff, and A. Hubbard, "Toward Understanding the Risk of Secondary Airborne Infection: Emission of Respirable Pathogens," *Journal of Occupationl and Environmental Hygiene* 2 (2005): 143–54.

32. Nicas et al. "Toward Understanding."

33. Lange, "SARS and Respiratory Protection," *Hong Kong Medical Journal* 10 (2004a): 71–72 (letter).

34. Tennessee Department of Health, "Tennessee Epi-News — Monkeypox: An Emerging Infectious Disease in North America," *Communicable and Environmental Disease Services,* July (2003): 00–00; S. M. Gordon and D. L. Longworth, "SARS: Here to Stay? Monkeypox: Beware of Exotic Pets," *Cleveland Clinic Journal of Medicine* 70 (2003): 889–95.

35. A. T. Fleischer et al., "Evaluation of Human-to-Human Transmission of Monkeypox from Infected Patients to Health Care Workers," *Clinical Infectious Disease* 40 (2005): 689–94; Gordon and Longworth, ibid.

36. J. J. Sejvar et al., "Human Monkeypox Infection: A Family Cluster in the Midwestern United States," *Journal of Infectious Disease,* 190 (2004): 1833–40.

37. Centers for Disease Control, "Update: Multistate Outbreak of Monkeypox — Illinois, Indiana, Kansas, Missouri, Ohio, and Wisconsin, 2003," *Mortality Morbidity Weekly Report* 52 (2003): 616.

38. Tennessee Department of Health, "Monkeypox," 2003.

39. J. Maskalyk, "Monkeypox Outbreak Among Pet Owners," *Canadian Medical Association Journal* 169 (2003): 44–45.

40. Ibid.

41. Fleischer et al. "Evaluation of Human-to-Human Transmission," 2005.

42. Gordon and Longworth, "SARS: Here to Stay?"

43. Maskalyk, "Monkeypox Outbreak Among Pet Owners"; B. C. Abrahams and D. M. Kaufman, "Anticipating Smallpox and Monkeypox Outbreaks: Complications of the Smallpox Vaccine," *Neurologist* 10: 265–74.

44. R. Connix et al., "Tuberculosis in Prisons in Countries with High Prevalence," *British Medical Journal* 320 (2000): 440–42.

45. D. Hillemann et al., "Use of the Genotype MTBDR Assay for Rapid Detection of Rifampin and Isoniazid Resistance in *Mycobacterium tuberculosis* Complex Isolates," *Journal of Clinical Microbiology* 43 (2005): 3699–3703.

46. R. M. Granich et al., "Multidrug Resistance Among Persons with Tuberculosis in California, 1994–2003," *Journal of the American Medical Association* 293 (2005): 2732–39.

47. Connix et al., "Tuberculosis in Prisons."

48. R. Laniado-Laborin, "Tuberculosis in Correctional Facilities: A Nightmare without End in Sight," *Chest* 119 (2001): 681–83.

49. Ibid.

50. Connix et al., "Tuberculosis in Prisons."

51. M. E. Kimerling et al., "Inadequacy of the Current WHO Re-Treatment Regimen in a Central Siberian Prison: Treatment Failure and MDR-TB," *International Journal of Tuberculosis and Lung Disease,* 3 (1999): 451–53.

52. Centers for Disease Control, "Influenza Virus," http://www.cdc.gov/flu/avian (accessed February 24, 2005a).

53. B. Petrini and S. Hoffner, "Drug-Resistant and Multidrug-Resistant Tubercule Bacilli," *International Journal of Antimicrobial Agents* 13 (1999): 93–97.

54. Centers for Disease Control, "Multidrug-Resistant Tuberculosis in a Hospital — Jersey City, New Jersey, 1990–1992," *Morbidity and Mortality Weekly Report* 43 (1993): 417–19.

55. Centers for Disease Control, "Influenza Virus," 2005a.

56. Ibid.

57. Monto, "Threat of Avian Influenza," 2005.

58. N. Williams, "Flu Pandemic Fears Continue," *Current Biology* 15 (2005): R313–14; J. Parry "WHO Confirms Avian Flu Outbreak in Hanoi," *British Medical Journal* 328 (2004a): 123.

59. Parry, ibid.

60. J. Parry, "Officials Report First Cambodian Case of Avian Flu," *British Journal of Medicine* 330 (2005): 273.

61. J. Parry, "WHO Investigates Possible Human to Human Transmission of Avian Flu," *British Medical Journal* 328 (2004b): 308.

62. F. Fleck "Avian Flu Virus Could Evolve into a Dangerous Human Pathogen Experts Fear," *Bulletin of the World Health Organization* 82 (2004): 236–37; Parry, "WHO Investigates," 2004b.

63. S. Jian et al., "Ventilation of Wards and Nosocomial Outbreak of Severe Acute Respiratory Syndrome Among Healthcare Workers, " *Chinese Medical Journal* (Eng.) 116 (2003): 1293–97.

64. Yassi et al., "Research Gaps," 2005.

65. Lange, "Respiratory Protection, " 2005a.

66. Ibid.
67. Koley, "Severe Acute Respiratory Syndrome," 2003; J. Kool, "Risk of Person-to-Person Transmission of Pneumonic Plague," *Clinical Infectious Disease* 15 (2005): 70–72; J. L. Derrick and C. D. Gromersall, "Protecting Healthcare Staff from Severe Acute Respiratory Syndrome: Filtration Capacity of Multiple Masks," *Journal of Hospital Infection* 59 (2005):365–68.
68. Lange, "Respiratory Protection," 2005a.
69. Koley, "Severe Acute Respiratory Syndrome," 2003.
70. Yassi et al., "Research Gaps," 2005.
71. J. L. Lange, "BIS/BTS SARS Guidelines," *Thorax* 59 (2004b): 725–27 (letter).
72. J. L. Lange, "A Questionnaire Survey During Asbestos Abatement Refresher Training for Frequency of Respirator Use, Respirator Fit Testing and Medical Surveillance," *Journal of Occupational Medicine and Toxicology* 2 (1993): 65–74.
73. J. L. Lange, "Clinics in Diagnosis Imaging: Hydatid Cyst of the Lung and Liver," *Singapore Medical Journal* 46 (2005c): 434 (letter); M. D. Nettleman, M. Fredrickson, R. L. Good, and S. A. Hunter, "Tuberculosis Control Strategies: The Cost of Particulate Respirators" 121 (1994): 37–40.
74. M. A. Rothstein et al., "Quarantine and Isolation: Lessons learned from SRAS — A Report to the Centers for Disease Control and Prevention," Institute for Bioethics, Healthy Policy and Law, University of Louisville School of Medicine, Louisville, KY 40292; http://www.instituteforbioethics.org
75. Ibid.
76. Ibid.
77. N. Ndayimirije and M. K. Kindhauser, "Marburg Hemorrhagic Fever in Angola — Fighting Fear and a Lethal Pathogen," *New England Journal of Medicine* 352 (2005): 2155–57.
78. Ibid.
79. Massachusetts Medical Society, "Good Health Is in Your Hands," May 11th (2001); U.S. Department of Labor, "Avian Flu Guidance," http://www.osha.gov/dsg/guidance/avian-flu.html (accessed February 24th, 2005).
80. Koley, "Severe Acute Respiratory Syndrome," 2003.
81. Lange, "Respiratory Protection," 2005a.
82. G. D. Thorne et al., "Using the Hierarchy of Control Technologies to Improve Healthcare Facility Infection Control: Lessons For Severe Acute Respiratory Syndrome," *Journal of Occupational and Environmental Medicine* 46 (2004): 613–22.
83. L. Ha et al., "Lack of SARS Transmission,"
84. D. A. Fisher et al. "Preventing Local Transmission of SARS: Lessons from Singapore," *Medical Journal of Australia* 178 (2003): 555–58; B. J. Tobis, "Respiratory Protection for Severe Acute Respiratory Syndrome (SARS)," *Disaster Management Response* 1 (2003): 91–92; Thorne et al., "Using the Hierarchy of Control Technologies,"; Derrick and Gomersall, "Protecting Healthcare Staff from Severe Acute Respiratory Syndrome: Filtration Capacity of Multiple Masks," *Journal of Hospital Infection* 59 (2005): 365–68.

85. M. Nicas et al., "Respiratory Protection," 2004a; Lange, "The Best Protection," 2003a; Lange, "SARS Respiratory Protection," 2003b; Lange, Respiratory Protection," 2005a.

86. Seto et al., "Advisors of Expert SARA Group."

87. Derrick and Gomersall, "Surgical Helmets."

88. Nicas et al., "Respiratory Protection."

89. Centers for Disease Control, "Guidelines for Preventing the Transmission of Mycobacterium tuberculosis in Health Care Facilities," *Morbidity and Mortality Weekly Report* 43 (1994): 1–27.

90. Centers for Disease Control, "Public Health Guidance for Community-Level Preparedness and Response to Severe Acute Respiratory Syndrome (SARS)," 2003; http://www.cdc.gov/ncidid/.sars/sarsprepplan.htm (accessed October 2003).

91. Canada Communicable Disease Report, "Cluster of Severe Acute Respiratory Syndrome Cases Among Protected Health Care Workers — Toronto," April (2003); Centers for Disease Control, "Interim Domestic Guidance On the Use of Respirators to Prevent Transmission of SARS," May 6, 2003; http://cdc.gove/ncidod/sars/rsepirators.htm, 2003c; Health Canada, 2003; Minnesota Department of Health, "N95 Disposable Respirators," St. Paul, MN http://www.health.state.mn.us/divs/idepc/dtopics/ (accessed April 26th, 2005).

92. J. L. Lange, "SARS Respiratory Protection," *Canadian Medical Association Journal* 169 (2003b): 541–42 (letter); Lange, ibid.; Lange, "SARS Respiratory Protection," (2004c).

93. Fisher et al., "Preventing Local Transmission," 2003.

94. Centers for Disease Control, "Preventing the Transmission of Mycobacterium Tuberculosis," 1994.

95. Lange, "Respiratory Protection," 2005a.

96. G. S. Rajhans and D. S. L. Blackwell, *Practical Guide to Respirator Usage in Industry* (Boston: Butterworth, 1985).

97. S. Grinshpun, "Respirator Performance with Infectious Agents (Study with Stimulant). http://www.cdc/gov/niosh/npptl/resources/pressrel/ (accessed April 26th, 2005).

98. Seto et al., "Effectiveness of Precautions," 2003.

99. Nicas, "Respiratory Protection."

100. Grinshpun, "Respirator Performance."

101. Centers for Disease Control, "Public Health Guidance," 2003b.

102. Grinshpun, "Respirator Performance."

103. Lange, "Best Protection," 2003a.

104. Lange, "Respiratory Protection," 2005a.

105. Minnesota Department of Health, "N95 Disposable Respirators."

106. J. L. Lange, "Health Effects of Respirator Use at Low Airborne Concentrations," *Medical Hypotheses* 54: 1005–07.

107. K. J. Davis et al., "Seroepidemiological Study of Respiratory Virus Infections Among Dental Surgeons," *British Dental Journal* 176 (1994): 262–65.

108. Koley, "Severe Acute Respiratory Syndrome."

109. K.L. Khoo, P.H. Leng, I.B. Ibrahim, T.K. Lim, "The changing face of healthcare workers perceptions on powered air-purifying respirators during the SARS outbreak," *Respiratory* 10 (2005): 107–110.

110. Derrick and Gromersall, "Surgical Helmets."

111. Ibid.

112. Lange, "Respiratory Protection," 2005a.

113. Rajans and Blackwell, "Practical Guide to Respirator Usage."

114. F. Lateef et al., "New Paradigm for Protection: The Emergency Ambulance Services in the Time of Severe Acute Respiratory Syndrome," *Prehospital Emergency Care* 8 (2004): 304–7.

115. Kool, "Risk of Person-to-Person Transmission"

116. Ibid.

117. Lange, "Respiratory Protection," 2005a.

118. Ibid.

119. Rajhans and Blackwell, "Practical Guide to Respirator Usage."

120. Lange, "SARS and Respiratory Protection," 2004a.

121. Ibid.; N. J. Bollinger and R. H. Schutz, *NIOSH Guide to Industrial Respiratory Protection* (Washington, D.C.: U.S. Department of Health and Human Services, 1987).

122. Yassi et al., "Research Gaps in Protecting Healthcare Workers"; S. Barnhart et al., "Tuberculosis in Health Care Settings and the Estimated Benefits of Engineering Controls and Respiratory Protection," *Journal of Occupational and Environmental Medicine* 39 (1997): 849–54.

123. B. A. Plog et al., Fundamentals of Industrial Hygiene (Itasca, IL: National Safety Council, 1996).

124. H. E. Mullins et al., "Development of New Qualitative Test for Fit Testing Respirators," *American Industrial Hygiene Association Journal* 56 (1995):

125. NIOSH 1994 Pocket Guide to Chemical Hazards. U.S. Department of Health and Human Services (DHHS). *NIOSH* 94–116. Washington, D.C.

126. Mullins et al., "Development of New Qualitative Test"; Clapham, "Comparison of N95 Disposable Filtering Facepiece Fits Using Bitrex Qualitative and TSI Portacount Quantitative Fit Testing," *International Journal of Occupational and Environmental Health* 6 (2000): 50–55.

127. J. R. Sibert and N. Frude, "Bittering Agents in the Prevention of Accidental Poisoning: Children's Reactions to Denatonium Benzoate (Bitrex)," *Archives of Emergency Medicine* 8 (1991): 1–7; M. E. Mullins and B. Zane-Horowitz, "Was It Necessary to Add Bitrex (Denatonium Benzoate) to Automative Products?" *Veterinarian and Human Toxicology* 46 (2004): 150–52.

128. B. Bjorkner, "Contact Urticaria and Asthma from Denatonium Benzoate (Bitrex)." *Contact Dermatitis* 6 (1980): 466–71; North Safety Products, "Material Safety Data Sheet for Denatonium Benzoate (Bitrex)," http://www.north safety.com/usa/en/images/MSDS/ (accessed May 8th, 2005); C-Tech Corporation, "Material Safety Data Sheet for Denatonium Benzoate," http://www.business.vsnl.com/whitestar/denatonium.html (accessed May 8th, 2005).

129. Clapham, "Comparison of N95 Disposable Filtering Facepiece Fits."

130. Derrick and Gromersall, "Protecting Healthcare Staff."

131. Lange, "A Questionnaire Survey."

132. Lange, ibid.; Lange, "Respiratory Protection," 2005a.

133. Lange, "SARS and Respiratory Protection," 2004a; Lange, "Respiratory Protection," 2005a; Lange, "SARS, Emerging Diseases," 2005b.

134. Oh and Linn, "Singapore Experience of SARS."

135. Yassi et al.,"Research Gaps in Protecting Healthcare Workers."

136. Ibid.

Chapter 4

1. J. O. Hendley and J. M. Gwaltney Jr., "Mechanisms of Transmission of Rhinovirus Infections," *Epidemiological Review* 10 (1988): 242–58.

2. W. H. Seto et al., "Effectiveness of Precautions against Droplets and Contact in Prevention of Nosocomial Transmission of Severe Acute Respiratory Syndrome (SARS)," *Lancet* 361 (2003):1519–20.

3. Y. Li et al. "Role of Air Distribution in SARS Transmission during the Largest Nosocomial Outbreak in Hong Kong," *Indoor Air* 15 (2004): 83–95.

4. C. N. Haas et al., *Quantitative Microbial Risk Assessment* (New York: John Wiley, 1999), 266–68, 275–77.

5. M. Nicas and A. Hubbard, "A Risk Analysis for Airborne Pathogens with Low Infectious Doses: Application to Respirator Selection against *Coccidioides immitis* Spores," *Risk Analysis* 22 (2002):1153–63.

6. Haas et al., *Quantitative Microbial Risk Assessment.*

7. Nicas M., "Use of a Probabilistic Infectious Dose Model for Estimating Airborne Pathogen Infection Risk: Application to Respiratory Protection," *Journal of International. Society for Respiratory Protection* 22 (2005): 24–37.

8. W. F. Wells, *Airborne Contagion and Air Hygiene* (New York: Cambridge University Press), 117–22. Experiments using a variety of small mammals indicate that the deposition of a single *M.tuberculosis* bacillus induces a focus of infection in lung tissue. Variable susceptibility to infection in the human population likely exists but has not been quantified.

9. R. M. Jones et al., "The Infectious Dose of *Coxiella burnetti* (Q fever). Submitted to *Applied Biosafety.*

10. R. M. Jones et al., "The Infectious Dose of *Francisella tularensis* (Tularemia)," *Applied Biosafety* 10 (2005): in press

11. M. Nicas and A. Hubbard, "A Risk Analysis for Airborne Pathogens with Low Infectious Doses: Application to Respirator Selection Against *Coccidioides immitis* Spores," *Risk Analysis* 22 (2002): 1153–63.

12. M. Nicas et al., "The Infectious Dose of Variola (Smallpox) Virus," *Applied Biosafety* 9 (2004): 118–27.

13. Nicas and Hubbard, "Risk Analysis."

14. NIOSH, *NIOSH Respirator Selection Logic 2004*, NIOSH Publication Number 2005-100, October 2004, http://www.cdc.gov/niosh/docs/2005-100/default.html#ack (accessed September 7th, 2005).

15. M. Nicas and J. Neuhaus, "Variability in Respiratory Protection and the Assigned Protection Factor," *Journal of Occupational and Environmental Hygiene* 1(2004): 99–109.
16. Ibid.
17. M. Nicas and J. Neuhaus, "Variability in Respiratory Protection and the Assigned Protection Factor," *Journal of Occupational and Environmental Hygiene* 1 (2004): 99–109.
18. NIOSH, *NIOSH Respirator Selection.*
19. R. Fairfax, OSHA, "Enforcement Policy Change for Respiratory Protection for Select Respirators Used in the Pharmaceutical Industry," memorandum to OSHA Regional Administrators, May 30th, 2002.
20. H. J. Cohen et al, "Simulated Workplace Protection Factor Study of Powered Air-Purifying and Supplied Air Respirators," *American Indian Hygiene Association Journal* 62 (2001): 595–604.
21. R. G. Loudon and R. M. Roberts, "Droplet Expulsion from the Respiratory Tract," *American Review of Respiratory Disease* 95 (1967):435–42.
22. M. Nicas et al., "Toward Understanding the Risk of Secondary Airborne Infection: Emission of Respirable Pathogens," *Journal of Occupational and Environmental Hygiene* 2 (2005):134–45.
23. Ibid.
24. Ibid.
25. R. G. Loudon and L. C. Brown, "Cough Frequency in Patients with Respiratory Disease," *American Review of Respiratory Disease* 96 (1967):1137–43.
26. H. Yeager et al., "Quantitative Studies of Mycobacterial Populations in Sputum and Saliva," *American Review of Respiratory Disease* 95 (1967):908–1004.
27. W. C. Hinds, *Aerosol Technology,* 2nd ed. (New York: John Wiley, 1999), 63–64, 235–42.
28. National Institute for Occupational Safety and Health, *A Guide to Industrial Respiratory Protection,* NIOSH Publication No. 76-189 (Cincinnati, Ohio: NIOSH, 1976), 13.
29. Fairfax, memorandum.

Chapter 5

1. L. M. Brousseau et al., "Mycobacterial Aerosol Collection Efficiency of Respirator and Surgical Mask Filters Under Varying Conditions of Flow and Humidity," *Applied Occupational & Environmental Hygiene,* 12, no.6 (1997): 435–45.
2. Y. Qian et al., "Particle Reentrainment from Fibrous Filters," *Aerosol Science and Technology,* 27, no. 3 (1997): 394–404.
3. K. Lee et al., "Respiratory Protection against Mycobacterium tuberculosis: Quantitative Fit Test Outcomes for Five Type N95 Filtering-Facepiece Respirators," *Journal of Occupational Environmental Hygiene,* 1 (2004): 22–28.
4. Brousseau et al., "Mycobacterial Aerosol Collection."

5. M. Lippmann et al., "Deposition, Retention, and Clearance of Inhaled Particles," *British Journal of Industrial Medicine*, 37, no. 4 (1980): 337–62.

6. M. Nicas, "Respiratory Protection and the Risk of Mycobacterium-Tuberculosis Infection," American Journal of Industrial Medicine, 27, no. 3 (1995): 317–33.

7. S. Sattar and M. Ijaz, "Spread of Viral Infections by Aerosols," *CRC Critical Reviews in Environmental Control*, 17, no. 2 (1987): 89–131.

8. B. O. Stuart, "Deposition and Clearance of Inhaled Particles," *Environmental Health Perspectives*, 16 (1976): 41–53.

9. J. H. Vincent, "The Fate of Inhaled Aerosols: A Review of Observed Trends and Some Generalizations," *Annals of Occupational Hygiene*, 34, no. 6 (1990): 623–37.

10. E. A. Nardell and J. M. Macher, "Respiratory Infections: Transmission and Environmental Control," in *Bioaerosols: Assessment and Control*, ed. J. Macher (Cincinnati, OH: American Conference of Governmental Industrial Hygienists, 1999).

11. B. L. Haagmans et al.. "Pegylated Interferon-α Protects Type 1 Pneumocytes against SARS Coronavirus Infection in Macaques," *Nature Medical* 10, no. 3 (2004): 290–93.

12. D. Evans, "Epidemiology and Etiology of Occupational Infectious Diseases," in *Occupational and Environmental Infectious Diseases: Epidemiology, Prevention and Clinical Management.*, ed. A Couturier (Beverley Farms, MA: OEM Press, 2000), 37–132.

13. C. Peters et al., "Patients Infected with High-Hazard Viruses: Scientific Basis for Infection Control," *Archives of Virology* 11, Suppl. (1996): 141–68.

14. Nardell and Macher, *Respiratory Infections.*

15. J. P. Duguid, "The Size and the Duration of Air Carriage of Respiratory Droplets and Droplet-Nuclei," *The Journal of Hygiene*, 44 (1946); 471–79.

16. R. S. Papineni and F. S. Rosenthal, "The Size Distribution of Droplets in the Exhaled Breath of Healthy Human Subjects," *Journal of Aerosol Medicine*, 10, no. 2 (1997): 105–16.

17. Ibid.

18. K. Fennelly et al., "Cough-Generated Aerosols of Mycobacterium Tuberculosis: A New Method to Study Infectiousness," *American Journal of Respiratory Critical Care Medicine* 169 (2004): 604–609.

19. N. J. Bollinger and R. M. Schutz, *NIOSH Guide to Industrial Respiratory Protection* (Washington, D.C.: NIOSH, 1987), 116.

20. L. Janssen, "Comparison of Five Methods for Fit-Testing N95 Filtering-Facepiece Respirators," *Applied Occupational & Environmental Hygiene*, 18, no. 10 (2003): 732–33.

21. K. W. Lee and B.Y.H. Liu, "On the Minimum Efficiency and the Most Penetrating Particle Size for Fibrous Filters," *Journal of the Air Pollution Control Association*, 30, no. 4 (1980): 377–81.

22. Ibid.

23. Brousseau et al., "Mycobacterial Aerosol Collection."

24. Bollinger and Schutz, *NIOSH Guide.*

25. Canadian Standards Association, *Selection, Use, and Care of Respirators* (Mississauga, ON: Canadian Standards Association, 2002).
26. Bollinger and Schutz, *NIOSH Guide.*
27. Canadian Standards Association, *Respirators.*
28. T. Hodous and C. Hodous, "The Role of Respiratory Protective Devices in the Control of Tuberculosis," *Occupational Medicine: State of the Art Reviews* 9, no. 14 (1994): 631–57.
29. Canadian Standards Association, *Respirators.*
30. Centers for Disease Control, *NIOSH TB Respiratory Protection Program in Health Care Facilities. Administator's Guide* (Atlanta, GA: NIOSH, 1999).
31. Canadian Standards Association, *Respirators.*
32. Centers for Disease Control, "Laboratory Performance Evaluation of N95 Filtering Facepiece Respirators, 1996," *Morbidity and Mortality Weekly Report,* 47, no. 48 (1998): 1045–49.
33. C. C. Coffey et al., "Simulated Workplace Performance of N95 Respirators," *American Industrial Hygiene Association Journal* 60, no. 5 (1999): 618–24.
34. C. Coffey et al., "Fitting Characteristics of Eighteen N95 Filtering-Face Respirators," *Journal of Occupational and Environmental Hygiene* 1 (2004): 262–71.
35. Lee et al., "Respiratory Protection."
36. R. T. McKay and E. Davies, "Capability of Respirator Wearers to Detect Aerosolized Qualitative Fit Test Agents (Sweetener and Bitrex) with Known Fixed Leaks," *Applied Occupational & Environmental Hygiene* 15, no. 6 (2000): 479–84.
37. Bollinger and Schutz, *NIOSH Guidelines.*
38. Lee and Liu, "Minimum Efficiency."
39. W. Hinds, *Aerosol Technology: Properties, Behavior, and Measurement of Airborne Particles.* (New York: Wiley, 1999), 191.
40. L. Janssen, "Principles of Physiology and Respirator Performance," *Occupational Health & Safety,* 72, no. 6 (2003): 73–78.
41. C. C. Chen et al., "Loading and Filtration Characteristics of Filtering Facepieces," *American Industrial Hygiene Association Journal* 54, no. 2 (1993): 51–60.
42. A.Weber et al., "Aerosol Penetration and Leakage Characteristics of Masks Used in the Health Care Industry," *AJIC: American Journal of Infection Control* 21, no. 4 (1993): 167–73.
43. L. Barrett and A. Rousseau, "Aerosol Loading Performance of Electret Filter Media," *American Industrial Hygiene Association Journal,* 59 (1998): 532–39.
44. Centers for Disease Control, "Guidelines for Preventing the Transmission of *Mycobacterium Tuberculosis* in Health-Care Facilities," *Morbidity and Mortality Weekly Report* 43, RR-13 (1994): –27.
45. NIOSH, *TB Respiratory Protection Program,* 1999.
46. Centers for Disease Control, "Guidelines," 1994.
47. Y. G. Qian et al., "Performance of N95 Respirators: Filtration Efficiency for Airborne Microbial and Inert Particles," *American Industrial Hygiene Association Journal* 59, no. 2 (1998): 128–32.

48. Centers for Disease Control, *NIOSH Guide to the Selection and Use of Particulate Respirators: Certified Under 42 CFR 84.* (Atlanta, GA: NIOSH, Centers for Disease Control and Prevention, 1996).

49. Lee and Liu, "Minimum Efficiency."

50. C.C. Chen and S. H. Huang, "The Effects of Particle Charge on the Performance of a Filtering Facepiece," *American Industrial Hygiene Association Journal* 59, no. 4 (1998): 227–33.

51. S. B. Martin, Jr. and E. S. Moyer, "Electrostatic Respirator Filter Media: Filter Efficiency and Most Penetrating Particle Size Effects," *Applied Occupational & Environmental Hygiene* 15, no. 8 (2000): 609–17.

52. Ibid.

53. E. S. Moyer and M. S. Bergman, "Electrostatic N-95 Respirator Filter Media Efficiency Degradation Resulting from Intermittent Sodium Chloride Aerosol Exposure," *Applied Occupational & Environmental Hygiene* 15, no. 8 (2000): 600–8.

54. Health Canada, "Routine Practices and Additional Precautions for Preventing the Transmission of Infection in Health Care," CCDR, 25, no. S4 (1999): 1–155.

55. N. L. Belkin, "The Evolution of the Surgical Mask: Filtering Efficiency versus Effectiveness," *Infection Control and Hospital Epidemiology* 18, no. 1 (1997): 49–57.

56. Hodous and Hodous, "Respiratory Protective Devices."

57. F. I. Gilmore, "The Tip of the Iceberg" 39, no. 10 (1994): 37–39.

58. C. C. Chen et al., "Aerosol Penetration Through Filtering Facepieces and Respirator Cartridges," *American Industrial Hygiene Association Journal*, 53, no. 9 (1992): 566–74.

59. C.C. Chen and K. Willeke, "Characteristics of Face Seal Leakage in Filtering Facepieces," *American Industrial Hygiene Association Journal* 53, no. 9 (1992): 533–39.

60. Weber et al., "Aerosol Penetration."

61. D. Wake et al., "Performance of Respirator Filters and Surgical Masks against Bacterial Aerosols, *Journal of Aerosol Science* 28, no. 7 (1997): 1311–29.

62. Brousseau et al., "Micobacterial Aerosol Collection."

63. MHRA, *Breathing System Filters: An Assessment of 104 Breathing System Filters,* Evaluation 04005, March 2004.

64. Belkin, "Evolution of the Surgical Mask."

65. Centers for Disease Control, *Understanding Respiratory Protection Against SARS* (Atlanta, GA: NIOSH/Centers for Disease Control and Prevention, 2003).

66. T. Tuomi, "Face Seal Leakage of Half Masks and Surgical Masks," *American Industrial.Hygiene Association Journal* 46, no. 6 (1985): 308–12.

67. Bollinger and Schutz, *NIOSH Guide,* 1987.

68. Canadian Standards Association, *Respirators,* 2002.

69. J. L. Derrick and C. D. Gomersall, "Surgical Helmets and SARS Infection," *Emerging Infectious Diseases* 10, no. 2 (2004):

70. NIOSH, *Respirator Decision Logic* (Atlanta, GA: Centers for Disease Control/NIOSH, 1987).

71. Quian et al., "Performance of N95 Respirator."

72. Nicas, "Respiratory Protection."

73. Nardell and Macher, "Respiratory Infection."

74. M. Nicas, "An Analytical Framework for Relating Dose, Risk, and Incidence: An Application to Occupational Tuberculosis Infection" *Risk Analysis* 16, no. 4 (1996): 527–38.

75. M. Nicas, "Refining a Risk Model for Occupational Tuberculosis Transmission," *American Industrial Hygiene Association Journal* 57, no. 1, (1996): 16–22.

76. M. Nicas, "Assessing the Relative Importance of the Components of an Occupational Tuberculosis Control Program," *Journal of Occupational and Environmental Medicine* 40, no. 7 (1998): 648–54.

77. Peters et al., "Patients Infected with High Hazard Viruses."

78. S. Barnhart et al., "Tuberculosis in Health Care Settings and the Estimated Benefits of Engineering Controls and Respiratory Protection," *Journal of Occupational and Environmental Medicine* 39, no. 9 (1997): 849–54.

80. L. Gammaitoni and M. C. Nucci, "Using a Mathematical Model to Evaluate the Efficacy of TB Control Measures," *Emerging Infectious Diseases*, 3, no. 3 (1997): 335–42.

81. Lee et al., "Respiratory Protection."

82. Brousseau et al., "Mycobacterial Aerosol Collection."

83. L. M. Brosseau, "Aerosol Penetration Behavior of Respirator Valves," *American Industrial Hygiene Association Journal*, 59, no. 3 (1998): 173–80.

84. N. V. McCullough et al., "Improved Methods for Generation, Sampling, and Recovery of Biological Aerosols in Filter Challenge Tests," *American Industrial Hygiene Association Journal*, 59, no. 4 (1998): 234–41.

85. C. C. Coffey et al., "Simulated Workplace Performance."

86. S. K. Chen et al., "Evaluation of Single-Use Masks and Respirators for Protection of Health Care Workers Against Mycobacterial Aerosols," *American Journal of Infection Control* 22, no. 2 (1994): 65–74.

87. C. C. Coffey et al., "Comparison of Five Methods for Fit-Testing N95 Filtering-Facepiece Respirators," *Applied Occupational & Environmental Hygiene*, 17, no. 10 (2002): 723–30.

88. R. D. Huff et al., "Personnel Protection During Aerosol Ventilation Studies Using Radioactive Technetium (Tc99m), *American Industrial Hygiene Association Journal*, 55, no. 12 (1994): 1144–48.

89. D. Hannum et al., "The Effect of Respirator Training on the Ability of Healthcare Workers to Pass a Qualitative Fit Test," *Infection Control & Hospital Epidemiology*, 17, no. 10 (1996): 636–40.

90. D. M. Bell, "Human Immunodeficiency Virus Transmission in Health Care Settings: Risk and Risk Reduction," *American Journal of Medicine* 91, no. 3B (1991): 294S–300S.

91. G. M. McCarthy et al., "Occupational Injuries and Exposures Among Canadian Dentists: The Results of a National Survey," *Infection Control and Hospital Epidemiology*, 20, no. 5 (1999): 331–36.

92. D. L. Kouri and J. M. Ernest, "Incidence of Perceived and Actual Face Shield Contamination During Vaginal and Cesarean Delivery," *American Journal of Obstetrics & Gynecology*, 169, no. 2 (1993): 1–5.

93. G. Kernbach-Wighton et al., "Bone-Dust in Autopsies: Reduction of Spreading," *Forensic Science International* 83, no. 2 (1996): 95–103.

94. T. Nighswonger, "How Much Eye Protection Is Enough? Too Many Workers Who Wear Eye Protection Still Suffer Injuries. Here's Help On How to Determine When More Protection Is Needed," *Occupational Hazards* 64 (2002): 40–44.

95. S. Lee, *American Industrial Hygiene Association Journal* 59, no. 1 (1988): A13–A14.

96. T. Finch et al., "Occupational Airborne Allergic Contact Dermatitis from Isoflurane Vapor," *Contact Dermatitis*, 42, no. 1 (2000): 46.

97. K C. Niven,et al., "Estimation of Exposure from Spilled Glutaraldehyde Solutions in a Hospital Setting," *Annals of Occupational Hygiene*, 41, no. 6 (1997): 691–98.

98. Kouri and Ernest, "Face Shield Contamination."

99. K. E. Leese et al., "Assessment of Blood-Splash Exposures of Medical-Waste Treatment Workers," *Journal of Environmental Health*, 61, no. 6 (1999): 8–11, 27–28.

100. A. P. Giachino et al., "Expected Contamination of the Orthopedic Surgeon's Conjunctiva," *Canadian Journal of Surgery* 31, no. 2 (1988): 51–52.

101. K. J. Davies et al., "Seroepidemiological Study of Respiratory Virus Infections Among Dental Surgeons," *British Dental Journal*, 176, no. 7 (1994): 262–65.

102. M. Basu et al., "A Survey of Aerosol-Related Symptoms in Dental Hygienists," *Journal of Social and Occupational Medicine* 38, no. 1/2 (1988): 23–25.

103. Centers for Disease Control, *Public Health Guidance for Community-Level Preparedness and Response to Severe Acute Respiratory Syndrome (SARS) Version 2 Supplement I: Infection Control in Healthcare, Home, and Community Settings* (Atlanta, GA: Centers for Disease Control and Prevention, 2004).

104. L. W. Green et al., *Health Education Planning: A Diagnostic Approach* (Palo Alto, CA: Mayfield, 1980).

105. D. DeJoy, "A Behavioral-Diagnostic Model for Fostering Self-Protective Behavior in the Workplace," in *Trends in Ergonomics/Human Factor,* vol. 3, ed. W. Karwowski (Amsterdam, The Netherlands: Elsevier Science,, 1986), 907–17.

106. D. M. DeJoy et al., "A Work Systems Analysis of Compliance with Universal Precautions Among Healthcare Workers," *Health Education Quarterly* 23, no. 2 (1996): 159–74.

107. D. M. DeJoy et al., "An Integrative Perspective On Worksite Health Promotion," *Journal of Occupational Medicine* 35 (1993): 1221–30.

108. N. Sheehy and C. AJ, "Industrial Accidents," in *International Review of Industrial and Organization Psychology*, ed. C. Cooper and I. Robertson (New York: John Wiley, 1987), 201–27.

109. M. Smith, "Human Factors in Occupational Injury, Evaluation and Control," in *Handbook of Human Factors*, ed. S. G. Editor. New York:, Wiley-Interscience, 1986).

110. B. Dalton and J. Harris, "A Comprehensive Approach to Health Management," *Journal of Occupational Medicine* 33 (1991): 338–48.

111. R. Winett et al., *Health Psychology and Public Health: An Integrative Approach* (New York: Pergamon, 1991).

112. A. S. Ho et al., "An Outbreak of Severe Acute Respiratory Syndrome Among Hospital Workers in a Community Hospital in Hong Kong," *Annals of Internal Medicine* 139, no. 7 (2003): 564–67.

113. D. C. Scales et al., "Illness in Intensive Care Staff after Brief Exposure to Severe Acute Respiratory Syndrome," *Emerging Infectious Diseases*, 9, no. 10 (2003): 1205–10.

114. M. Ofner et al., "Cluster of Severe Acute Respiratory Syndrome Cases among Protected Health-Care Workers — Toronto, Canada, April 2003," *Journal of the American Medical Association*, 289, no. 21 (2003): 2788–89.

115. M. D. Christian et al., "Possible SARS Coronavirus Transmission during Cardiopulmonary Resuscitation," *Emerging Infectious Diseases*, 10, no. 2 (2004):

116. H. A. Dwosh et al., "Identification and Containment of an Outbreak of SARS in a Community Hospital," *Canadian Medical Association Journal* 168, no. 11 (2003): 1415–20.

117. W. H. Seto et al., "Effectiveness of Precautions Against Droplets and Contact in Prevention of Nosocomial Transmission of Severe Acute Respiratory Syndrome (SARS)," *Lancet* 361, no. 9368 (2003): 1519–20.

118. M. Varia et al., "Investigation of a Nosocomial Outbreak of Severe Acute Respiratory Syndrome (SARS) in Toronto, Canada," *Canadian Medical Association Journal* 169, no. 4 (2003): 285–92.

119. M. Loeb et al., "SARS among Critical Care Nurses, Toronto," *Emerging Infectious Diseases*, 10, no. 2 (2004):

120. W. Wong et al., "Cluster of SARS among Medical Students Exposed to Single Patient, Hong Kong," *Emerging Infectious Diseases*, 10, no. 2 (2004):

121. J. T. F. Lau et al., "SARS Transmission among Hospital Workers in Hong Kong," *Emerging Infectious Diseases* 10, no. 2 (2004):.

122. CDC, *Public Health Guidance,* 2004.

123. Health Canada, *Infection Control Guidance for Health Care Workers in Health Care Facilities and Other Institutional Settings — Severe Acute Respiratory Syndrome. (SARS).* WHO (2003).

124. WHO, *Hospital Infection Control Guidance for Severe Acute Respiratory Syndrome (SARS)* (Geneva, Switzerland: World Health Organization, 2003).

125. Lau et al., "SARS Transmission."

126. Scales, "Illness in Intensive Care Staff."

127. Ofner et al., "Cluster."

128. "Minor Breach, Major Problem: Toronto Medical Workers Find SARS 'Unforgiving': CDC Sends Team of Investigators to Canada," *Hospital Infection Control,* 30, no. 6 (2003): 73–75, 77.

129. Dwosh et al., "Cluster,"2003.

130. Ibid.

131. Ibid.

132. Ibid.

133. A. Cooper et al., "A Practical Approach to Airway Management in Patients with SARS," *Canadian Medical Association Journal,* 169, no. 8 (2003): 785–87.

134. B. J. Park et al., "Lack of SARS Transmission among Healthcare Workers, United States," *Emerging Infectious Diseases* 10, no. 2 (2004): 106–110.

135. R. Garcia et al., "Nosocomial Respiratory Syncytial Virus Infections: Prevention and Control in Bone Marrow Transplant Patients," *Infection Control Hospital Epidemiology* 18, no. 6 (1997): 412–16.

136. P. Madge et al., "Prospective Controlled Study of Four Infection-Control Procedures to Prevent Nosocomial Infection with Respiratory Syncytial Virus," *Lancet,* 340, no. 8827 (1992): 1079–83.

137. S. E. Beekmann et al., "Rapid Identification of Respiratory Viruses: Impact on Isolation Practices and Transmission among Immunocompromised Pediatric Patients," *Infection Control Hospital Epidemiology* 17, no. 9 (1996): 581–86.

138. M. Zambon et al., "Molecular Epidemiology of Two Consecutive Outbreaks of Parainfluenza 3 in a Bone Marrow Transplant Unit," *Journal of Clinical Microbiology* 36, no. 8 (1998): 2289–93.

139. S. E. Moisiuk et al., "Outbreak of Parainfluenza Virus Type 3 in an Intermediate Care Neonatal Nursery," *Pediatric Infectious Disease Journal,* 17, no. 1 (1998): 49–53.

140. N. Singh-Naz et al., "Outbreak of Parainfluenza Virus Type 3 in a Neonatal Nursery," *Pediatric Infectious Disease Journal* 9, no. 1 (1990): 31–33.

141. R. Haley et al., "The Efficacy of Infection Surveillance and Control Programs in Preventing Nosocomial Infections in U.S. Hospitals," *American Journal of Epidemiology* 121 (1985): 182–205.

142. R. Haley, *Managing Hospital Infection Control for Cost-Effectiveness: A Strategy for Reducing Infectious Complications* (Chicago: American Hospital Publishing, 1986).

143. D. Zoutman et al., "The State of Infection Surveillance and Control in Canadian Acute Care Hospitals," *American Journal of Infection Control* 31 (2003): 266–73.

144. American Medical Association, *Occupational Health Services A Practical Approach* (Chicago: American Medical Association, 1989).

145. D. M. DeJoy et al., "Influence of Employee, Job Task, and Organizational Factors on Adherence to Universal Precautions Among Nurses," *International Journal Industrial Ergonomics* 16, no. 1 (1995): 43–55.

146. R. R. M. Gershon et al., "Compliance with Universal Precautions Among Health Care Workers at Three Regional Hospitals," *AJIC: American Journal of Infection Control* 23, no. 4 (1995): 225–36.

147. R. R. M. Gershon et al., "Compliance with Universal Precautions in Correctional Health Care Facilities," *Journal of Occupational and Environmental Medicine* 41, no. 3 (1999): 181–89.

148. J. C. DiGiacomo et al., "Barrier Precautions in Trauma Resuscitation: Real-Time Analysis Utilizing Videotape Review," *American Journal of Emergency Medicine*, 15, no. 1(1997): 34–39.
149. L. E. Kim et al., "Improved Compliance with Universal Precautions in the Operating Room Following an Educational Intervention," *Infection Control & Hospital Epidemiology* 22, no. 8 (2001): 522–24.
150. DeJoy et al., "Influence of Employee, Job/Task."
151. W. Moongtui et al., "Using Peer Feedback to Improve Handwashing and Glove Usage Among Thai Healthcare Workers," *American Journal of Infection Control*, 28, no. 5 (2000): 365–69.
152. J. Salyer, "Environmental Turbulence: Impact in Nurse Performance," *Journal of Nursing Administration* 25 (1995): 12–20.
153. C. Boylan and G. Russell, "Beyond Restructuring," *Journal of Nursing Administration* 27 (1997): 13–20.
154. L. Hall and G. Donner, "The Changing Role of Hospital Nurse Managers: A Literature Review," *Canadian Journal of Nursing Administration* 10 (1997): 14–39.
155. C. Curran, "Changing the Way We Do Business," *Nursing Economics*, 9 (1991): 296–97.
156. J. Shindal-Rothschild and M. Duffy, "The Impact of Restructuring and Work Design On Nursing Practice and Patient Care," *Best Practice Benchmarking Healthcare* 1 (1996): 271–82.
157. DeJoy et al., "Work Systems Analysis."
158. R. R. Gershon et al., "Hospital Safety Climate and Its Relationship with Safe Work Practices and Workplace Exposure Incidents," *AJIC: American Journal of Infection Control* 28, no. 3 (2000): 211–21.
159. Brousseau et al., "Mycobacterial Aerosol Collection."
160. Fennelly et al. "Cough-Generated Aerosols."
161. DeJoy et al., "Work Systems Analysis."
162. D. Zohar, Safety Climate in Industrial Organizaations: Theoretical and Applied Implications," *Journal of Applied Psychology* 65 (1980): 96–102.
163. R. Brown and H. Holmes, "The Use of a Factor-Analytic Proceedure for Assessing the Validity of an Employee Safety Climate Model," *Accident Analysis and Prevention* 18 (1986): 455–70.
164. R Simonds and Y. Shafai-Sahrai, "Factors Apparently Afffecting the Injury Frequency in Eleven Matched Pairs of Companies," *Journal of Safety Research* 9 (1977): 120–27.
165. Smith et al., "Characteristics of Successful Safety Programs," *Journal of Safety Research* 10 (1978): 5–15.
166. H. Cohen and R. Cleveland, "Safety Program Practices in Recording-Holding Plants," *Professional Safety* 28 (1983): 26–33.
167. Gershon et al., "Hospital Safety Climate."
168. T. Cox and M. Leiter, "The Health of Health Care Organizations," *Work & Stress* 6 (1992): 219–27.
169. C. White and M. Berger, "Using Force-Field Analysis to Promote Use of Personal Protective Equipment," *Infection Control Hospital Epidemiology* 13 (1992): 752–55.

170. J. Ford and S. Fisher, "The Transfer of Safety Training in Work Organizations: A Systems Perspective to Continuous Learning," *Occupational Medicine* 9 (1994): 241–59.

171. I. Goldstein, "Training in Work Organizations," in *Handbook of Industrial and Organizational Psychology,* ed. M. Dunnette and L. Hough (Palo Alto, CA: Consulting Psychologists Press, 1991), 506–619.

172. White and Berger, "Using Force-Field Analysis."

173. Gershon et al., "Compliance with Universal Precautions."

174. DeJoy et al., "Influence of Employee, Job/Task."

175. DeJoy et al., "Behavioral-Diagnostic Analysis of Compliance with Universal Precautions among Nurses," *Journal of Occupational Health Psychology* 5, no. 1 (2000): 127–41.

176. DeJoy et al., "Work Systems Analysis."

177. DeJoy et al., "Influence of Employee, Job/Task."

178. Gershon et al., "Hospital Safety Climate."

179. D. L. Rivers et al., "Predictors of Nurses' Acceptance of an Intravenous Catheter Safety Device," *Nursing Research* 52, no. 4 (2003): 249–55.

180. Gershon et al., "Hospital Safety Climate."

181. Ho et al., "An Outbreak of Severe Acute Respiratory Syndrome," 2003; Scales et al., "Illness in Intensive Care Staff"; Ofner et al., "Cluster of Severe Acute Respsiratory Syndrome"; Christian et al., "Possible SARS Coronavirus Transmission"; Dwosh et al., "Identification and Containment of an Outbreak of SARS"; Seto et al., "Effectiveness of Precautions against Droplets"; Varia et al., "Investigation of a Nosocomial Outbreak"; Loeb et al., "SARS among Critical Care Nurses"; Wong et al., "Cluster of SARS among Medical Students"; Lau et al., "SARS Transmission.

182. World Health Organization, *Consensus Document On the Epidemiology of Severe Acute Respiratory Syndrome* (Geneva, Switzerland: WHO, 2003).

183. Dwosh et al., "Identification and Containment."

184. A. S. M. Abdullah et al., "Lessons From the Severe Acute Respiratory Syndrome Outbreak in Hong Kong," *Emerging Infectious Diseases* 9, no. 9 (2003): 1042–45.

185. L. D. Ha et al., "Lack of SARS Transmission among Public Hospital Workers, Vietnam," *Emerging Infectious Diseases,* 10, no. 2 (2004): 1204–1206.

186. Varia et al., "Investigation of a Nosocomial Outbreak."

187. World Health Organization, *Consensus Document.*

188. Park et al., "Lack of SARS Transmission."

189. P. K. H. Chow et al., "Healthcare Worker Seroconversion in SARS Outbreak," *Emerging Infectious Diseases,* 10, no. 2 (2004): 1202–1204.

190. Health Canada, *Infection Control Guidance,* 2003.

191. World Health Organization, *Hospital Infection Control Guidance,* 2003.

192. S. Riley et al., "Transmission Dynamics of the Etiological Agent of SARS in Hong Kong: Impact of Public Health Interventions," *Science* 300, no. 5627 (2003): 1961–66.

193. Ofner et al., "Cluster of Severe Acute Respiratory Syndrome."

194. Christian et al., "Possible SARS Coronavirus Transmission."

195. Cooper et al., "A Practical Approach to Airway Management."

196. Loeb et al., "SARS Among Critical Care Nurses."
197. Ha et al., "Lack of SARS Transmission."
198. Wong et al., "Cluster of SARS Among Medical Students."
199. Varia et al., "Investigation of a Nosocomial Outbreak."
200. MHRA, "Breathing System Filters."
201. WHO, *Consensus Document*, 2003.
202. Ibid.
203. Ho et al., "An Outbreak of Severe Acute Respiratory Syndrome."
204. Seto et al., "Effectiveness of Precautions Against Droplets."
205. Ibid.
206. Ibid.
207. Loeb et al., "SARS Among Critical Care Nurses."
208. Seto et al., "Effectiveness of Precautions Against Droplets."
209. Loeb et al., "SARS Among Critical Care Nurses."
210. Ha et al., "Lack of SARS Transmission."
211. Seto et al., "Effectiveness of Precautions Against Droplets."
212. Lau et al. "Transmission Among Hospital Workers."
213. Ibid.
214. Christian et al., "Possible SARS Coronavirus Transmission."
215. A. Gagneur et al., "Coronavirus-Related Nosocomial Viral Respiratory Infections in a Neonatal and Paediatric Intensive Care Unit: A Prospective Study" *Journal of Hospital Infections* 51, no. 1 (2002): 59–64.
216. A. R. Falsey et al., "The 'Common Cold' in Frail Older Persons: Impact of Rhinovirus and Coronavirus in a Senior Daycare Center," *Journal of the American Geriatric Society*, 45, no. 6 (1997): 706–11.
217. National Advisory Committee on Immunization, "Statement on Influenza Vaccination for the 2003–2004 Season," *CCDR* 29, no. ACS-4 (2003): 1–20.
218. A. Yassi et al., "Morbidity, Cost and Role of Health Care Worker Transmission in an Influenza Outbreak in a Tertiary Care Hospital," *Canadian Journal of Infectious Disease* 4, no. 1 (1993): 52–56.
219. Ibid.
220. Health Canada, "Routine Practices," 1999.
221. C. B. Hall and R. G. Douglas, Jr., "Nosocomial Respiratory Syncytial Viral Infections. Should Gowns and Masks Be Used?" *American Journal of Diseases of Children* 135, no. 6 (1981): 512–15.
222. Garcia et al., "Nosocomial Respiratory Syncytial Virus Infections"; Madge et al., "Prospective Controlled Study Four Infection-Control Procedures"; Beekman et al. "Rapid Identification of Respiratory Viruses."
223. K. K. Macartney et al., "Nosocomial Respiratory Syncytial Virus Infections: The Cost-Effectiveness and Cost-Benefit of Infection Control," *Pediatrics* 106, no. 3 (2000): 520–26.
224. B. Jones et al., "Control of an Outbreak of Respiratory Syncytial Virus in Immunocompromised Adults," *Journal of Hospital Infections* 44 (2000): 53–57.
225. J. M. Langley et al., "Nosocomial Respiratory Syncytial Virus Infection in Canadian Pediatric Hospitals: A Pediatric Investigators Collaborative Network on Infections in Canada Study," *Pediatrics* 100, no. 6 (1997): 943–46.

226. C. B.Hall et al., "Possible Transmission by Fomites of Respiratory Syncytial Virus," *Journal of Infectious Disease* 141, no. 1 (1980): 98–102.

227. M. T. Brady et al., "Survival and Disinfection of Parainfluenza Viruses on Environmental Surfaces," *American Journal of Infection Control* 18, no. 1 (1990): 18–23.

228. Zambon et al., "Molecular Epidemiology"; Moisiuk et al., "Outbreak of Parainfluenza"; Sing-Naz et al., "Outbreak of Parainfluenza," 1990.

229. D. Menzies et al., "Hospital Ventilation and Risk for Tuberculous Infection in Canadian Health Care Workers," *Annals of Internal Medicine* 133, no. 10 (2000): 779–89.

230. Centers for Disease Control, "Guidelines for Preventing the Transmission of *Mycobacterium tuberculosis.*"

231. Health Canada, *Guidelines for Preventing the Transmission of Tuberculosis in Canadian Health Care Facilities and Other Institutional Settings* (: Health Canada, 1996), Ottawa, Canada.

232. Menzies et al., "Hospital Ventilation."

233. S. Segalmaurer and G. E. Kalkut, "Environmental-Control of Tuberculosis — Continuing Controversy," *Clinical Infectious Diseases* 19, no. 2 (1994): 299–308.

234. M. M. Hannan et al., "Investigation and Control of a Large Outbreak of Multi-Drug Resistant Tuberculosis at a Central Lisbon Hospital," *Journal of Hospital Infections* 47, no. 2 (2001): 91–97.

235. Segalmaurer and Kalkut, "Environmental-Control of Tuberculosis."

236. P. M. Sutton et al., "Tuberculosis Isolation: Comparison of Written Procedures and Actual Practices in Three California Hospitals," *Infection Control & Hospital Epidemiology* 21, no. 1 (2000): 28–32.

237. Menzies et al., "Hospital Ventilation."

238. Segalmaurer and Kalkut, "Environmental-Control of Tuberculosis."

239. J. I. Tokars et al., "Use and Efficacy of Tuberculosis Infection Control Practices at Hospitals with Previous Outbreaks of Multidrug-Resistant Tuberculosis," *Infection Control & Hospital Epidemiology* 22, no. 7 (2001): 449–55.

240. A. W. Helfgott et al., "Compliance with Universal Precautions: Knowledge and Behavior of Residents and Students in a Department of Obstetrics and Gynecology," *Infectious Diseases in Obstetrics & Gynecology* 6, no. 3 (1998): 123–28.

241. Gershon et al., "Compliance with Universal Precautions."

242. B. Evanoff et al., "Compliance with Universal Precautions among Emergency Department Personnel Caring for Trauma Patients," *Annals of Emergency Medicine* 33, no. 2 (1999): 160–65.

243. DeJoy et al., "Behavioral–Diagnostic Analysis of Compliance."

244. M.Cooke, "House Staff Attitudes toward the Acquired Immunodeficiency Virus," *AIDS and Public Policy* 3 (1988): 59–60.

245. M. Becker et al., "Non-Compliance with Universal Precautions Policy: Why Do Physicians and Nurses Recap Needles?" *American Journal of Infection Control* 18 (1990): 232–39.

246. M. Hoffman-Terry et al., "Impact of Human Immunodeficiency Virus on Medical and Surgical Residents," *Archives of Internal Medicine* 152 (1992):1788–96.

247. Gershon et al., "Compliance with Universal Precautions."

248. E. W. Young et al., "Rural Nurses' Use of Universal Precautions in Relation to Perceived Knowledge of Patient's HIV Status," *International Journal of Nursing Studies*, 33, no. 3 (1996): 249–58.

249. Kim et al., "Improved Compliance with Universal Precautions."

250. W. Afif et al., "Compliance with Methicillin-Resistant Staphylococcus Aureus Precautions in a Teaching Hospital," *AJIC: American Journal of Infection Control* 30, no. 7 (2002): 430–33.

251. Gershon et al., "Compliance with Universal Precautions."

252. T. Angtuaco et al., "Universal Precautions Guideline: Self-Reported Compliance by Gastroenterologists and Gastrointestinal Endoscopy Nurses — A Decade's Lack of Progress," *American Journal of Gastroenterology* 98, no. 11 (2003): 2420–23.

253. J. Prieto and J. Clark, "Infection Control. Dazed and Confused ... the Implementation of Infection Control Policies and Guidelines," *Nursing Times* 95, no. 28 (1999): 49–50, 53.

254. D. B. Jeffe et al., "Healthcare Workers' Attitudes and Compliance with Universal Precautions: Gender, Occupation, and Speciality Differences," *Infection Control and Hospital Epidemiology* 18, no. 10 (1997): 710–12.

255. Gershon et al., "Compliance with Universal Precautions."

256. DeJoy et al., "Behavioral–Diagnostic Analysis of Compliance."

257. G. Godin et al., "Determinants of Nurses' Adherence to Universal Precautions for Venipunctures," *American Journal of Infection Control* 28, no. 5 (2000): 359–64.

258. Ibid.

259. Gershon et al., "Hospital Safety Climate."

260. Helfgott et al., "Compliance with Universal Precautions."

261. DeJoy et al., "Behavioral–Diagnostic Analysis of Compliance."

262. L. Nickell et al., "Psychosocial Effects of SARS On Hospital Staff: Survey of a Large Tertiary Institution," *Canadian Medical Association Journal* 170, no. 5 (2004): 793–98.

263. DeJoy et al., "Behavioral–Diagnostic Analysis of Compliance."

264. K. P. Fennelly, "Personal Respiratory Protection against Mycobacterium Tuberculosis," *Clinics in Chest Medicine* 18, no. 1 (1997): 1–17.

265. Prieto and Clark, "Infection Control."

266. DeJoy et al., "Behavioral–Diagnostic Analysis of Compliance."

267. M. Willy et al., "Adverse Exposures and Universal Precautions Practices among a Group of Highly Exposed Health Professionals," *Infection Control Hospital Epidemiology* 11 (1990): 351–56.

268. G. Kelen et al., "Adherence to Universal Precautions during Interventions on Critically Ill and Injured Emergency Department Patients," *Journal of AIDS* 3 (1990): 987–94.

269. L. Linn et al., "Physicians' Perceptions about Increased Glove-Wearing in Response to Risk of HIV Infection," *Infection Control Hospital Epidemiology* 11 (1990): 248–54.

270. S. Osborne, "Influences On Compliance with Standard Precautions Among Operating Room Nurses," *American Journal of Infection Control* 7 (2003): 415–23.

271. Nickell et al., "Psychosocial Effect of SARS.."

272. L. Bero et al., "Closing the Gap between Research and Practice: An Over of Systematic Reviews of Interventions to Promote the Implementation of Research Findings," *British Medical Journal* 317 (1998): 465–68.

273. R. Grol et al., "Attributes of Clinical Guidelines the Influence Use of Clinical Guidelines in General Study: Observational Study," *British Medical Journal* 317 (1998): 858–61.

Chapter 6

1. Becki Jenkins, "Standards of Practice — Historical Voices Cry out, 'Patient First,'"*Infection Control Today* 9 no. 2 (2005): 42.

2. Eric Toner, CBN Weekly Bulletin, Oct. 31, 2005, www.upmc-CBN.ong/brief.

3. Joint Commission Hospital Accreditation, Environment of Care, Washington, D.C., 2003.

4. Jenkins, "Standards of Practice."

5. Ibid.

6. Centers for Disease Control, "Guidelines for Environmental Infection Control in Health-Care Facilities," *Morbidity and Mortality Weekly Report* 54, RR-10 (June 6, 2003): Centers for Disease Control, Special Pathogens Branch, "Viral Hemorrhagic Fevers," August 23, 2004; Occupational Safety and Health Administration, *Respsiratory Protection.* 29 CFR 1910. 34. (Atlanta, GA: OSHA).

7. A. Streifel, "Airborne Infectious Disease: Best Practices For Ventilation Management," *HPAC Engineering* (September 2003):

8. University of Medicine and Dentistry of New Jersey Magazine, Spring, 2004, Vol. 1, No. 1.

9. CDC MMWR, May 11, 1990, 39(RR-7); 1–15, Prevention and Control of Influenza. http://iier.iscii.es/mmwr/preview/mmwrhtml/00001644.htm.

10. R. B. Couch, "Viruses and Indoor Pollution," *Bulletin of the New York Academy of Medicine* 57 (1981): 907–21.

11. American Institute of Architects, Academy of Architecture for Health (AIA), *Guidelines for Design and Construction of Hospital and Health Care Facilities* (Washington, D.C.: American Institute of Architects, 2001).

12. American Society of Heating, Refrigerating, and Air Conditioning Engineers, *Proposed New Standard 170, Ventilation of Healthcare Facilities* (Atlanta, GA: ASHRAE, 2005).

13. AIA, *Guidelines,* 2001.

14. Streifel, "Airborne Infectious Disease," 2003.

15. P. Sutton, "California Department of Health Study," in *Handbook of Modern Hospital Safety*, ed. William Charney (Boca Raton, FL: CRC Press, 1999).

16. Centers for Disease Control, "Guidelines," 2003.

17. Julia Garner, *Guidelines for Isolation Precautions in Hospitals — Hospital Control Advisory Committee Rationale For Isolation Precautions in Hospitals*. (Atlanta, GA: Centers for Disease Control, 1996).

18. Sutton, "California Dept. of Health Study."

19. S. Cody and M. Fenstersheib, "Preparedness and Response in Healthcare Facilities" Santa Clara County, California, May 24th, 2004.

20. Centers for Disease Control, "Guidelines."

21. Cody and Fenstersheib, "Preparedness."

22. Centers for Disease Control, "Guidelines."

23. AIA *Guidelines*, 2001; Sutton, "California Department of Health Study," 1999.

24. Garner, *Guidelines*, 1996.

25. ASHRAE, "Standard 170," 2005.

26. Ibid.

27. William Charney, "Engineering Controls Options for TB in a Hospital Setting," in *Handbook of Modern Hospital Safety*, ed. William Charney (Boca Raton, FL: CRC Press, 1998), 216–22.

28. J. Mead, "Challenge Testing of Portable HEPA Filter Units," *Journal of the American College of Emergency Physicians* 44 (2005): no. 6 (): 635–45.

Chapter 7

1. D. A. Henderson, "Surveillance Systems and Intergovernmental Cooperation," in *Emerging Viruses*, ed. S. S. Morse (New York: Oxford University Press; 1993), 283–89.

2. A. Benenson, "Infectious Diseases," in *Epidemiology and Health Policy*, ed. S. Levine and A. M. Lilienfield (New York: Tavistock, 1987), 207–26.

3. M. J. Roseneau, "The Uses of Fear in Preventive Medicine," *Boston Medical and Surgical Journal* 162, no. 10 (1910):305–07.

4. B. Frist, "Manhattan Project for the 21st Century," Paper presented at Harvard Medical School Health Care Policy Seidman Lecture, June 1, 2005, Cambridge, MA.

5. R. L. Berkelman et al., "Infectious Disease Surveillance: A Crumbling Foundation" *Science* 264, no. 5157 (1994):368–70.

6. E. Fee and T. M. Brown, "The Unfulfilled Promise of Public Health: Deja Vu All Over Again," *Health Affairs* 21, no. 6 (2002):31–43.

7. M. Drexler, *Secret Agents: The Menace of Emerging Infections* (Washington, D.C.: Joseph Henry Press, 2002).

8. Committee for the Study of the Future of Public Health, *The Future of Public Health* (Washington, D.C.: National Academy Press, 1988).

9. L. J. Letgers et al., "Are We Prepared for a Viral Pandemic Emergency?" in *Emerging Viruses*, ed. S. S. Morse (New York: Oxford University Press, 1993).

10. W. R. Dowdle "The Future of the Public Health Laboratory," *Annual Review of Public Health* 14 (1993): 649–64.

11. Berkelman et al., "Infectious Disease Surveillance."

12. J. Lederberg ed. *Microbial Threats to Health in the United States* (Washington, D.C.: National Academy Press, 1992).

13. U. Desselberger, "Emerging and Re-emerging Infectious Diseases," *Journal of Infection* 40, no. 1(2000):3–15.

14. H. Feldman et al., "Emerging and Re-Emerging Infectious Diseases," *Medical Microbiology and Immunology* 191 (2002):63–74.

15. D. G. Maki, "SARS Revisited: The Challenge of Controlling Emerging Infectious Diseases at the Local, Regional, Federal, and Global Levels," *Mayo Clinic Proceedings* 79, no. 11 (2004):1359–66.

16. Berkelman et al., "Infectious Disease Surveillance."

17. V. Sidel, H. Cohen, and R. Gould, "Good Intentions and the Road to Bioterrorism Preparedness," American Journal of Public Health 91, no. 5 (2001):716–18.

18. J. D. Mayer "Geography, Ecology, and Emerging Infectious Disease,"*Social Science and Medicine* 50 (2000):937–52.

19. P. R. Epstein, "Climate Change and Emerging Infectious Diseases," *Microbes and Infection* 3 (2001): 747–54.

20. S. S. Morse, "Examining the Origins of Emerging Viruses," in *Emerging Viruses*, ed. S. S. Morse (New York: Oxford University Press, 1993).

21. L. Garrett, *The Coming Plague: Newly Emerging Diseases in a World Out of Balance* (New York: Farrar, Strauss, and Giroux, 1994).

22. M. Shnayerson and M. J. Plotkin, *The Killers Within: The Deadly Rise of Drug Resistant Bacteria* (Boston: Back Bay Books, 2002).

23. R. H. Kahn et al., "Syphilis Outbreaks Among Men Who Have Sex With Men," *Sexually Transmitted Diseases* 29, no. 5 (2002):285–87.

24. A. Trampuz et al., "Avian Influenza: A New Pandemic Threat?" *Mayo Clinic Proceedings* 79, no. 4 (2004):523–30.

25. M.-J. Earls and S. A. Hearne. *Facing the Flu: From the Bird Flu to a Possible Pandemic, Why Isn't America Ready?* (Washington, D.C.: Trust for America's Health, 2004).

26. H. G. Stiver "The Threats and Prospects for Control of an Influenza Pandemic," *Future Drugs* 3, no. 1 (2004):35–42.

27. M. A. Hamburg, "Public Health Preparedness," *Science* 295, no. 5559 (2002).

28. A. Fine and M. Layton, "Lessons From the West Nile Viral Encephalitis Outbreak in New York City, 1999: Implications for Bioterrorism Preparedness," *Clinical Infectious Diseases* 32, no. 2 (2001):277–82.

29. H. L. Hinton, *Combating Terrorism: Considerations for Investing Resources in Chemical and Biological Preparedness* (Washington, D.C.: General Accounting Office, 2001).

30. A.E. Smithson and L-A. Levy, *Ataxia: The Chemical and Biological Terrorism Threat and the US Response* (Washington, D.C.: The Henry L. Stimson Center, 2000).

31. E. L. Baker et al., "The Public Health Infrastructure and Our Nation's Health," *Annual Reviews in Public Health* 26 (2005):203–318.

32. M. L. Boulton et al., "Assessment of Epidemiologic Capacity in State and Territorial Health Departments — United States, 2001," *Morbidity and Mortality Weekly Report* 52, no. 43 (2003):1049–51.

33. R. E. Hoffman et al., "Capacity of State and territorial health Agencies to Prevent Foodborne Illness," *Emerging Infectious Diseases* 11, no. 1 (2005):11–16.

34. S. Smith, "Anthrax vs. the Flu," *International Journal of Health Services* 34, no. 1 (2004):169–72.

35. K. Eban "Waiting for Bioterror: Is Our Public Health System Ready?" *The Nation* 275, no. 20 (2002):11.

36. C. Keane et al., "Services Privatized in Local Health Departments: A National Survey of Practices and Perceptions," *American Journal of Public Health* 92, no. 8 (2002):1250–53.

37. J. Heinrich, *Bioterrorism: Review of Public Health Preparedness Programs* (Washington, D.C.: General Accounting Office, 2001).

38. General Accounting Office, *West Nile Virus Outbreak: Lessons for Public Health Preparedness* (Washington, D.C.: General Accounting Office, 2000).

39. J. L. Pressman and A. B. Wildavsky, *Implementation: How Great Expectations in Washington Are Dashed in Oakland; Or, Why It's Amazing that Federal Programs Work at All, This Being a Saga of the Economic Development Administration as Told by Two Sympathetic Observers Who Seek to Build Morals on a Foundation of Ruined Hopes* (Berkeley: University of California Press, 1973).

40. J. G. March and J. P. Olsen, "Garbage-Can Models of Decision-making in Organizations," in *Ambiguity and Command: Organizational Perspectives on Military Decision-Making*, ed. J. G. Marsh and S. Weissinger-Baylon (Marshfield, MA: Pittman; 1986).

41. S. Weissinger-Baylon, "Garbage Can Processes in Naval Warfare," in *Ambiguity and Command: Organizational Perspectives on Military Decision-Making*, ed. J. G. March and S. Weissinger-Baylon (Marshfield, MA: Pittman, 1986).

42. March et al, "Garbage-Can Models."

43. J. G. March and J. P. Olsen, *Rediscovering Institutions: The Organizational Basis of Politics* (New York: The Free Press, 1989), 21–38.

44. B. T. Pentland and H. R. Rueter, "Organizational Routines as Grammars of Action," *Administrative Science Quarterly* 39 (1994):484–510.

45. S. A. Kauffman, *The Origins of Order: Self Organization and Selection in Evolution* (New York: Oxford University Press, 1993).

46. J. G. March and H. A. Simon, *Organizations* (New York: John Wiley,1967).

47. J. Von Neumann and O. Morgenstern, *Theory of Games and Economic Behaviour* (Princeton, NJ: Princeton University Press, 1944).

48. W. B. Arthur "Inductive Reasoning and Bounded Rationality,"*American Economic Review* 84, no. 2 (1994):406–11.

49. L. Cosmides and J. Tooby, "Better Than Rational: Evolutionary Psychology and the Invisible Hand," *American Economic Review* 84, no. 2 (1994):327–32.

50. P. t'Hart, U. Rosenthal, and A. Kouzmin, "Crisis Decision-Making: The Centralization Theory Revisited," *Administration and Society* 25, no. 1 (1993):12–45.

51. G. T. Allison, "Conceptual Models and the Cuban Missile Crisis," *American Political Science Review* 63 (1969): 689–718.

52. F. D. Scutchfield, E. A. Knight, A. V. Kelly, M. W. Bhandari, and I. P. Vasilescu, "Local Public Health Agency Capacity and Its Relationship to Public Health System Performance," *Journal of Public Health Management Practice* 10, no. 3 (2004):204–15.

53. S. A. McCann, "View from the Hill: Congressional Efforts to Address Bioterrorism,"*Emerging Infectious Diseases* 5, no. 4 (1999):496.

54. Committee for the Study of Public Health, 1988, 83–86.

55. M. A, Hamburg and S. A. Hearne, *SARS and Its Implications for U.S. Public Health Policy - "We've Been Lucky"* (Washington, D.C.: Trust for America's Health, 2003).

56. L. D. Weiss, *Private Medicine and Public Health* (Boulder, CO: Westview Press,1997).

57. L. O. Gostin, *Public Health Law: Power, Duty, and Restraint* (Berkeley: University of California Press, 2000).

58. J. W. Kingdon, *Agendas, Alternatives, and Public Policies*, 2nd ed. (New York: Longman, 2002).

59. C. H. Foreman, *Plagues, Products, and Politics: Emergent Public Health Hazards and National Policymaking* (Washington, D.C.: Brookings Institute, 1994).

60. Centers for Disease Control, *Preventing Emerging Infections: A Strategy for the 21st Century* (Atlanta, GA: Centers for Disease Control and Prevention, 1998).

61. Centers for Disease Control, "The Centers for Disease Control and Prevention on Emerging Infectious Disease Threats," *Population and Development Review* 20, no. 3 (1994):687–90.

62. M. T. Osterholm, "Bioterrorism: Media Hype or Real Potential Nightmare?" *American Journal of Infection Control* 27, no. 6 (1999):461–62.

63. R. Katz, "Public Health Preparedness: The Best Defense Against Biological Weapons," *The Washington Quarterly* 25, no. 3 (2002):69–82.

64. J. L. Bryan and H. F. Fields, "An Ounce of Prevention is Worth a Pound of Cure - Shoring Up the Public Health Infrastructure to Respond to Bioterrrorist Attacks," *American Journal of Infection Control* 27, no. 6 (1999):465–67.

65. J. E. McDade, "Addressing the Potential Threat of Bioterrorism — Value Added to an Improved Public Health Infrastructure," *Emerging Infectious Diseases* 5, no. 4 (1999):591–92.

66. G. Avery, "Bioterrorism, Fear, and Public Health Reform: Matching a Policy Solution to the Wrong Window," *Public Administration Review* 64, no. 3 (2004):275–88.

67. McCann, "View from the Hill."

68. Sidel et al., "Good Intentions."

69. H. J. Geiger, "Terrorism, Biological Weapons, and Bonanzas: Assessing the Real Threat to Public Health," *American Journal of Public Health* 91, no. 5 (2001):708–09.

70. V. W. Sidel et al. "Bioterrorism Preparedness: Cooptation of Public Health?" *Medicine and Global Survival* 7, no. 2 (2002):82–89.

71. Sidel et al., "Good Intentions."

72. Geiger, "Terrorism, Biological Weapons."

73. N. Ward et al., "Policy Framing and Learning the Lessons from the UK's Foot And Mouth Disease Crisis," *Environmental and Planning C: Government and Policy* 22 (2004): 291–306.

74. Drexler, *Secret Agents,* 267–68.

75. M. T. Osterholm and J. Schwartz, *Living Terrors: What America Needs to Know to Survive the Coming Bioterrorism Catastrophe* (New York: Delacorte Press, 2000).

76. L. Garrett, "Responding to the Nightmare of Bioterrorism," *The Responsive Community* 12, no.1 (2002):88–93.

77. Geiger, "Terrorism, Biological Weapons."

78. S. A. Hearne et al., *Ready or Not? Protecting the Public's Health in the Age of Bioterrorism* (Washington, D.C.: Trust for America's Health, 2004).

79. Garrett, "Responding to the Nightmare."

80. Sidel et al., "Bioterrorism Preparedness," 29.

81. W. L. Waugh and R. T. Sylves, "Organizing the War on Terror," *Public Administration Review* 62, Suppl. 1 (2002):145–53.

82. D. P. Moyhnihan, *Secrecy* (New Haven, CT: Yale University Press, 1998).

83. t'Hart et al., "Crisis Decision-Making."

84. M. Siegel, "The Anthrax Fumble: Bureaucratic Timididty and Turf Battles Needlessly Put Many Americans at Risk." *The Nation* 274, no.10 (2002):14.

85. P. K. Dewan et al., "Inhalational Anthrax Outbreak Among Postal Workers, Washington, DC: 2001," *Emerging Infectious Diseases* 8, no. 10 (2002): 1066–72.

86. R. A. Falkenrath et al., *America's Achilles Heel: Nuclear, Biological, and Chemical Terrorism and Covert Attack* (Cambridge, MA: MIT Press, 1998).

87. Drexler, *Secret Agents,* 266.

88. S. Brownlee "Under Control — Why America Isn't Ready for Bioterrorism," *The New Republic* (2001):22–24.

89. M. K. Wynia and L. O. Gostin, "Ethical Challenges in Preparing for Bioterrorism: Barriers Within the Health Care System," *American Journal of Public Health* 94, no.7 (2004):1096–1102.

90. Hamburg, "Public Health Preparedness."

91. M. K. Wynia and L. Gostin, "The Bioterrorist Threat and Access to Healthcare," *Science* 296, no.5573 (2002).

92. J. Lederberg, "Concluding Remarks," in *Managed Care Systems and Emerging Infections: Challenges and Opportunities for Strengthening Surveillence, Research, and Prevention, Workshop Summary*, ed. J. R. Davis (Washington, D.C.: National Academy Press; 2000), 76–77.

93. N. Pourat et al., "Medicaid Managed Care and STDs: Missed Opportunities to Control the Epidemic," *Health Affairs* 21, no. 3 (2002):228–39.

94. H. E. Frech, "Physician Fees and Price Controls," in *American Health Care: Government, Market Processes, and Public Health*, ed. R. D. Feldman (Oakland, CA: The Independent Institute; 2000), 347–63.

95. W. Higgins et al., "Assessing Hospital Preparedness Using an Instrument Based on the Mass Casualty Disaster Plan Checklist: Results of a Statewide Survey" *American Journal of Infection Control* 32 (2004):327–32.

96. J. R. Richards et al., "Survey of Directors of Emergency Departments in California on Overcrowding,"*Western Journal of Medicine* 172 (2000):385–88.

97. Ibid.

98. R. W. Derlet et al., "Frequent Overcrowding in US Emergency Departments," *Academic Emergency Medicine* 8, no. 2 (2001):151–55.

99. J. Heinrich et al., *Infectious Disease Outbreaks: Bioterrorism Preparedness Efforts Have Improved Public Health Response Capacity, But Gaps Remain* (Washington, D.C.: General Accounting Office, 2003).

100. J. G. Bartlett, "Decline in Microbial Studies for Patients With Pulmonary Infections," *Clinical Infectious Diseases* 39 (2004):170–72.

101. D. Hanflinget al., "Making Healthcare Preparedness a Part of the Homeland Security Equation,"*Topics in Emergency Medicine* 26, no. 2(2004):128–42.

102. D. C. Wetteret al., "Hospital Preparedness for Victims of Chemical or Biological Terrorism," *American Journal of Public Health* 91, no. 5 (2001):710–16.

103. W. R. Jarvis, "Infection Control and Changing Health-Care Delivery Systems," *Emerging Infectious Diseases* 7, no.2 (2001):170–73.

104. N. Lurie, "Perspective: The Public Health Infrastructure: Rebuild Or Redesign?" *Health Affairs*21, no. 6 (2002):28–30.

105. J. G. Wheeler et al., "Barriers to Public Health Management of a Pertussis Outbreak in Arkansas,"*Archives of Pediatric and Adolescent Medicine* 158, no. 2 (2004):146–52.

106. S. C., Alder et al., "Physician Preparedness for Bioterrorism Recognition and Response: A Utah-Based Needs Assessment," *Disaster Management and Response*2, no. 3 (2004):69–74.

107. N. C. Mann et al., "Public Health Preparedness for Mass Casualty Events: A 2002 State-by-State Assessment," *Prehospital and Disaster Medicine* 19, no. 3(2004):245–55.

108. D. N. Fox, "From AIDS to TB: Value Conflicts in Reporting Diseases," *Hastings Center Reports* 11 (1986):L11–L16.

109. Ibid.

110. Wynia et al., "Bioterrorist Threat."

111. C. R. Wise and R. Nader, "Organizing the Federal System for Homeland Security: Problems, Issues, and Dilemmas," *Public Administration Review* 62, Suppl. 1 (2002):44–58.

112. Fine and Layton, "Lessons from West Nile."

113. Baker et al., "Public Health Infrastructure."

114. Ibid.

115. J. Kayyem, *US Preparations for Biological Terrorism: Legal Limitations and the Need for National Planning* Cambridge, MA: Harvard University, John F. Kennedy School of Government, 2001.

116. L. O. Gostin et al., "The Model State Emergency Health Powers Act: Planning for and Response to Bioterrorism and Naturally Occurring Infections," *Journal of the American Medical Association* 288, no. 5 (2002):622–28.

117. J. Kincaid and R. L.Cole, "Issues of Federalism in Response to Terror,"*Public Administration Review* 62, Suppl. 1(2002):181–92.

118. J. Colgrove and R. Bayer, "Manifold Restraints: Liberty, Public Health, and the Legacy of Jacobsen v. Massachusetts," *American Journal of Public Health* 95, no. 4 (2005): 571–76.

119. W. K. Mariner et al., "Jacobsen v Massachusetts: It's Not Your Great-Great-Grandfather's Public Health Law," *American Journal of Public Health* 95, no. 4 (2005):581–90.

120. L. O. Gostin, "Influenza Pandemic Preparedness: Legal and Ethical Dimensions," *Hastings Center Report* 34, no. 5 (2004):10–11.

121. G. J. Annas, "Perspective: Bioterrorism, Public Health, and Human Rights," *Health Affairs*; 21, no.6 (2002) 94–97.

122. A. Etzioni, "Perspective: Public Health Law: A Communitarian Perspective," *Health Affairs*; 21, no. 6 (2002):102–04.

123. R. Addlakha, "State Legitimacy and Social Suffering in a Modern Epidemic: A Case Study of Dengue Hemmorhagic Fever in Delhi," *Indian Sociology*; 35, no. 2 (2001):151–79.

124. B. D. Stein et al., "Emotional and Behavioral Consequences of Bioterrorism: Planning a Public Health Response," *The Millbank Quarterly* 82, no. 3 (2004):413–56.

125. R. J. Wray et al., "Theoretical Perspectives on Public Communication Preparedness for Terrorist Attacks," *Family and Community Health*; 27, no. 3 (2004):232–41.

126. Fine and Layton, "Lessons from West Nile," 105.

127. A. S. Khan and D. A. Ashford, "Ready or Not — Preparedness for Bioterrorism," *New* England Journal of Medicine; 345, no. 4 (2001): 287–89.

128. Adler et al., "Physician Preparedness."

129. J. W. Buehler, "Surveillence," in Rothman KJ, Greenland S, editors. *Modern Epidemiology*, 2nd ed., K. J. Rothman and S. Greenland (Philadelphia, PA: Lippincott, Williams, and Wilkins, 1998), 435–57.

130. Foreman et al., "Plagues, Products, Politics," 146–49.

131. D. L. Heymann and G. R. Rodier, "The WHO Operational Support Team to the Global Outbreak Alert and Response Network. Hot Spots in a Wired World: WHO Surveillence of Emerging and Re-Emerging Infectious Diseases," *The Lancet Infectious Diseases*; 1 (2001): 345–53.

132. Berkelman et al., "Infectious Disease Surveillance."

133. Henderson, "Surveillance Systems."

134. J. W. Buehler, "Review of the 2004 National Syndromic Surveillence Conference - Lessons Learned and Questions To Be Answered," *Morbidity and Mortality Weekly Report*; 53, Suppl. (2004):18–22.

135. J. A. Pavin et al., "Innovative Surveillence Methods for Rapid Detection of Disease Outbreaks and Bioterrorism: Results of an Interagency Workshop on Health Indicator Surveillence," *American Journal of Public Health*; 93, no. 8 (2003):1230–35.

136. J. M. Townes et al., "Investigation of an Electronic Emergency Department Information System as a Data Source for Respiratory Syndrome Surveillance," *Journal of Public Health Management Practice*; 10, no. 4 (2004):299–307.

137. C. V. Broome and J. Loonsk, "Public Health Information Network Improving Early Detection by Using a Standards-Based Approach to Connecting Public Health and Clinical Medicine" *Morbidity and Mortality Weekly Report*; 53, Suppl. (2004):199–202.

138. D. M. Bravata et al., "Systematic Review: Surveillance Systems for Early Detection of Bioterrorism-Related Diseases," *Annals of Internal Medicine*; 140, no. 11 (2004):910–22.

139. D. Koo, "Managed Care and Infectious Disease Surveillance: Opportunities for Collaboration," in *Managed Care Systems and Emerging Infections: Challenges and Opportunities for Strengthening Surveillence, Research, and Prevention, Workshop Summary*, ed. J. R. David (Washington, D.C.: National Academy Press, 2000).

140. R. Platt, "Collaborative Surveillance Efforts and Monitoring of Data," in *Managed Care Systems and Emerging Infections: Challenges and Opportunities for Strengthening Surveillence, Research, and Prevention, Workshop Summary*, ed. J. R. Davis (Washington, D.C.: National Academy Press, 2000).

141. D. F. Gordon and D. G. F. Noah, *The Global Infectious Disease Threat and Its Implications for the United States* (Langley, VA: Central Intelligence Agency, 2000).

142. J. Heinrich et al., *Gaps Remain in Surveillance Capabilities of State and Local* (Washington, D.C.: General Accounting Office, 2003).

143. D. Drociuk et al., "Health Information Privacy and Syndromic Surveillance Systems," *Morbidity and Mortality Weekly Report*; 53, Suppl. (2004): 221–25.

144. D. M. Sosin and J. DeThomasis, "Evaluation Challenges for Syndromic Surveillance," *Morbidity and Mortality Weekly Report* 53, Suppl. (2004): 125–29.

145. T. Zwilich, "Lawmakers Urge Bioterror, Disease Surveillance Law," Reuters (2003).

146. M. Lipsky, "Street Level Bureaucracy and the Analysis of Urban Reform," *Urban Affairs Quarterly* 6 (1971):391–409.

147. B. Hjern, "Implementation Research: The Link Gone Missing," *Journal of Public Policy* 2 (1982):301–08.

148. E. Ostrom, "The Danger of Self-Evident Truths," *PS — Political Science and Politics*; 33, no. 1 (2000):33–44.

149. McHugh et al., "How Prepared are Americans for Public Health Emergencies? Twelve Communities Weigh In," *Health Affairs*; 23, no. 3 (2004):201–09.

150. Z. Bashir et al., "Local and State Collaboration for Effective Preparedness Planning," *Journal of Public Health Management Practice* 9, no.5 (2003): 344–51.
151. "Smallpox Fiasco," *The Washington Post,* 2003. p. A20.
152. S. A. Hearne et al., *Public Health Laboratories: Unprepared and Overwhelmed.* (Washington, D.C.: Trust for America's Health, 2003).
153. L. A. Cole, "Risks of Publicity about Bioterrorism: Anthrax Hoaxes and Hype," *American Journal of Infection Control*;27, no. 6 (1999):470–73.
154. W. A. Orenstein et al., "Immunizations in the United States: Success, Structure, and Stress," *Health Affairs*; 24, no. 3 (2005): 599–610.
155. Benenson, "Infectious Diseases."
156. National Vaccine Advisory Committee, "Strengthening the Supply of Routinely Recommended Vaccines in the United States: Recommendations From the National Vaccine Advisory Committee," Journal of the American Medical Association; 290 no. 23 (2003): 3122–28.
157. Earls and Hearne, *Facing the Flu.*
158. Ibid.
159. P. A. Offit, "Why Are Pharmacueetical Companies Gradually Abandoning Vaccines?" *Health Affairs*;24, no. 3 (2005):622–30.
160. I. D. Gus et al., "Planning for the Next Pandemic of Influenza," *Reviews in Medical Virology* 11 (2001): 59–70.
161. J. E. Calfee, *Bioterrorism and Pharmacueticals: The Influence of Secretary Thompson's Cipro Negotiations* (Washington, D.C.: American Enterprise Institute, 2001).
162. Gust et al., "Planning for the Next Pandemic."
163. P. A. Offit and R. K. Jew, "Addressing Parents' Concerns: Do Vaccines Contain Harmful Preservatives, Adjuvants, Additives, or Residuals?" *Pediatrics*; 112, no. 6 (2003): 1394–401.
164. Stiver, "Threats and Prospects."
165. Gust et al., "Planning for the Next Pandemic."
166. R. J. Webby and R. G. Webster, "Are We Ready for Pandemic Influenza?" *Science*; 302, no. 5650(2003): 1519-22.
167. D. Carpenter, "The Political Economy of FDA Drug Review: Processing, Politics, and Lessons for Policy," *Health Affairs*; 23, no. 1 (2004):52–63.
168. S. A. Plotkin, "Why Certain Vaccines Have Been Delayed or Not Developed at All," *Health Affairs*; 24, no. 3 (2005): 631–34.
169. P. Ritvo et al., "Vaccines in the Public Eye," *Nature Medicine*; 11, 4 Suppl. (2005): S20–S24.
170. P. F. Harrison and J. Lederberg, eds., *Antimicrobial Resistance: Issues and Options* (Washington, D.C.: National Academy Press, 1998).
171. Alder et al., "Preparedness for Bioterrorism Recognition and Response."
172. A. S. Fauci et al., "Emerging Infectious Diseases: A 10-Year Perspective From the National Institute of Allergy and Infectious Diseases," *Emerging Infectious Diseases* 11, no. 4 (2005): 519–25.
173. Smithson and Levy, *Ataxia.*
174. Ibid.

175. R. G. Ridley, "Research on Infectious Disease Requires Better Coordination," *Nature Medicine*;10, 12 Suppl. (2004): S137–S40.

176. J. Heinrich, *Bioterrorism: Public Health and Medical Preparedness* (Washington, D.C.: General Accounting Office, 2001).

177. L. P. Cohen, "Safe and Effective: Many Medicines Remain Potent for Years Past Their Expiration Dates," *Wall Street Journal*.

178. Smithson and Levia, *Atazia*, 261–62, 296–98.

179. G. Avery, "Public Health System Has Foot in Mouth,"*Washington Times*.

Chapter 8

1. H. W. Cohen et al., "Bioterrorism Initiatives: Public Health in Reverse?" *American Journal of Public Health;* 89 (1999): 1629–31.[ISI][Medline]

2. M. R. Fraser and D. L. Brown, "Bioterrorism Preparedness and Local Public Health Agencies: Building Response Capacity," *Public Health Report* 115 (2000): 326–30.[ISI][Medline]

3. H. W. Cohen et al., "Bioterrorism Preparedness: Dual Use or Poor Excuse?" *Public Health Rep*ort 115 (2000): 403–405.[ISI][Medline]

4. B. S. Levy and V. W. Sidel, eds., *Terrorism and Public Health* (New York: Oxford University Press, 2003).

5. "Update: Investigation of Bioterrorism-Related Inhalational Anthrax —Connecticut, 2001," *Morbidity and Mortality Weekly Report*, 50 (2001):1049–51.[Medline]

6. A. Regaldo et al., "FBI Makes Military Labs Key Focus on Anthrax," *Wall Street Journal*, February 12th, 2002.

7. W. J. Broad et al., "Subject of Anthrax Inquiry Tied to Anti-Germ Training," *New York Times*, July 2nd, 2003.

8. Cohen et al., "Bioterrorism Initiatives."

9. U.S. General Accounting Office, *Combating Terrorism: Need for Comprehensive Threat and Risk Assessments of Chemical and Biological Attacks: Report to Congressional Requesters* (Washington, D.C.: U.S. General Accounting Office, 1999),1–35. GAO publication NSIAD-99-163.

10. V. W. Sidel et al., "Good Intentions and the Road to Bioterrorism Preparedness. *American Journal of Public Health* 91(2001): 716–18.

11. J. B. Tucker, "Bioterrorism Is the Least of Our Worries," *New York Times,* October 16th, 1999.

12. J. Miller et al., "U.S. Germ Warfare Research Pushes Treaty Limits," *New York Times*, September 4th, 2001.

13. P. Gorner, "U.S. War on Anthrax Has Its Risks: Rush to Stock New Vaccine Has Scientists Wary," *Chicago Tribune* March 28th, 2004.

14. G. W. Bush, "President Bush Discusses Iraq with Congressional Leaders." http://www.whitehouse.gov/news/releases/2002/09/20020926-7.html (accessed September 9th, 2003).

15. Associated Press, "U.S. Weapons Hunters Find No Evidence Iraq Had Smallpox," *USA Today*, September 18th, 2003, http://www.usatoday.com/news/world/iraq/2003-09-18-iraq-smallpox_x.htm (accessed October 29th, 2003).
16. "Bush's Comments On His Plan For Smallpox Vaccinations across the U.S.," *New York Times*, December 14th, 2002.
17. G. H. Brundtland, "World Health Organization Announces Updated Guidance On Smallpox Vaccination," http://www.who.int/inf-pr-2001/en/state2001-16.html (accessed September 9th, 2003).
18. American Public Health Association Executive Board, "APHA Policy On Smallpox Vaccinations, http://www.apha.org/legislative/ policy/smallpox.pdf (accessed September 9th, 2003).
19. D. Linzer, "No Trace Found of Reputed Smallpox in Iraq," *Miami Herald*, September 19th, 2003.
20. California Nurses Association, "CNA Adds Voice to Opposition to Smallpox Vaccination Plan," http://www.calnurse.org/cna/calnursejanfeb03/smallpox.html (accessed July 15th, 2004).
21. H. W. Cohen and S. Eolis, "Smallpox Vaccine: Don't Do it," *American Journal of Nursing* 103 (2003):13.[Medline]
22. Centers for Disease Control and Prevention, "Update: Adverse Events Following Civilian Smallpox Vaccination — United States, 2003," *Morbidity and Mortality Weekly Report* 53 (2003): 106–107.
23. Ibid.
24. Institute of Medicine, Committee on Smallpox Vaccination Program Implementation, Board on Health Promotion and Disease Prevention, *Review of the Centers for Disease Control and Prevention's Smallpox Vaccination Program Implementation* (Washington, D.C.: National Academy Press, 2003).
25. Ibid.
26. Elliott vs. Public Health Funding: Feds Giveth but the States Taketh Away," http://www.ama-assn.org/sci-pubs/amnews/2002/10/28/hIl21028.htm (accessed August 31st, 2003), 18–19.
27. S. Smith, "Anthrax vs. the flu," *Boston Globe*, July 29th, 2003.
28. M. Eserink, "New Biodefense Splurge Creates Hotbeds, Shatters Dreams," *Science* 302 (2003): 206–207.
29. Sunshine Project, "Map of High Containment and Other Facilities of the U.S. Biodefense Program, http://www.sunshine-project.org/biodefense/ (accessed May 24th, 2004).
30. J. Miller, "New Biolabs Stir Debate Over Secrecy and Safety," *New York Times*, February 10th, 2004.
31. R. Weiss and D. Snyder, "Second Leak of Anthrax Found at Army Lab,"*Washington Post*, April 24th, 2002.
32. E. Williamson, "Ft. Detrick Unearths Hazardous Surprises," *Washington Post*, May 27th, 2003.
33. U.S. General Accounting Office, "Combating Bioterrorism: Actions Needed to Improve Security at Plum Island Animal Disease Center," http://www.gao.gov/atext/d03847.txt (accessed October 29th, 2003).
34. Sunshine Project, "Map of High Containment."

35. V. W. Sidel et al., "Bioterrorism Preparedness: Cooptation of Public Health?" *Medical Global Survival* 7 (2002):82–89.

36. M. Kelley and J. Coghlan, "Mixing Bugs and Bombs," *Bulletin of Atomic Scientists* 59, no. 5 (2003): 24–31.

37. O. A. Divis and N. M..Horrock, "Living Terror: Lab Secrets in Dispute," http://www.upi.com/view. efm? StoryID=20030806-061348-4757r (accessed September 9th, 2003).

38. V. W. Sidel and B. Levy, "Security and Public Health," in "Global Threats to Security," ed. R. M. Gould and P. Sutton, special issue, *Social Justice* 29, no. 3 (2002):108–119.

Chapter 9

1. Centers for Disease Control and Prevention, "Questions and Answers: The Disease," <http://www.cdc.gov/flu/about/qa/disease.htm> (accessed May 10, 2005).

2. Associated Press, "Flu Vaccine Shortage Could Cost U.S. $20 Billion," *USA Today*, October 21st, 2004, <http://www.usatoday.com/news/health/2004-10-21-flu-vaccine-cost_x.htm> (accessed May 10th, 2005).

3. Madeline Drexler, *Secret Agents, The Menace of Emerging Infections* (Washington, D.C.: Joseph Henry Press, 2002), 190.

4. Centers for Disease Control and Prevention, "Information about Influenza Pandemics," <http://www.cdc.gov/flu/avian/gen-info/pamkmics.htm.> (accessed May 10th, 2005).

5. Scientists have issued a range of projections about the potential impact of a pandemic flu outbreak, based on different assumptions about the severity of the strain. This projection is based on using a formula developed by the CDC called FluAid, using assumptions about predictions about the potential severity of the virus.

6. Martin I. Meltzer et al., "The Economic Impact of Pandemic Influenza in the United States: Priorities for Intervention," *Emerging Infectious Diseases*, 5, no. 5 (1999): 659–71, <http://www.cdc.gov/ncidod/EID/vol5 no5/pdf/meltzer.pdf> (accessed May 5th, 2005).

7. Michael T. Osterholm, "Preparing for the Next Pandemic," *New England Journal of Medicine* 18, no.352: (2005): 1839–42.

8. Centers for Disease Control and Prevention, "Questions and Answers."

9. World Health Organization "Fact Sheet: Avian Influenza," (February 5th, 2004) <http://www.who.int/csr/don/2004_01_15/en/.6> (accessed May 16th, 2004).

10. Associated Press, "Bird Flu Called Global Human Threat," *Washington Post,* February 24th, 2005, http://www.washingtonpost.com/wp-dyn/articles/ A46424-2005Feb 23.html> (accessed May 10th, 2005).

11. World Health Organization, "Cumulative Number of Confirmed Human Cases of Avian InfluenzaA/(H5N1) since 28 January 2004." June 17th, 2005,<http://www.who.int/csr/disease/avian_influenza/country/cases_ table_2005_06_17/en/index.html> (accessed June 21st, 2005).

12. Keith Bradsher, "Some Asian Bankers Worry About the Economic Toll From Bird Flu," *New York Times*, April 5th, 2005.

13. Mike Leavitt, "U.S. Health Secretary Calls Bird Flu Outbreak Urgent Challenge," remarks of U.S. Health and Human Services Secretary at World Health Assembly, U.S. Department of State, May 16th, 2005, http://usinfo.state.gov/gi/Archive/2005/May/16-413291.htm1?chanlid= globalissues> (accessed May 23rd, 2005).

14. Centers for Disease Control and Prevention, "Basic Information about Avian Influenza (Bird Flu)," February 6th, 2004, http://www.cdc.gov/flu/ avian/facts.htm (accessed May 16th, 2005).

15. World Health Organization, "International Response to the Distribution of H2N2 Influenza Virus for Laboratory Testing: Risk Considered Low for Laboratory Workers and the Public," April 12th, 2005, htltp://www.who.int/ csr/disease/inf1uenza/h2n2_2005_04_12/en/ (accessed May 2nd, 2005).

16. World Health Organization, "FactSheet: Avian Influenza," February 5th, 2004, http://www. who.int/csr/don/2004_01_15/en/.6 (accessed May 16th, 2005.

17. World Health Organization, "Strengthening Pandemic Influenza Preparedness and Response: Report by the Secretariat," April 7th, 2005, http://www.who.int/csr/disease/inf1uenza/preparedness/report.pdf> (accessed May 19th, 2005).

18. Anthony S. Fauci "The Role of NIH Biomedical Research in Pandemic Influenza Preparedness," Testimony Before the House Committee on Appropriations Subcommittee on Labor, HHS, and Education, United States House of Representatives, <http://appropriations.house.gov/_files/AnthonyFauciTestimony.pdf> (accessed at May 18th, 2005).

19. Julie L Gerberding, "Influenza and Influenza Vaccine Safety and Supply," Testimony before The Committee on Appropriations Subcommittee on Labor, Health and Human Services, & Education U.S. House of Representatives, October 5th, 2004, http://www. hhs.gov/asl/testify/t041005c.html (accessed May 18th, 2005).

20. Keiji Fukuda,, Chief, Epidemiology and Surveillance Section, Influenza Branch, Division of Viral and Rickettsial Diseases, National Center for Infectious Diseases, Centers for Disease Control and Prevention, Power Point Presentation: "Influenza and Pandemics: What You Need to Know," Congressional Briefing, Washington, D.C., March 8th, 2005.

21. Centers for Disease Control and Prevention National Vaccine Program Office, "FluAid Home," July 25th, 2000, http://www2a.cdc.gov/od/fluaid/ Default.htm (accessed April 27th, 2005).

22. Keiji Fukuda, "Influenza and Pandemics,"

23. American Hospital Association Resource Center, "Fast Facts on U.S. Hospitals from AHA Hospital Statistics," http://www.hospitalconnect.com/aha/ resource_center/fastfacts/fast_facts_US_hospitals.htm1#community (accessed June 13th, 2005).

24. Osterholm "Preparing for the Next Pandemic."

25. U.S. Department of Health and Human Services, "Draft Pandemic lnf1uenza Preparedness and Response Plan." August 2004, http://www.hhs.gov/ nvpo/pandemic-plan/final-pandemiccore.pdf (accessed March 16th, 2005).

26. Summaries of public comments on the pandemic plan were found at http://www.hhs.gov/ nvpo/pandemicplan/ (accessed March 2005).

27. U.S. House of Representative·"The Threat of and Planning for Pandemic Flu," Transcript of the Hearing of the Health Subcommittee of the House Energy and Commerce Committee, May 26th, 2005, http://energycom-merce.house.gov/108/Hearings/0526 2005hearing1530/hearing.htm#Tran-script (accessed June 10th, 2005).

28. USAID Health "Avian Influenza Response." May 25th, 2005, http://www.usaid.gov/our_work/global_health/home/News/news_items/avian_influenza.html (accessed May 31st, 2005).

29. Barak Obama, "S.969: A Bill to Amend the Public Health Service Act With Respect to Preparation for an Influenza Pandemic, Including an Avian Influenza Pandemic, and for Other Purposes," introduced April 28th, 2005, http://thomas.loc.gov/cgi-bin/bdquery/D?dl09:3:./temp/~bdbj0S:: (accessed May 31st, 2005).

30. "Executive Order: Amendment to E.O. 13295 Relating to Certain Influenza Viruses and Quarantinable Communicable Diseases," White House News Releases, April 1st, 2005, http://www.whitehouse.gov/news/releases/2005/04/20050401-6.html (accessed May 23rd, 2005).

31. U.S. Department of State, "Avian Flu Fact Sheet." April 8th, 2005, http://travel.state. gov/travel/tips/health/health_2126.html (accessed May 24th, 2005).

32. U.S. Congressional Research Service, "State Department FY2006-2007 Authorization Bill," introduced March 10th, 2005, http://www.congress.gov/cgi-bin/query/z?c109:S.600: (accessed May 24th, 2005).

33. "Flu Pandemic? We're Really Not Ready, Folks." *CQ Healthbeat,* May 26th, 2005, http://www.cq.com/healthbeatnews.html (accessed May 27th, 2005).

34. Trust for America's Health, *Ready or Not? Protecting the Public's Health in the Age of Bioterrorism* (Washington. D.C.: Trust for America's Health, 2004), 33.

35. Summaries of public comments on the pandemic plan, see note 26.

36. Frank D. Roylance, "Fears of Flu Pandemic Spurring Preparations; The Threat of Global Influenza Prompts Research, but Critics Say the Efforts Fall Short," *Baltimore Sun,* June 12th, 2005.

37. Paul A. Offit, "Why Are Pharmaceutical Companies Gradually Abandoning Vaccines?" *Health Affairs,* 24, no.3 (2005): 622.

38. Gerberding "U.S. Influenza Supply," 2.

39. "Health Secretary: No Flu Vaccine Crisis," CNN.com, October 18th, 2004, http://www.cnn.com/2004/ HEALTH/10/18/flu/ (accessed May 16th, 2005).

40. World Health Organization, "WHO Guidelines on the Use of Vaccines and Antivirals during Influenza Pandemics," 2002, http://www.who.int/emc/diseases/flu/annex3.htm (accessed May 16th, 2005).

41. Gerberding, "U.S. Influenza Supply," 2.

42. U.S. House of Representatives, "The Threat of and Planning for Pandemic Flu." see note 27.

43. Ibid.

44. British Department of Health, "News Release: Improving Preparedness for Possible Flu Pandemic: Purchase of Antiviral Drugs and Publication of Plan," March 1st, 2005, http://www.dh.gov.uk/PublicationsAndStatistics/PressReleases/PressReleasesNotices (accessed April 27th, 2005).

45. U.S. House of Representatives, "The Threat of and Planning for Pandemic Flu," see note 27.

46. David S. Fedson, "Preparing for Pandemic Vaccination: An International Policy Agenda for Vaccine Development," *Journal of Public Health Policy*, 26, no.4 (2005):29. http://www.palgrave-journals.com/jphp/fedson,pdf (accessed May 31st, 2005.

47. "Bird Flu Spreads Among Java's Pigs." *Nature*, 435, no. 7041 (2005):390.

48. Ibid.

49. Mike Leavitt, Opening Remarks of the U.S. Secretary of Health and Human Services, at the Ministerial Meeting on Avian Influenza, May 16th, 2005, http://usinfo.state.gov/gi/Archive/2005/May/16-413291.html?chanlid=globalissues (accessed May 24th, 2005).

50. "On a Wing and a Prayer," *Nature*, 435, no.7041 (2005): 390.

Chapter 10

1. Centers for Disease Control and Prevention, "Key Facts about Influenza and Influenza Vaccine, Department of Health and Human Services, 2005, http://www.cdc.gov/flu/keyfacts.htm (accessed August 22nd, 2005).

2. World Health Organization. Influenza. WHO; March 2003. Available at http://www.who.int/mediacentre/factsheets/fs211/en/print.html. Last accessed August 22nd, 2005.

3. N. P. Johnson and J. Mueller, "Updating the Accounts: Global Mortality of the 1918–1920 'Spanish' Influenza Pandemic," *Bulletin of the History of Medicine* 76 (2002):105–15.

4. Centers for Disease Control, "Key Facts about Influenza."

5. World Health Organization, "Influenza."

6. Centers for Disease Control, "Estimated Influenza Vaccination Coverage among Adults and Children — United States, September 1, 2004–January 31, 2005," Morbidity and Mortality Weekly Report 54 (2005):304–307.

7. U.S. Department of Health and Human Services, *Healthy People 2010: With Understanding and Improving Health and Objectives for Improving Health*, 2nd ed., 2 vols. (Washington, D.C.: U.S. Department of Health and Human Services; November 2000).

8. Centers for Disease Control, "Tiered Use of Inactivated Influenza Vaccine in the Event of a Vaccine Shortage," Morbidity and Mortality Weekly Report 54 (2005): 749–50.

9. G. Frankel and G. Cooper, "Britain: U.S. Told of Vaccine Shortage," *Washington Post*, October 9th, 2004.

10. Centers for Disease Control, "Estimated Vaccination Coverage."

11. D. Brown, "World Not Set to Deal with Flu," *Washington Post*, July 31st, 2005.

12. B. A. Cunha, "Influenza: Historical Aspects of Epidemics and Pandemics," *Infectious Disease Clinics of North America* 18 (2004):141–55.

13. G. Kolata, *Flu: The Story of the Great Influenza Pandemic of 1918 and the Search for the Virus That Caused It* (New York:. Farrar, Straus and Giroux, 1999).

14. Johnson and Mueller, "Updating the Accounts."

15. J. K. Taubenberger et al., "Capturing a Killer Flu Virus," *Scientific American* 292 (2005):62–71.

16. Ibid.

17. Centers for Disease Control, "Information about Influenza Pandemics," Department of Health and Human Services, 2005, http://www.cdc.gov/flu/avian/gen-info/pandemics.htm

18. Ibid.

19. R. E. Neustadt and H. V. Fineberg, *The Epidemic that Never Was* (New York: Vintage, 1983).

20. N. M. Ferguson et al., "Strategies For Containing an Emerging Influenza Pandemic in Southeast Asia," *Nature Online* August 3, 2005:1–6.

21. I. M. Longini Jr. et al., "Containing Pandemic Influenza at the Source," *Science Express*, August 3, 2005: 1–9.

22. Kolata, "Flu."

23. Taubenberger et al., "Capturing a Killer Flu Virus."

24. J. Miller et al., "U.S. Germ Warfare Research Pushes Treaty Limits," *New York Times*, September 4, 2001.

25. L. R. Ember "Testing the Limits," *Chemical and Engineering News*, 83 (2005): 26–32.

26. A. Regaldo et al., "FBI Makes Military Labs Key Focus On Anthrax," *Wall Street Journal*. February 12, 2002.

27. J. Miller, "New Biolabs Stir Debate Over Secrecy and Safety," *New York Times*, February 10, 2004.

28. V. W. Sidel et al., "Bioterrorism Preparedness: Cooptation of Public Health? *Medical Global Survival* 7 (2002): 82–89.

29. M. Kelley and J. Coghlan, "Mixing Bugs and Bombs," *Bulletin of Atomic Scientists* 59, no. 5 (2003): 24–31.

30. A. Srinivasan et al., "Glanders In a Military Research Microbiologist," *New England Journal of Medicine* 345 (2001): 256–58.

31. F. Kunkle, "Md. Lab Ships Live Anthrax in Error," *Washington Post*, June 12th, 2004.

32. R. Stein and S. Vedantam, "Officials Race to Destroy Samples," *Washington Post*, April 13th, 2005.

33. H. W. Cohen et al., "Bioterrorism — Scare Stories Can Be Dangerous to Our Health," http://www.thedoctorwillseeyounow.com/articles/other/biotb_13/ (accessed August 25th, 2005).

34. P. Harrison, "U.S. Seeks Massive Stock of Smallpox Vaccine," *Reuters Health*, August 16th, 2005, http://www.nlm.nih.gov/medlineplus/news/fullstory_26322.html (accessed August 25th, 2005).

35. CNN, "Bush Military Bird Flu Role Slammed," CNN.com, http://www.cnn.com/2005/POLITICS/10/05/bush.reax/ (accessed October 7th, 2005).

36. U.S. Department of Homeland Security, "National Preparedness Month," http://www.ready.gov/npm/ (accessed September 18th, 2005).

37. H. W. Cohen et al., "The Pitfalls of Bioterrorism Preparedness: The Anthrax and Smallpox Experiences," *American Journal of Public Health* 94, no. 10 (2004):1667–71.

38. S. Borenstein, "Federal Government Wasn't Ready for Katrina, Disaster Experts Say," Knight Ridder Newspapers, http://www.realcities.com/mld/krwashington/12528233.htm (accessed September 18th, 2005).

39. Wall Street Article 2004 hurricane

40. CNN, "Bush Military Bird Flu Role Slammed."

41. One of the references regarding diversion of funds from public health to bioterrorism?

42. J. J. Treanor, "Influenza Virus," in *Principles and Practices of Infectious Disease*, ed. G. L. Mandell et al. (Philadelphia: Churchill Livingstone; 2004), 2060–85.

43. S. S. Chiu et al., "Influenza-Related Hospitalizations Among Children in Hong Kong," *New England Journal Medicine* 347 (2002):2097-2103.

44. A. S.. Monto and N. H. "Arden Implications of Viral Resistance to Amantadine in Control of Influenza A, *Clinical Infectious Disease* 15 (1992):362–67.

45. F. G. Hayden et al., "Efficacy and Safety of the Oral Neuraminidase Inhibitor Oseltamivir in Treating Acute Influenza: A Randomized, Controlled Trial," *Journal of the American Medical Association* 283 (2000): 1016–24.

46. Monto and Arden, "Implications of Viral Resistance."

47. F. L. Ruben, "Prevention and Control of Influenza: Role of Vaccine," *American Journal of Medicine* 82 (1987):31–33.

48. K. L. Nichol et al., "The Effectiveness of Vaccination Against Influenza in Healthy, Working Adults: A Randomized Controlled Trial, *New England Journal of Medicine* 333 (1995): 889–93.

Chapter 11

1. H. Small, *Florence Nightingale. Avenging Angel* (London: Constable, 1999).

2. S. J. Dancer, "Mopping Up Hospital Infection," *Journal of Hospital Infections* 43 (1999): 85–100.

3. G. A. J. Ayliffe and M. P. English, "Hospital Infection," in *From Miasmas to MRSA,* 1st ed. Cambridge, UK: Cambridge University Press, 2003.

4. R. J. Mullan and T. M. Frazier, "The Authors Reply" (letter), *American Journal of Epidemiology* 362 (1986): 150–155.

5. J-L. Vincent, "Nosocomial Infections in Adult Intensive-Care Units," *Lancet* 361 (2003): 2068–77.

6. Ayliffe and English, "Hospital Infection."

7. Centers for Disease Control, *Guidelines for Environmental Infection Control in Health-Care Facilities (GIDAC)* (Atlanta, GA: Centers for Disease Control, 2003).

8. CCDR., *Supplemental Infection Control Guidelines: Handwashing, Cleaning, Disinfecting and Sterilization in Health Care* (Health Canada, 1998).

9. L. T. Kohn et al., *To Err Is Human: Building a Safer Health System* (Washington, D.C.: National Academy Press, 2000).

10. V. Curtis, "Hygiene: How Myths, Monsters and Mothers-in-Law Can Promote Behaviour Change," *Journal of Infections* 43 (2001): 75–59.

11. M. Thompson and P. Hempshall, "Dirt Alert," *Nursing Times* 94 (1998): 66–69.

12. S. Patel, "The Impact of Environmental Cleanliness On Infection Rates (Clinical Evidence)," *Nursing Times* 100 (2004): 32–34.

13. UK Department of Health, Towards Cleaner Hospitals and Lower Rates of Infection: A Summary of Action," July 2004, htpp://www.nhsestates.gov.uk.

14. R. E. Malik et al., "Use of Audit Tools to Evaluate the Efficacy of Cleaning systems in Hospitals," *American Journal of Infection Control* 31 (2003): 181–87.

15. C. J. Griffith et al., "An Evaluation of Hospital Cleaning Regimens and Standards," *Journal of Hospital Infection* 45 (2000): 19–28.

16. V. J. Fraser and M. A. Olsen, "The Business of Health Care Epidemiology: Creating a Vision for Service Excellence," *American Journal of Infection Control* 30 (2002): 77–85.

17. J. P. Burke, "Infection Control — A Problem for Patient Safety," *New England Journal of Medicine* 348 (2003): 651–56; 654.

18. U.S. Department of Health and Human Services Conference, *Healthy People 2010.* (Washington, D.C.: U.S. Department of Health and Human Services, 2000).

19. W. R. Jarvis, "Infection Control and Changing Health-Care Delivery Systems," *Emerging Infectious Diseases* 7 (2001): 170–73.

20. Burke, "Infection Control."

21. R. B. Wenzel and M. B. Edmond, "The Impact of Hospital-Acquired Bloodstream Infections," *Emerging Infectious Diseases* 7 (2001): 174–77.

22. D. E. Zoutman et al., "The State of Infection Surveillance and Control in Canadian Hospitals," *American Journal of Infection Control* 312 (2003): 266–73.

23. Centers for Disease Control and Prevention, "Public Health Focus: Surveillance, Prevention, and Control of Nosocomial Infections," *Morbidity and Mortality Weekly Report* 41 (1992): 783–87.

24. P. W. Stone et al., "A Systematic Audit of Economic Evidence Linking Nosocomial Infections and Infection Control Interventions: 1990–2000," *American Journal of Infection Control* 30 (2002): 145–52.

25. A. S. Benenson, *Control of Communicable Diseases in Man,* 16th ed. (Atlanta, GA: American Public Health Association, 1995).

26. Ayliffe and English, "Hospital Infection."

27. N. N. Damani, *Manual of Infection Control Procedures,* 2nd ed. (London: Greenwich Medical Media, 2003).

28. C.Beck-Sague et al., "Outbreak Investigations," *Infection and Control Hospital Epidemiology* 18 (1997): 138.

29. Damani, *Manual of Infection Control Procedures.*

30. Centers for Disease Control, "Nososcomial Enterococci Resistant to Vancomycin — United States 1989–1993," *Morbidity and Mortality Weekly Report* 42 (1993): 597–99.

31. C. Wendt et al., "Survival of Acinetobacter Baumanii On Dry Surfaces," *Journal of Clinical Microbiology* 35 (1997): 1394–97.

32. Damani, *Manual of Infection Control Procedures.*

33. C. Beck-sague et al., "Outbreak Investigations."

34. S. Engelhart et al., "Pseudomonas Aeruginosa Outbreak in a Haematology–Oncology Unit Associated with Contaminated Surface Cleaning Equipment," *Journal of Hospital Infection* 52 (2002): 93–98.

35. Centers for Disease Control, *Guidelines for Environmental Infection Control.*

36. Health Canada, "Supplement Infection Control Guidelines."

37. R. A. Hailey et al., "Nosocomial Infections in U.S. Hospitals, 1975–1976: Estimated Frequency by Selected Characteristics of Patients," *American Journal of Medicine* 70 (1981): 947–59.

38. W. R. Jarvis, Commentary, *Canadian Medical Association Journal* 31 (2003): 272–73.

39. Burke, "Infection Control."

40. J. L. Gerberding, "Hospital-Onset Infections: A Patient Safety Issue," *Annals of Internal Medicine* 137 (2002): 665–70.

41. J. Giesbrecht, *Modern Infectious Disease Epidemiology*, 2nd ed. (London: Arnold Publishers, 2002).

42. Burke, "Infection Control."

43. P. W. Stone et al., A Systematic audit."

44. Burke, "Infection Control."

45. F. Daschner and A. Schuster, "Disinfection and the Prevention of Infectious Disease: No Adverse Effects?" *American Journal of Infection Control* 32 (2004): 224–25.

46. M. Kennedy, "Overuse of Superdrugs Leads to "Super Bugs,". WMJ 99 (2000): 22–23, 28.

47. D. Leaper, "Nosocomial Infection," *British Journal of Surgery* 91 (2004): 526–27.

48. N. Crowcroft and M. Catchpole, "Mortality from Methicillin-Resistant *Staphylococcus aureus* in England and Wales: Analysis of Death Certificates," British Medical Journal 325 (2002): 1390–91.

49. S. Harbarth and D. Pittet, "Methicillin-Resistant Staphylococcus Aureus," *Lancet Infectious Disease* 5 (2005): 653–63.

50. J. M. Boyce, "Understanding and Controlling Methicillin-Resistant Staphylococcus Infections," *Infection Control Hospital Epidemiology* 23 (2002): 485–87.

51. J. A. Martinez et al., "Role of Environmental Contamination as a Risk Factor for Acquisition of Vancomycin-Resistant Enterococci in Patients Treated in a Medical Intensive Care Unit," *Archives of Internal Medicine* 163 (2003): 1905–12.

52. W. A. Rutala et al., "Studies On the Disinfection of VRE-Contaminated Surfaces," *Infection Control Hospital Epidemiology* 21 (2000): 548.

53. A. Rampling et al., "Evidence that Hospital Hygiene Is Important in the Control of Methicillin-Resistant Staphylococcus Aureus," *Journal of Hospital Infections* 49 (2001): 109–16.

54. C. Muto et al., "SHEA Guideline for Preventing the Nosocomial Transmission of Multidrug-Resistant Strains of Staphylococcus Aureus and Enterococcus," *Infection Control Hospital Epidemiology* 24 (2003): 362–86.

55. J. S. Cheesborough et al., "Possible Prolonged Environmental Survival of Small Round Structured Viruses," *Journal of Hospital Infections* 35 (1997): 325–26.

56. L. A. Sawyer et al., "25- to 30-NM Virus Particle Associated with a Hospital Outbreak of Acute Gastroenteritis with Evidence for Airborne Transmission," *American Journal of Epidemiology* 127 (1988): 1261–71.

57. E. O. Caul, "Viral Gastroenteritis: Small Structured Round Virus, Calciviruses and Astroviruses" (Part 1), *Journal of Clinical Pathology* 49 (1996): 874–80.

58. J. Barker et al., "Effects of Cleaning and Disinfection in Reducing the Spread of Norovirus Contamination via Environmental Surfaces" *Journal of Hospital Infections* 58 (2004): 42–49.

59. B.Christie, "Vomiting Virus Closes Scottish Hospital," *British Medical Journal* 324 (2002): 258–59.

60. L. Valiquette et al., "Clostridium Difficile Infections in Hospitals: A Brewing Storm," *Canadian Medical Association Journal* 171 (2004): 27–29.

61. M. A. Miller et al., *Infection Control Hospital Epidemiology* (2002): 137–40.

62. A. E. Simor et al., "Clostridium Difficile in Long-Term Care Facilities for the Elderly," *Infection Control Hospital Epidemiology* (2002): 696–703.

63. J. Pepin et al., "Clostridium Difficile-Associated Diarrhea in a Region of Quebec from 1991–2003: A Changing Pattern of Disease Severity," *Canadian Medical Association Journal* 171 (2004): online1– 7.

64. M. H. Samore et al., "Clinical and Molecular Epidemiology of Sporadic and Clustered Cases of Nosocomial Clostridium Difficile Diarrhea," *American Journal of Medicine* 100 (1996): 32–40.

65. L. C. Eggerston, "Difficile Hits Sherbrooke, Quebec Hospital: 100 Deaths," *Canadian Medical Association Journal* (2004); 171 online 1.

66. L. Eggerston and B. Sibbald, "Hospitals Battling Outbreaks of C. Difficile," *Canadian Medical Association Journal* 17 (2004): 19–21.

67. T. V. Riley, "Nosocomial Diarrhea Due to Clostridium Difficile," *Current Opinion on Infectious Disease* 17 (2004): 323–27.

68. S. M. Poutanen and A. E. Simor, "Clostridium Difficile-Associated Diarrhea in Adults" (review). *Canadian Medical Association Journal* 171 (2005): 51–58.

69. J. L. Mayfield et al., "Environmental Control to Reduce Clostridium Difficile," *Clinical Infectious Disease* 31 (2000): 995–1000.

70. M. H. Samore et al., "Clinical and Molecular Epidemiology of Sporadic and Clustered Cases of Nosocomial Clostridium Difficile Diarrhea,"

71. Eggerston and Sibbald, "Hospitals Battling Outbreaks of C. Difficile."

72. C. M. Pellowe et al., "Evidence-Based Guidelines for Preventing Healthcare-Associated Infections in Primary and Community Care in England," *Journal of Hospital Infections* 55 (2003): S3–S4.

73. D. Pittet et al., "Bacterial Contamination of the Hands of Hospital Staff During Routine Patient Care," *Archives of Internal Medicine* 159 (1999): 821–26.

74. A. E. Aiello and E. L. Larson, "What Is the Evidence of a Causal Link between Hygiene and Infections?" *Lancet Infectious Disease* 2 (2002): 103–110.

75. B. S. Cooper et al., "Preliminary Analysis of the Transmission Dynamics of Nosocomial Infections: Stochastic and Management Effects," *Journal of Hospital Infection* 43 (1999): 131–47.

76. J. M. Boyce et al., "Environmental Contamination Due to Methicillin-Resistant Staphylococcus Aureus: Possible Infection Control Implications," *Infection Control Hospital Epidemiology* 18 (1997): 622–27.

77. A. E. Aiello and E. L. Larson, "Causal Inference: The Case of Hygiene and Health," *American Journal of Infection Control* 30 (2002): 503–10.

78. E. A. Jenner et al., "Infection Control — Evidence into Practice," *Journal of Hospital Infections* 42 (1999): 91–104.

79. E. Larson and E. K..Kretzer, "Compliance with Handwashing and Barrier Precautions," *Journal of Hospital Infections* 30, Suppl. (1995): 88–106.

80. L. Buchanan et al., "Microbial Risk Assessment: Dose-Response Relations and Risk Characterization," *International Journal of Food Microbiology* 58 (2000): 159–72.

81. Health Canada, "A Framework for the Application of Precaution in Science-Based Decision Making about Risk," http://www.hc-sc.gc.ca/sab-ccs/jun2000_precautionary_approach_e.html

82. Centers for Disease Control,, *Guidelines for Environmental Infection.*

83. C. M. Pellowe et al., "Evidence-Based Guidelines."

84. Rampling et al., "Evidence that Hospital Hygiene Is Important."

85. R. J Pratt et al., "The Epic Project: Developing National Evidence-Based Guidelines for Preventing Healthcare-Associated Infections. Phase 1: Guidelines for Preventing Hospital-Acquired Infections," *Journal Hospital Infection* 47, Suppl. (2001): S1–S82.

86. C. M. Pellowe et al., "The Epic Project. Updating the Evidence-Base for National Evidence-Based Guidelines for Preventing Healthcare-Associated Infections in NHS Hospitals in England: A Report with Recommendations," *British Journal of Infection Control* 5 (2004): 10–16.

87. S. Dancer "How Do We Assess Hospital Cleaning? A Proposal for Microbiological Standards for Surface Hygiene in Hospitals" (Review). *Journal of Hospital Infections* 56 (2004):10–5.

88. C. J. Griffith et al., "Environmental Surface Cleanliness and the Potential for Contamination During Handwashing," *American Journal of Infection Control* 31 (2003): 93–96.

89. C. J. Griffith et al., "An Evaluation of Hospital Cleaning Regimes and Standards," *Journal of Hospital Infections* 45 (2000): 19–28.

90. A. Feeney et al., "Socioeconomic and Sex Differentials in Reason for Sickness Absence from the Whitehall II Study," *Occupational and Environmental Medicine* 55 (1998): 91–98.

91. F. M. North et al., "Explaining Socioeconomic Differences in Sickness Absence: The Whitehall II Study," *British Medical Journal* 306 (1993): 361–66.

92. T. S. Kristensen, "Sickness Absence and Work Strain among Danish Slaughterhouse Workers: An Analysis of Absence from Work Regarded as Coping Behaviour," *Social Science and Medicine* 32 (1991): 15–27.

93. J. V. Johnson et al., "Long-Term Psychosocial Work Environment and Cardiovascular Mortality among Swedish Men,"*American Journal of Public Health* 86 (1996): 324–31.

94. M. Marmot, "Importance of the Psychosocial Environment in Epidemiological Studies," *Scandinavian Journal of Work and Environmental Health* 25, Suppl 4 (1999): 49–53.

95. S. Stansfield et al., *Work Related Factors and Ill Health. The Whitehall II Study. Health and Safety Executive.* Contract Research Report 266/2000. (2000)

96. C. A. Woodward et al., "The Impact of Re-Engineering and Other Cost Reduction Strategies On the Staff of a Large Teaching Hospital," *Medical Care* 37 (1999): 556–69.

97. K. Messing, *One-Eyed Science: Occupational Health and Women Workers* (Philadelphia: Temple University Press, 1998).

98. M. Gampiere et al., "Duration of Employment Is Not a Predictor of Disability in Cleaners: A Longitudinal Study," *Scandinavian Journal of Public Health* 31 (2003): 63–68.

99. S. Salminen et al., "Stress Factors Predicting Injuries of Hospital Personnel," *American Journal of Indian Medicine* 44 (2003: 32–36.

100. P. J. Carrivick et al., "Effectiveness of a Workplace Risk Assessment Team in Reducing the Rate, Cost, and Duration of Occupational Injury," *Journal of Occupational and Environmental Medicine* 44 (2002): 155–59.

101. J. P. Zock, "World at Work: Cleaners," *Occupational and Environmental Medicine* 62 (2005): 581– 84.

102. M. Kivimaki et al., "Organisational Justice and Health of Employees: Prospective Cohort Study,"*Occupational and Environmental Medicine* 60 (2003): 27–34.

103. M. Elovainio et al., "Organizational Justice: Evidence of a New Psychosocial Predictor of Health," *American Journal of Public Health* 92 (2002): 105–8.

104. K. Messing et al., "Light" and "Heavy" Work in the Housekeeping Service of a Hospital," *Applied Ergonomics* 29 (1998): 451–59.

105. K. Messing "Hospital Trash: Cleaners Speak of Their Role in Disease Prevention," *Medical Anthropology Quarterly* 12 (1998): 168–87.

106. M. Gervais, *Bilan de santé des travailleurs Québecois* (Montréal: Institut de recherche en santé et sécurité du travail au Québec, 2004).

107. D. M. Berwick, "Improvement, Trust and the Healthcare Workforce," *Quality and Safety in Health Care* 12 (2003): 448–52.
108. K. D. Dieckhaus and B. W. Cooper, "Infection Control Concepts in Critical Care," *Critical Care Clinic* 14 (1998): 55–70.
109. J. S. Garner and The Hospitals Infection Control Practices Advisory Committee, "Guidelines for Isolation Precautions in Hospitals," *Infection Control Hospital Epidemiology* 17 (1996): 53–80.
110. H. Humphreys, "Infection Control Team in the Operating Room: Separating Aspiration from Reality," *Journal of Hospital Infections* 42 (1999): 265–67.
111. J. S. Garner and The Hospitals Infection Control Practices Advisory Committee, "Guidelines for Isolation."
112. M. Rutherford, "Tackling the Problem of Poor Ward Cleanliness," *Professional Nursing* 16 (2001):1148–52; 1150.
113. M. Dettenkofer et al., "Does Disinfection of Environmental Surfaces Influence Nosocomial Infection Rates? A Systematic Review," *American Journal of Infection Control* 32 (2004): 84–89.
114. F. Daschner, "The Hospital and Pollution: Role of the Hospital Epidemiologist in Protecting the Environment," in *Infectious Disease Epidemiology,* ed.. K. E Nelson, et al. (Gaithersburg, MD: Aspen, 2001).
115. C. J. Griffith et al., "Environmental Surface Cleanliness."
116. Griffith, Cooper et al., "An Evaluation of Hospital Cleaning Regimes and Standards."
117. Thompson and Hempshall, "Dirt Alert."
118. UK Department of Health, "A Matron's Charter: An Action Plan for Cleaner Hospitals," http://www.dh.gov.uk/PublicationsAndStatistics/Publications/PublicationsPolicyAndGuidance/PublicationsPolicyAndGuidanceArticle/fs/en?CONTENT_ID=4091506&chk=L2Id5d
119. D. M. Berwick, "Improvement, Trust and the Healthcare Workforce."
120. CCDR, "Supplement Infection Control Guidelines."
121. D. J. Weber and W. A. Rutala, "The Emerging Nosocomial Pathogens Cryptosporidium, Escheria Coli 0157Z:H7, Helicobacter Pylori, and Hepatitis C: Epidemiology, Environmental Survival, Efficacy of Disinfection and Control Measures," *Infection Control Hospital Epidemiology* (2001); 306–15.
122. M. J. Terpstra, "The Correlation between Sustainable Development and Home Hygiene," *American Journal of Infection Control* 29 (2001): 211–17.
123. S. King, "Decontamination of Equipment and Environment," *Nursing Study* 12 (1998): 57–60.
124. Dieckhaus and Cooper, "Infection Control Concepts in Critical Care."
125. D. J. Weber et al., "The Effect of Blood on the Antiviral Activity of Sodium Hypochlorite, a Phenolic and a Quaternary Compound," *Infection Control Hospital Epidemiology* 20 (1999): 821–27.
126. G. L. French et al., "Tackling Contamination of the Hospital Environment by Methicillin-Resistant Staphylococcus Aureus (MRSA): A Comparison between Conventional Terminal Leaning and Hydrogen Vapour Decontamination," *Journal of Hospital Infection* 57 (2004): 31–37.
127. Garner and The Hospitals Infection Control Practices Advisory Committee, "Guidelines for Isolation Precautions."

128. French et al., "Tackling Contamination of the Hospital Environment."
129. A. Voss et al., "Should We Routinely Disinfect Floors?" (Letter) *Journal of Hospital Infection* 53 (2003): 150.
130. S. Bloomfield et al., "Disinfection and the Prevention of Infectious Disease," *American Journal of Infection Control* 32 (2004): 312– 312.
131. S. King, "Decontamination of Equipment and the Environment," *Nursing Standard* 12 (1998): 57–60.
132. F. Fitzpatrick et al., "A Purpose Built MRSA Unit." *Journal of Hospital Infection* 46 (2000): 271–79.
133. S. Dharan et al., "Routine Disinfection of Patients' Environmental Surfaces. Myth or Reality?" *Journal of Hospital Infection* 42 (1999):113–17.
134. W. A. Rutala and D. J. Weber, "The Benefits of Surface Disinfection," *American Journal of Infection Control* 32 (2004): 226–31.
135. M. Rutherford, "Tackling the Problem of Poor Ward Cleanliness."
136. Rutala and Weber, "The Benefits of Surface Disinfection."
137. Rutherford, "Tackling the Problem of Poor Ward Cleanliness."
138. Centers for Disease Control, *Guidelines for Environmental Infection Control.*
139. CCDR, "Supplement Infection Control Guidelines."
140. Rutala and Weber,"The Benefits of Surface Disinfection."
141. F. Daschner and A. Schuster, "Disinfection and the Prevention of Infectious Disease: No Adverse Effects?" *American Journal of Infection Control* 32 (2004): 224–25.
142. Rutala and Weber. "The Benefits of Surface Disinfection."
143. Auditor General for Scotland, "A Clean Bill of Health? A Review of Domestic Services in Scottish Hospitals. Audit Scotland, 2000." http://www.audit-scotland.gov.uk/index/00h01ag.asp
144. H. S. Gold and R. C. Moellering, "Antimicrobial-Drug Resistance,"*New England Journal of Medicine* 335 (1996); 1445–48.
145. D. E. Zoutman and B. D. Ford, "The Relationship between Hospital Infection Surveillance and Control Activities and Antibiotic-Resistant Pathogen Rates," *American Journal of Infection Control* 33 (2005): 1–5.
146. Rutala and Weber, "The Benefits of Surface Disinfection."
147. Garner and The Hospitals Infection Control Practices Advisory Committee, "Guidelines for Isolation."
148. Muto et al., "SHEA Guidelines for Preventing the Nosocomial Transmission of Multidrug-Resistant Strains."
149. A Simor, et al., "The Canadian Nosocomial Infection Surveillance Program: Results of the First 18 Months of Surveillance for Methicillin-Resistant Staphylococcus Aureus in Canadian Hospitals," *Canadian Community Disease Report* 23 (1997): 41–45.
150. Z. Filetothe, *Hospital Acquired Infection. Causes and Control.* London: Whurr, 2003).
151. Damani, *Manual of Infection Control Procedures.*
152. Rutala and Weber, "The Benefits of Surface Disinfection."
153. A. Rampling et al., "Evidence that Hospital Hygiene Is Important in the Control of Methicillin-Resistant Staphylococcus Aureus."

154. P. Toynbee, "Quality Care Means Valuing Care Assistants, Porters and Cleaners too," *The Guardian,*
155. Auditor General for Scotland, "A Clean Bill of Health?"
156. "Audit Tools for Monitoring Infection Control Standards 2004," http://www.dh.gov.uk/PublicationsAndStatistics/Publications/Publications PolicyAndGuidance/PublicationsPolicyAndGuidanceArticle/fs/en? CONTENT_ID=4090854&chk=MNt4Af.
157. Malik et al., "Use of Audit Tools to Evaluate the Efficacy of Cleaning Systems in Hospitals."

Chapter 12

1. D. Butler-Jones, Chief Public Health Officer. Public Health Agency of Canada. Ottawa Ontario. Personal correspondence, September 1st, 2005.
2. World Health Organization, note to come: WHO Status Report May 20, 2003.
3. C. W. Tyler and J. M. Last, "Epidemiology," in *Public Health and Preventive Medicine,* ed. R. B. Wallace (Stamford, CT: Appleton and Lange,1998).
4. J. O'Grady, "Joint Health and Safety Committees," in *Injury and the New World of Work,* ed. T. Sullivan (Vancouver: UBC Press, 2000).
5. U.S. National Research Council, National Academy of Sciences, Washington, D.C.
6. *Morbidity and Mortality Weekly Report* 52(182003);405-411.
7. T. Svoboda et al., Public Health Measures to Control the Spread of Severe Acute Respiratory Syndrome during the Outbreak in Toronto," *New England Journal of Medicine,* 350 (2004):2352-61.
8. A. Campbell, Commissioner, *The SARS Commission Interim Report: SARS and Public Health In Ontario* (April 15, 2004 and April 5, 2005): 15-16.
9. OPSEU/ONA submission to the SARS Commission Hearings, November 2003.
10. H, Checkoway et al., Research Methods in Occupational Epidemiology. Oxford University Press, Oxford, 1989.
11. R. M. Park, "Hazard Identification in Occupational Injury: Reflections on Standard Epidemiologic Methods," *International Journal of Occupational and Environmental Health* 8, no 4 (2002): 354-62.
12. International Labour Office, *SARS Practical and Administrative Responses to an Infectious Disease in the Workplace* (Geneva, Switzerland: International Labour Office, March 2004).
13. International Labour Office, Occupational Safety and Health Convention, 1982, No.155.
14. Public Health Agency of Canada, "Learning from SARS," in *Canada: Anatomy of an Outbreak.* Report of the National Advisory Committee on SARS, October, 2003.
15. M. Kirby, "Reforming Health Protection and Promotion in Canada: Time to Act," 14th Report of the Standing Senate Committee on Social Affairs, Science and Technology, Report, November 5th, 2003.

16. D. Koh et al., "Occupational Health Response to SARS" (letter). *Emerging Infectious Diseases* [serial on the Internet], January 2005, http://www.cdc.gov/ncidod/EID/vol11no01/04-0637.htm

Epilogue

William Charney

This book is meant to be a warning: "Code Red" with red being the highest alert status. Being prepared for an emerging, naturally occurring pandemic is not an easy task. International cooperation at the level still not achieved would need to take place. Public health would have to be refunded at the same level as we are funding the war in Iraq. Hospitals would need to change "business as usual," with special hospitals designated and designed for the treatment of contagious flu, being a more efficient design model as well as providing a higher level of public safety. Communities would need to be trained and more involved. Police and fire fighters and all first responders would need retraining on the risk controls. Mechanical ventilation systems for receiving hospitals would need retrofitting, respiratory protection models for healthcare workers would need upgrading, hospitals would need to have surge model training, and healthcare workers would need specialized training in protecting against cross-contamination — and this is just a partial list.

Even as recently as November 2005, the Bush Pandemic Plan did not learn the class lessons of Katrina or what Hillel Cohen refers to as the socioeconomic divide. The Bush Plan still, according to Cohen, does not take into consideration the various economic obstacles that poor people would face, like paying for prescription medications during a pandemic. Or, to take another example: Bush referred to a website for further explanation of the plan, forgetting that most poor people do not have a home computer.

Both the public health systems and the occupational health response systems are not ready for pandemic flu and need repair as quickly as the levees in New Orleans. In the occupational health arena it is interesting to note that the Department of Health and Human Services has issued an influenza plan entitled, "HHS Pandemic Influenza Plan" with a subsection titled "Infection Control." After a reading of this section it is apparent that the national agency in charge of protecting Americans is still not assimilating the newer science on transmission paradigms. This document, by downplaying the airborne vector route of transmission, recommends either surgical masks or N95 respirators, neither of which will be protective if the virus is airborne, and there is enough evidence that this could be the case. Triage protocols, as mentioned in the Flu Plan, of patients just being "3 feet away from each other" does not reflect modern standards of patient developmental flow or separation, nor is there mention of negative pressure isolation in this section on infection control.

It is still the overwhelming opinion of many emergency room directors and physicians in the United States that most emergency rooms could not handle a major plane crash let alone a surge of patients with influenza. "It is a struggle to meet the nightly demand of 911 calls," said an emergency room physician in a major trauma center in Seattle, Washington, "but somehow we are supposed to deal with a new strain of influenza." The recent trend in the United States of scaling down trauma units, along with a growing U.S. population, is making it very difficult to imagine how U.S. trauma centers are going to cope with a surge of pandemic patients. The disconnect is not being granted enough attention. Major trauma centers around the country have been complaining for years of overcrowding, lack of funding, and a lack of attention from all public agencies for redress of their growing grievances.

One such grievance is the Bush Administration's willingness to spend billions of dollars on stockpiling of Tamiflu — even though it is not clear whether this medicine would be effective — while showing no interest in investing in emergency rooms that would have to handle the flu cases during a pandemic.

Dealing with surge numbers of patients requires a lot of coordination between local, state, and federal governments, the National Guard, schools, religious organizations, food suppliers, and, of course neighboring hospitals. Providing space for many extra beds in an environment where lack of space is a prime issue will pose a challenge, and the purchase of cots, mattresses, and other resources, and storage of these resources will also challenge many hospital administrations. The example of the earthquake in Pakistan in late October 2005 revealed that there were not enough tents or temporary housing in that whole country to now assist all the homeless who then had to protect themselves against the Himalayan winter. Using churches, arenas, and other community resources means stockpiling of materials and coordination and upgrading of ventilation systems within the chosen auxiliary facilities to prevent cross-transmissions.

This book has dealt with many of the contradictions that now exist in our public health and occupational health readiness for pandemic scenarios. Some macro issues such as global warming, the lack of a universal health plan in the United States, class biases, and the differences between health systems in the first world as compared to the third world, will all contribute to the pandemic paradigm that has been discussed on some level in this volume. Micro issues, such as lack of staffing and proper funding of our public health systems, and lack of coordination among different public health entities, as stated by Avery, leave us vulnerable. And monies spent on bioterror, as explained by Cohen, do not cross over and protect the public health from naturally occurring infections. Inner city and community hospitals are unprepared for surge patients, and the questionable protocols of occupational protection for healthcare workers could tend to accelerate the pandemic.

One clear message after 9/11 was that the public health system in the United States was broken. Since then the repairs, according to the evidence supplied here, have not been made. A similar message was sent after the SARS outbreak in Asia and Canada, namely that many of the infection control protocols recommended by guideline agencies did not work for this mutating corona virus (in Canada, 50% of cases were healthcare workers) However, since then, according to different sources, many of the oversights and mistakes that were identified have not been corrected. The American Flu Plan, by downplaying the airborne transmission potential, and against existing scientific evidence, puts both occupational and public safety into a more vulnerable position. After Katrina, the postmortems on the class system in the United States revealed the lack of planning for the poor, but planning since Katrina is not reflecting lessons learned.

High level training of healthcare workers is crucial to prepare for a pandemic. If the healthcare workers become ill, there will be fewer and fewer people to care for the sick. It was apparent during the SARS outbreak in Canada that healthcare workers had multiple problems preventing cross-contamination of themselves and other healthcare workers, solely for surface removable contamination. The chapters on the Canadian Commission report included discussion of these issues to draw attention to the need to train healthcare workers on a model that rises to the level of risk and that incorporates applicable skills needed during a virulent outbreak. This updated training and skill testing still has yet to been done.

The book devotes three chapters to respiratory protection because this is one of the more contentious issues facing the occupational health of the healthcare community. It is the interpretation of this volume that compromises were made in the recommendation of N95 respirators and that this choice of respirator will not protect healthcare workers from potential exposures, either during incidental patient room contact or during aerosolizing procedures. The book argues that the recommendation of N95 respirators was a balance between cost ($1.00 per mask) and comfort, but also transgresses some important existing rules in the OHSA Respirator Protection Guidelines, especially the dose–response clauses. (Dr. Nicas even argues that the recommendation of the N95 respirator is made without any consideration for present uncertainties regarding airborne exposure intensity, a healthcare worker's duration of exposure, the infectious dose of the virus, and the potential consequences of the infection (i.e., a high case fatality proportion).

However, the recommendation of an N95 is nonproductive economically since the downstream cost of treating a healthcare worker who contracts the disease is far greater than providing a more protective respirator. The new Flu Plan downplays airborne transmission, preferring instead to issue protective guidelines for surface removable contamination. This was a fallacious analysis for SARS and it will be for any upcoming virus.

Dr. Nicas argues the following about the recently released CDC avian flu plan.

The recommendation "of a surgical mask or procedure mask for close contact with infectious patients" as stated in the Plan has multiple problems, one of which is that the surgical mask's minimum level of resistance to virus penetration is unspecified; the infectious dose of a future flu virus is unknown at present; and the concentrations of the avian flu virus in saliva and nasal secretions (number per ml) of future sick patients is also unknown. Therefore, the CDC cannot know whether a surgical mask would provide sufficient resistance to the penetration of a high challenge load of avian flu virus to prevent infection in the healthcare worker.

Nicas goes on to explain that the document implies that the use of an actual respirator (as opposed to a surgical mask) is not needed unless an aerosol-generating procedure is being performed on the patient. The associated assumption in the Plan is that inhalation exposure will not be a normal transmission route of avian flu. This assumption is entirely unwarranted given that other influenza virus infections can be transmitted by inhalation and the document itself acknowledges that the "relative contributions and clinical importance of different modes of transmission are currently unknown" (section S4-11.A page S4-3). If transmission routes are unknown the higher level of precautions would apply. Nicas states, "These circumstances argue for routine use of powered air-purifying respirators (PAPRs) at the beginning of an avian flu outbreak. In the event that evidence accumulates that airborne transmission is not important the use of PAPRs can be discontinued."

If, as we suspect, airborne transmission will be a vector, then hospital ventilation systems need to be ready for this contingency. Derman in his chapter dealt specifically with the need for mechanical ventilation system capability in hospitals to both isolate and scrub the air, two scientifically proven methods to reduce transmission rates. His assessment is that this capability still remains an open question, especially for producing ade-quate air exchanges per hour to reduce the number of toxic particles in the air.

We attempted in this book to draw attention to existing needs using a critical methodology. All too often bureaucrats in various governmental guideline agencies want the public to feel safe and to assume that they are working hard and protections are in place. In some areas of the pandemic flu paradigm this is the case. There are many people hard at work providing plans, coverage scenarios, and exercises. However, for many of the needed details in emerging infectious diseases, as raised in this volume, there remains much work to be done, tasks that must be followed up, and a scientific rigor applied. There are hundreds of public health and occupational health officers and staff all over North America with growing concerns about readiness for the next potential pandemic. Many of these concerns are not being addressed, especially issues of budgets, targeting of resources, vaccines that will not work, and vaccine shelf lives that may exceed expiration dates, labor issues, and communi-cation issues between public health agencies. Congress is finally taking an oversight role of the agencies responsible for providing functional plans and responses, such as the National Institutes of Health, the Centers for Disease Control, and others. Congressional committee debates on pan-demic flu are being televised on cable networks. However, there still remains a sense that the dots are not yet connected and that consensus has become a compromise between political parties rather than scientific solutions and that millions of lives may be lost as a cost of doing or not doing business.

The depletion of the treasury and the huge budget deficits are now coming into play as money is desperately needed to fix many systems. "Less government" or "smaller government" at a time when real federal leadership and dollars are needed has left holes in our response capability. Every state is developing its own plan and there is lack of clarity as to how they will all interact with the federal agencies. The debriefing after Katrina revealed the "blame game" approach of state blaming federal, federal blaming state, and both blaming local governments for mistakes made. Cronyism has depleted the expertise and morale of leading responding public health agencies and departments. One federal politician close to all these dynamics was quoted as saying, "it will be every man for himself."

The good news is there is still time to correct many of the problems pointed out in this volume. A real disaster plan provides for the worst case scenario and is protective of every stratum of society.

APPENDICES

Appendix A

American Federation of Labor and Congress of Industrial Organizations

EXECUTIVE COUNCIL

April 11, 2005

Julie Louise Gerberding, MD, MPH Director
Centers for Disease Control and Prevention
1600 Clifton Road, NE
MS-014
Atlanta, Georgia 30333

Dear Dr. Gerberding:

The terrorist attacks of September 11, 2001, dealt a staggering blow to the American people and workers in particular. As we have already seen, all healthcare providers, firefighters, first responders, first receivers, and emergency medical personnel will be among the first victims of terrorist acts. Emergency response/receiver personnel, our members, are on the front lines of homeland defense as we respond to these attacks. As

America's domestic defenders, we must be equipped, trained, and supported equally as well as the soldiers who wage war on foreign soil.

The AFL-CIO and undersigned unions are, therefore, greatly disturbed by the CDC's interim guidance on protecting healthcare workers caring for patients in the event of a bioterrorist plague attack released on April 4, 2005. This document, Interim Guidance for Protecting Health Care Workers Caring for Patients Potentially Exposed to Aerosolized Yersinia pestis from a Bioterrorism Event, advises that a surgical mask is sufficient to protect healthcare workers caring for patients exposed to plague. It further states that in the event of a bioterrorism incident, "exigent circumstances may require the suspension of some of the respiratory protection requirements found in the Occupational Safety and Health Administration Respiratory Standards (29 CFR 1910.134), such as fit testing and medical clearance."

This guidance is completely at odds with scientific evidence, legal requirements under the Occupational Safety and Health Act, existing NIOSH and OSHA guidance and recommendations for protecting responders, and the Worker Safety and Health Support Annex of the National Response Plan issued in December 2004. This guidance, if followed, would put healthcare workers at risk of serious and potentially deadly exposure. We ask that this inaccurate and harmful document be withdrawn immediately.

Protection offered by any respiratory protective device is contingent upon the employer of the respirator user adhering to complete program requirements (including all requirements of federal OSHA, e.g., 29CFR1910.134), the use of NIOSH-certified respirators in their approved configuration, individual respirator fit testing to rule out those respirators that cannot achieve a good fit on individual workers; and proper training of employees that must wear or have the potential to wear respirators.

There is NO scientific or medical evidence, historical or contemporary, that has demonstrated that a surgical mask provides adequate protection. Surgical masks are not respirators. They are simple devices, introduced without much change over a century ago to protect patients during surgical procedures. It was believed then, and unfortunately by some now that the cloth surgical mask would prevent harmful microorganisms hosted by the breathing passages of the healthcare provider from migrating to the patient's open surgical site and/or wound. Unless the device is an air purifying respirator certified by NIOSH, it does not afford protection to the provider and it does not comply with federal workplace standards.

Furthermore, there is a widely held misperception within the infection control community and perpetuated by the CDC that many biological agents are exclusively droplet transmitted; not airborne. However, as we learned at the CDC Workshop on Respiratory Protection for Airborne Infectious Agents held on November 30 —December 1, 2004, this is simply a false and misleading dichotomy. Professor Eugene C. Cole, DrPH, from Brigham Young University, clarified this fact during the first plenary session of the conference in his talk entitled: "Aerobiology of Infectious Agents." He stated that as soon as a droplet is expelled from a patient in a sneeze or a cough, it is only a matter of nanoseconds before the droplet begins to dessicate to convert to airborne droplet nuclei. These droplet nuclei in turn exhibit Brownian motion — meaning to float in the air largely unaffected by the forces of gravity. As such, these now airborne particles have been shown to easily bypass the significant gaps that exist between the face of a healthcare worker and a typical surgical mask due to the fact that surgical masks are incapable of achieving a proper face seal.

In October 2001, NIOSH issued Interim Recommendations for the Selection and Use of Protective Clothing and Respirators Against Biological Agents (DHHS (NIOSH) Publication Number 2002-1 09; found at http://www.cdc.gov/niosh/unp-intrecppe.htm) This document specifies the minimum recommendations for selection and use of protective clothing and respirators to protect against biological exposures associated with a suspected act of biological terrorism. The minimum level of protection recommended for responders is either a powered air purifying respirator, or in the case of exposure to unknown agents, a NIOSH approved, pressure demand SCBA. This minimum level of protection is similar to the level of respiratory protection recommended by the OSHA in the OSHA Best Practices for Hospital Based Receivers of Victims from Mass Casualty Incidents Involving the Release of Hazardous Substances issued in January 2005 (http://www.osha.gov/dts/osta/bestpractices/firstreceivers _hospital.pdf)

Moreover, the Worker Safety and Health Support Annex of the National Response Plan issued in 2004, to which the Department of Health and Human Services is a signatory, clearly states that "developing, implement-ing, and monitoring an incident personal protective equipment program, including the selection, use, and decontamination of PPE; implementation of a respiratory fit-test program; and distribution of PPE," is a key part of a response plan to protect workers in the event of a major incident, including a bioterrorist attack (http://www.dhs.gov/interweb/assetlibrary/ NRP_FullText.pdf).

The CDC interim guidance on plague states that the recommended measures are not for emergency responder or emergency receivers. How-ever, it is unclear as to whom these guidelines apply.

After a mass casualty bioterrorism event, all medical personnel would be first receivers. As defined by the source cited in this guidance statement, federal OSHA states that all healthcare workers at a hospital receiving contaminated victims for treatment may be termed first receivers *and* include personnel in the following roles: clinicians and other hospital staff who have a role in receiving and treating contaminated victims (e.g., triage, decontamination, medical treatment, and security) and those whose roles support these functions (e.g., set up and patient tracking). (See: http://www.osha.gov/dts/osta/bestpractices/html/hospital_firstreceivers html#a2).

The need for adequate personal protective equipment in the event of an emergency incident has been clearly demonstrated by the aftermath of September 11th. We have seen extensive and serious adverse respiratory effects among many of the firefighters and other workers who responded to the 9-11 attacks and who worked in the recovery operation at ground zero. Much of this disease is a result of responders not being provided adequate respiratory protection equipment, proper fit testing, or training. CDC and others should learn from this experience, that adequate respiratory protection is key for protecting workers in the event of such an emergency.

Healthcare workers, firefighters, police, skilled support personnel and other responders will be on the front line in the event of a bioterrorist incident or other attack. Providing these individuals with the highest level of protection is the government's responsibility. Absent the provision of such protection, healthcare workers cannot be expected to participate and respond in the event of such an incident. And as was clearly demonstrated during the failed smallpox vaccination program, absent adequate protection, participation will not occur. Failure to provide healthcare workers and other responders adequate protection not only endangers these workers, it undermines the effectiveness of any emergency response and thus endangers the public as well.

CDC's interim guidance on protecting healthcare workers caring for patients in the event of a bioterrorist plague attack is irresponsible and puts workers and the public in danger. It should be immediately withdrawn.

Our members are facing these threats now and they need protection today.

Sincerely,

Peg Seminario
Director
Departrnent of Occupational
Safety and Health, AFL-CIO

On behalf of:

American Federation of Government Employees
American Federation of State, County and Municipal Employees
American Federation of Teachers
Building and Construction Trades Department, AFL-CIO
Communications Workers of America
International Association of Firefighters
International Brotherhood of Teamsters
International Union, UAW
Service Employees International Union
United American Nurses
United Food and Commercial Workers
United Steelworkers of America

cc: The Honorable Michael O. Leavitt
    Secretary of Health and Human Services
    John Howard, MD, Director, NIOSH
    Jonathan L. Snare, Acting Assistant Secretary of
    Labor for Occupational Safety and Health

Interim Guidance for Protecting Health Care Workers Caring for Patients Potentially Exposed to Aerosolized *Yersinia pestis* from a Bioterrorism Event

PLAGUE

April 4, 2005

This interim guidance provides safety and health recommendations for workers In health care settings who may treat patients infected, or suspected of being infected, with Yersinia pestis (the bacterium that causes plague) as a result of a bioterrorism event. It Is NOT intended for emergency responders (e.g., EMTs at the scene of an event or transporting potentially contaminated victims) or first receivers (covered under other guidance: www.osho.gov/dts/osta/bestpractices/html/hospital_ firstreceivers.html) Nor is it intended for health care workers caring for patients with naturally occurring plague.

Naturally occurring plague is uncommon In the United States and occurs after infection with Y. pestis, most commonly transmitted to humans from infected rodents via fleas. Bubonic plague is most common. Pneumonic plague occurs when Y. pestis infects the lungs, either as a primary or secondary (spread of infection to the lungs in a patient with bubonic or septicemic plague) infection. The first signs of primary pneumonic plague are fever, headache, weakness, and rapidly developing pneumonia with shortness of breath, chest pain, cough, and, in a later stage, sometimes bloody sputum. The pneumonia progresses rapidly, may cause respiratory failure and shock, and without early treatment is frequently fatal. Plague is treated with antibiotics.

As soon as a diagnosis of suspected plague is made, the patient should be hospitalized under droplet precautions. Local and state health departments should be notified, confirmatory laboratory work should be initiated, and antibiotic therapy should be started as soon as possible after the laboratory specimens are taken. Contacts of pneumonic plague patients should be placed under observation or given preventive antibiotic therapy, depending on the degree and timing of contact.

Patients (or animals) with pneumonic plague can spread the Infection to other persons through the air, particularly during advanced disease when bloody sputum is present. Case reports describing occurrences of pneumonic plague have found that transmission of Y. pestis from persons (or animals) with pneumonic plague usually occurs among persons in direct and close contact with the ill person (or animal). Health care workers uncommonly may become infected during the care of patients with naturally occurring infection who develop advanced plague pneumonia. Therefore, droplet precautions have been recommended for health care workers caring for patients with pneumonic plague.

Plague may be encountered during a bioterrorism event, in which case transmission of the infection may occur by breathing in aerosolized bacteria, and a likely clinical manifestation would be primary pneumonic plague. A bioterrorism event with Y. pestis may be suspected if several cases of plague are identified within a short period of time (particularly in nonendemic areas) and the cases are not associated with transmission by infected fleas. A bioterrorism event can introduce initial uncertainty about the agent (e.g., drug resistance, persistence in the environment, infectiousness) and can place considerable stress on the health care system. Complete health care worker compliance with antimicrobial prophylaxis cannot be guaranteed during a bioterrorism event and the organism may be genetically altered to be antibiotic resistant. Plague vaccine, which is not currently available in the United States, has demonstrated efficacy only against bubonic plague and does not prevent the development of primary pneumonic plague.

When there is a suspicion of a bioterrorism event, infection control practice should include droplet and contact precautions. Historical and contemporary epidemiological evidence from naturally occurring pneumonic plague outbreaks indicates that the infection is not easily transmitted from person to person and that a surgical mask in combination with other droplet precautions provides adequate protection for health care workers. However, given the initial uncertainties associated with a bioterrorism event, additional precautions may be prudent, and the use of an N95 filtering face piece respirator will offer an additional degree of protection. In a large-scale bioterrorism event, exigent circumstances may require the suspension of some of the respiratory protection requirements found in the Occupational Safety and Health Administration Respiratory Standards (29 CFR 1910.134), such as fit testing and medical clearance.

Other complementary strategies should also be in place, including temperature monitoring of unprotected close contacts. In addition, the use of antibiotic prophylaxis may be considered. These precautions should be continued at least until a definitive diagnosis is established, antimicrobial sensitivity of the agent is known, and the presence of other agents is ruled out.

If the isolation capacity of the health care facility is adequate, patients should be isolated under droplet precautions for at least 48 hours of effective antibiotic therapy and until clinical improvement has taken place. If strict isolation is not possible due to the great number of patients, then patients with pneumonic plague may be cohorted while receiving treatment.

Health care faciilties should avoid surgery or other aerosol-generating procedures on known or suspected plague cases (including autopsies), unless deemed medically necessary (see "Plague as a Biological Weapon: Medical and Public Health Management" at http://jama.ama-assn.org/cgi/content/short/283/17/2281). If such procedures are performed, a higher level of respiratory protection may be necessary. If such procedures are performed on a patient with known or suspected pneumonic plague, airborne precautions should be implemented regardless of agent sensitivity or source of exposure (suspected bioterrorism versus naturally occurring).

For more information, visit www.bt.cdc.gov/agent/plague or call CDC at 800-CDC-INFO (English and Spanish) or 888-232-6348 (TTY).

Index

417

scientific literature, 314–316
teamwork, need for, 318
training, 318–319
working conditions, 317–318
Clostridium difficile diarrhea, 312–313
Colorado
    antiviral availability, 258
    effect of flu pandemic, 256, 275
Commission to Investigate Introduction and
        Spread of SARS, 31–38
    nurses' union officials reports, 3–38
Communication
    effect on infection control guideline
        compliance, 165–167
    importance of, 134–136
    planning of, 269
Compliance with infection control
        guidelines, 3–38, 96–98,
        149–151, 153–177
    anterooms, 169
    attitudes, impact of, 171–172
    availability of protective equipment,
        170–171
    beliefs, impact of, 172–173
    communication, 165–167
    consistency with safety instructions, lack
        of, 158
    content analysis, 156–168
    discomfort, 174–175
    enforcement by regulatory agencies, 159
    environmental decontamination, 170
    environmental factors, 98, 168–170
    evidenced-based policies, 161–163
    exhaustion, 176
    fatigue, 176
    fit testing, 167–168
    individual factors, 98
    isolation rooms, 168–169
    knowledge, 171
    negative pressure rooms, 169–170
    organizational factors, 97–98, 158,
        168–177
    others', attitudes of, impact of, 176–177
    peer environment, 175–176
    physical space separation, 168
    time constraints, 173–174
    training, 164–165
    workload, increase in, 174
    workplace attitudes, 159–161
Connecticut
    antiviral availability, 258

effect of flu pandemic, 256, 275
Cooperative synergy, agencies,
        departments, 325–338
    hurricane Katrina, 335–337
    Public Health Agency of Canada, 326
    SARS case study, 329–335
Coordination, importance of, 223–225
Cost containment, 229–230
Coughing patient, 78–86
    airborne concentration/inhaled dose,
        83–84
    cough particles, 78–81
    exposure intensity estimation, 78–86
    hypothetical example of, 83–84

D

Decontamination, scientific literature
        coverage, 94
Delaware
    antiviral availability, 258
    effect of flu pandemic, 256, 275
Department of Homeland Security, 252, 295
Department of State, advisory statement
        about avian flu, 262
Deterioration of public health system,
        219–246
    bioterrorism, 239–246
    public health responses, 221–238
DHS. *See* U.S. Department of Homeland
        Security
Diagnosis of influenza, 300–302
Discomfort, effect on infection control
        guideline compliance, 174–175
District of Columbia
    antiviral availability, 258
    effect of flu pandemic, 256, 275
Droplet-spread respiratory infections,
        144–145, 407–412. *See also
        under* specific infection
Drug availability issues, 233–235, 258–259,
        261, 263–266, 268–269, 290–292

E

Engineering controls, in healthcare work
        protection, 140–141, 146–147

J

Job control of hospital
cleaners/housekeepers, 317
Joint health and safety committees, role of,
19–21

K

Kansas
antiviral availability, 258
effect of flu pandemic, 256, 275
Katrina disaster, 293–298, 335–337
American Red Cross in, 295
armed troops, use of, 296
Bush Administration, failings of, 296
class, impact of, 294–295
Federal Emergency Management
Agency in, 295
lessons from, 293–298
limitations of preparedness, 295–297
race, impact of, 294–295
U.S. Department of Homeland Security,
295
Kentucky
antiviral availability, 258
effect of flu pandemic, 256, 275
Knowledge, effect on infection control
guideline compliance, 171
Knowledge gaps in occupationally acquired
respiratory infectious disease
control, 89–204

L

Laboratory accident, 84–86
airborne concentration/inhaled dose,
84–85
hypothetical example, 85–86
Leadership in public health arena,
importance of, 230–231
Legal issues, public health, 230–231
Lessons from hurricane Katrina disaster,
293–298
Louisiana
antiviral availability, 258
effect of flu pandemic, 256, 275

M

Maine
antiviral availability, 258
effect of flu pandemic, 256, 275
Maryland
antiviral availability, 258
effect of flu pandemic, 256, 275
Massachusetts
antiviral availability, 258
effect of flu pandemic, 256, 275
Medication availability issues, 233–235,
258–259, 261, 263–266,
268–269, 290–292
Michigan
antiviral availability, 258
effect of flu pandemic, 256, 275
Ministry of Labour, Canada
chronology of events involving, 23–29
events that should have triggered
enforcement, 22–23
role of, 21–22
Minnesota
antiviral availability, 258
effect of flu pandemic, 256, 275
Mississippi
antiviral availability, 258
effect of flu pandemic, 256, 275
Missouri
antiviral availability, 258
effect of flu pandemic, 256, 275
Monkeypox, 45–46
Montana
antiviral availability, 258
effect of flu pandemic, 256, 275
Mount Sinai Hospital, 15

N

National Institute for Occupational Safety
and Health, 410
air-purifying respirators tested, certified
by, 75
Nebraska
antiviral availability, 258
effect of flu pandemic, 256, 275
Negative pressure rooms, effect on
infection control guideline
compliance, 169–170

Nevada
  antiviral availability, 258
  effect of flu pandemic, 256, 275
New Hampshire
  antiviral availability, 258
  effect of flu pandemic, 256, 275
New Jersey
  antiviral availability, 258
  effect of flu pandemic, 256, 275
New Mexico
  antiviral availability, 258
  effect of flu pandemic, 256, 275
New Orleans devastation. *See* Hurricane
      Katrina
New York
  antiviral availability issues, 258
  effect of flu pandemic, 256, 275
Nightingale, Florence, 306
NIOSH. *See* National Institute for
      Occupational Safety and Health
North Carolina
  antiviral availability, 258
  effect of flu pandemic, 256, 276
North Dakota
  antiviral availability, 258
  effect of flu pandemic, 256, 276
Norwalk virus, 312
Nosocomial infection, 133–139, 305–324.
      See also under specific infection
Nursing shortage, effect of, 36–37

O

Obama, Barack, Senator, introduction of
      AVIAN Act of 2005, 261
Occupational Health and Safety Act, 17–19
Ohio
  antiviral availability, 258
  effect of flu pandemic, 256, 276
Oklahoma
  antiviral availability, 258
  effect of flu pandemic, 256, 276
ONA. *See* Ontario Nurses Association
Ontario Nurses Association, 3–38
Ontario Public Services Employees Union,
      3–30
OPSEU. *See* Ontario Public Services
      Employees Union

Oregon
  antiviral availability, 258
  effect of flu pandemic, 256, 276
Oseltamivir, availability issues, 258–259,
      265–266

P

Pandemic, flu, 249–276, 281–282
  antivirals/vaccine, 233–235, 263–266,
      269
  bird transmission, 267
  breaks of twentieth century, 251–253
  dangers of, 250–251
  model estimates, 253–254
  mortality, 256–257, 275–276
  pig transmission to humans, 267
  recommendations, 270–276
  Southeast Asia, 267
  state by state analysis, 254–259
  state level, 275–276
  state readiness, 262
  stockpiling of medications, 265–267
  stopping, 285–287
"Pandemic Preparedness and Influenza
      Vaccine Supply--CDC, NIAID,
      and Office of Secretary of HHS,"
      261
PAPR. *See* Powered air-purifying respirator
Pathogen emission rate, 81–82
Pathogen removal pathways, 82–83
Patient load per nurse, increase in, 36–37
Peer environment, effect on infection
      control guideline compliance,
      175–176
Pennsylvania
  antiviral availability, 259
  effect of flu pandemic, 257, 276
"The Perplexing Shift from Shortage to
      Surplus: Managing This Season's
      Flu Shot Supply and Preparing
      for Future," 261
Personal protective equipment, 39–86,
      142–144
  airborne pathogens, 71–86
  fit testing, 62–68
  respirators, 41–70
  scientific literature coverage, 94–95
PHAC. *See* Public Health Agency of Canada